Springer Series in Optical Sciences Volume 2

Edited by David L. MacAdam

# Springer Series in Optical Sciences

Edited by David L. MacAdam



1 **Solid-State Laser Engineering**
By W. Koechner

2 **Table of Laser Lines in Gases and Vapors**
3rd Edition
By R. Beck, W. Englisch, and K. Gürs

3 **Tunable Lasers and Applications**
Editors: A. Mooradian, T. Jaeger, and
P. Stokseth

4 **Nonlinear Laser Spectroscopy**
By V. S. Letokhov and V. P. Chebotayev

5 **Optics and Lasers**
An Engineering Physics Approach
By M. Young

6 **Photoelectron Statistics**
With Applications to Spectroscopy and
Optical Communication
By B. Saleh

7 **Laser Spectroscopy III**
Editors: J. L. Hall and J. L. Carlsten

8 **Frontiers in Visual Science**
Editors: S. J. Cool and E. J. Smith III

9 **High-Power Lasers and Applications**
2nd Printing
Editors: K.-L. Kompa and H. Walther

10 **Detection of Optical and Infrared Radiation**
2nd Printing
By R. H. Kingston

11 **Matrix Theory of Photoelasticity**
By P. S. Theocaris and E. E. Gdoutos

12 **The Monte Carlo Method in Atmospheric Optics**
By G. I. Marchuk, G. A. Mikhailov,
M. A. Nazaraliev, R. A. Darbinian, B. A. Kargin,
and B. S. Elepov

13 **Physiological Optics**
By Y. Le Grand and S. G. El Hage

14 **Laser Crystals** Physics and Properties
By A. A. Kaminskii

15 **X-Ray Spectroscopy**
By B. K. Agarwal

16 **Holographic Interferometry**
From the Scope of Deformation Analysis of
Opaque Bodies
By W. Schumann and M. Dubas

17 **Nonlinear Optics of Free Atoms and Molecules**
By D. C. Hanna, M. A. Yuratich, D. Cotter

18 **Holography in Medicine and Biology**
Editor: G. von Bally

19 **Color Theory and Its Application in Art and Design**
By G. A. Agoston

20 **Interferometry by Holography**
By Yu. I. Ostrovsky, M. M. Butusov,
G. V. Ostrovskaya

21 **Laser Spectroscopy IV**
Editors: H. Walther, K. W. Rothe

22 **Lasers in Photomedicine and Photobiology**
Editors: R. Pratesi and C. A. Sacchi

23 **Vertebrate Photoreceptor Optics**
Editors: J. M. Enoch and F. L. Tobey, Jr.

24 **Optical Fiber Systems and their Components**
An Introduction
By A. B. Sharma, S. J. Halme, and
M. M. Butusov

R. Beck · W. Englisch · K. Gürs



# Table of Laser Lines in Gases and Vapors

Third Revised and Enlarged Edition

Springer-Verlag Berlin Heidelberg New York 1980

Rasmus Beck, Ph. D., Wolfgang Englisch, Ph. D., and
Professor Karl Gürs, Ph. D.

Batelle-Institut e.V.
Am Römerhof 35, 6000 Frankfurt/M., Fed. Rep. of Germany



ISBN 3-540-10347-3 3. Auflage Springer-Verlag Berlin Heidelberg New York
ISBN 0-387-10347-3 3rd edition Springer-Verlag New York Heidelberg Berlin

ISBN 3-540-08603-X 2. Auflage Springer-Verlag Berlin Heidelberg New York
ISBN 0-387-08603-X 2nd edition Springer-Verlag New York Heidelberg Berlin

Library of Congress Cataloging in Publication Data. Beck, Rasmus, 1939 – Table of laser lines in gases and vapors. (Springer series in optical sciences ; v. 2) Bibliography: p. Includes index. 1. Laser spectroscopy–Tables. 2. Gases–Spectra–Tables. 3. Vapors–Spectra–Tables. I. Englisch, W., joint author. II. Gürs, Karl, joint author. III. Title. QC454.L3B4 1980 535.5′8 80-21379



Offset printing and bookbinding: Konrad Triltsch, Graphischer Betrieb, 8700 Würzburg.

2153/3130-543210

# Introductory Remarks

Numerous applications of lasers require use of specific wavelengths (gas analysis including remote sensing, Raman spectroscopy, optical pumping, laser chemistry and isotope separation). Scientists active in these fields have been compelled to search, in addition to the available, mostly obsolete, laser-line tables, the entire recent literature in order to find suitable laser transitions.

This book is intended to facilitate such search. It is a computer print-out of laser spectroscopic data that are stored on punched cards.

The third edition of this book contains over 6100 laser transitions, 1000 more than the second (1978) edition. An additional list of the lines arranged in order of wavelength should greatly facilitate the search for a laser material that generates a specific wavelength. Further information has also been supplied by listing the pump transition for each of the FIR lines obtained with the optically pumped organic vapors.

In addition to the laser lines, the operating conditions under which emission has been achieved are briefly specified at the top of the list for each active medium. They have been optimized only for specific technologically or scientifically important laser types. For this reason, the output power is not indicated for the individual lines. Strong lines are marked by >.

Storage of the data on punched cards and listing of the tables by computer, ready for photoreproduction, made it possible to up-date them until immediately before they went to press. This, however, entails some restrictions because of the limited number of type styles (only capital letters, no subscripts or superscripts). For example, the symbol N2+ stands either for doubly ionized atomic nitrogen or singly ionized molecular nitrogen. However, confusion should not occur because atomic and molecular media are listed separately. In other cases, also, the meaning is clear from the context.

The order in which the atomic laser media are listed is based on the periodic system, beginning with the noble gases, continuing with hydrogen and the alkalies to the halogens and the rare earths. The molecular laser media are arranged in order of chemical composition, beginning with the compounds of noble gases (the excimers), then other diatomic molecules, triatomic molecules, and ending with the more complex molecules of organic vapors. In addition to the table of contents, a subject index appears at the end of the book.

Numerous suggestions and contributions from the laser community have been included in the third edition. Further active participation in the up-dating of the tables is welcome.

A special problem has been brought up by the development of optically pumped dimer lasers (especially the $I_2$ laser). They make it possible to generate so many lines (~$10^6$) that it is prohibitive to list them all separately. In this case the reader is compelled to calculate the wave numbers from the molecular constants, of which very accurate values are available from the cited references.

Frankfurt, July 1980

*R. Beck · W. Englisch · K. Gürs*

# Contents

LEGEND:
> STRONG LINE
? THIS LINE HAS NOT BEEN IDENTIFIED
@ THIS LINE HAS NOT YET BEEN OBSERVED
[...] REFERENCES
TOTAL NUMBER OF OBSERVED LINES: 6145

# Table of Laser Lines Arranged by Active Medium

ACTIVE MEDIUM : HELIUM
SYMBOL : HE

OPERATING CONDITIONS : PULSED AND CONTINUOUS DISCHARGE IN PURE HELIUM 0.2-0.8 TORR.
PULSED DISCHARGE (120 KV) IN 11-15 TORR HE AND 3 TORR H2.

| WAVELENGTH IN AIR [MICROMETER] | | | WAVELENGTH IN AIR [MICROMETER] | | |
|---|---|---|---|---|---|
| HE | 0.706517 | [471] | HE | 4.60535 | [006,011] |
| HE | 0.706521 | [008,471] | HE[3] | 4.60567 | [011] |
| HE | 1.8685 | [002] | HE | 8.53 | [011] |
| HE | 1.9543 | [003,004] | HE | 95.763 | [007,009,428] |
| HE | 2.05813 | [005] | HE | 216.12 | [009,428] |
| HE | 2.0603 | [003,004] | | | |

NUMBER OF LINES IN HELIUM 11

---

ACTIVE MEDIUM : NEON
SYMBOL : NE

OPERATING CONDITIONS : OPERATION AS ION LASER WITH HIGH CURRENT DENSITIES (PULSED 1 KA/CM2. SOME LINES CONTINUOUS 300-800 A/CM2). TYPICAL PRESSURE 0.001-0.01 TORR [101]. WITH ATOMIC NEON, MOST OF THE LINES ARE GENERATED IN A MIXTURE OF 0.01-0.1 TORR NE AND 0.5 TORR HE. MANY LINES CAN BE OBTAINED IN PURE NEON, SOME OF THESE AS PULSED SUPERRADIANCE. IN PULSED DISCHARGE AT HIGH PRESSURE (10-40 TORR NE, 20-80 TORR HE) THREE LINES IN THE NEAR INFRARED HAVE BEEN OBTAINED [108].

| WAVELENGTH IN AIR [MICROMETER] | | | WAVELENGTH IN AIR [MICROMETER] | | |
|---|---|---|---|---|---|
| NE3+ | 0.2018424 | [371] | NE2+ | 0.26134 | [371] |
| NE3+ > | 0.2022186 | [371] | NE2+ > | 0.2677918 | [012,013,371] |
| NE3+ > | 0.2065304 | [371] | NE2+ > | 0.2678690 | [012,013,371] |
| NE2+ | 0.2177705 | [371] | NE2+ | 0.2777634 | [013,371] |
| NE2+ | 0.2180858 | [371] | NE2+ | 0.2866726 | [013,012,371] |
| NE4+ | 0.22657 | [371] | NE+ | 0.3319745 | [012,014,371] |
| NE3+ | 0.2285793 | [371] | NE+ > | 0.3323745 | [013,012,371] |
| NE3+ > | 0.2357980 | [012,371] | NE+ | 0.332717 | [013] |
| NE3+ ? | 0.2373200 | [371] | NE+ | 0.332923 | [014] |
| NE2+ > | 0.2473398 | [012,371] | NE2+ | 0.333114 | [014] |
| NE2+ | 0.2609982 | [371] | NE+ | 0.3345446 | [012,014,371] |

| | | WAVELENGTH IN AIR [MICROMETER] | |
|---|---|---|---|
| NE+ | > | 0.3378256 | [013,012,371] |
| NE+ | > | 0.3392799 | [013,012,371] |
| NE+ | | 0.339320 | [014] |
| NE+ | | 0.348195 | [516] |
| NE+ | | 0.371309 | [015] |
| NE | | 0.54006 | [016,017,102] |
| NE | | 0.58525 | [003,013] |
| NE | | 0.59448 | [018,016] |
| NE | | 0.59393 | [003,019] |
| NE | | 0.60461 | [003,019] |
| NE | | 0.61180 | [003,019] |
| NE | | 0.61431 | [020,016] |
| NE | > | 0.63282 | [003,021,004] |
| NE | | 0.63518 | [003,019] |
| NE | | 0.64011 | [003,022,019] |
| NE | | 0.73048 | [003,019] |
| NE | | 0.84634 | [326] |
| NE | | 0.86353 | [326] |
| NE | | 0.87717 | [326] |
| NE | | 0.88653 | [003,023] |
| NE+ | | 0.92674 | [326] |
| NE | | 0.89886 | [003,023] |
| NE | | 1.0295 | [003,023] |
| NE | | 1.0621 | [003,023] |
| NE | | 1.0798 | [024,023,004] |
| NE | | 1.0844 | [024,023,004] |
| NE | | 1.11441 | [024,023,112] |
| NE | | 1.1177 | [025,023,004] |
| NE | | 1.1390 | [024,023,004] |
| NE | | 1.1409 | [026,023,004] |
| NE | | 1.15235 | [025,023,112] |
| NE | | 1.15259014 | [003,155] |
| NE | | 1.1601 | [026,023,004] |
| NE | | 1.1614 | [025,023,004] |
| NE | | 1.17673 | [024,023,112] |
| NE | | 1.1789 | [003,023] |
| NE | | 1.1985 | [025,023,004] |
| NE | | 1.2066 | [025,023,028] |
| NE | | 1.2460 | [003,023] |
| NE | | 1.2689 | [003,029,030] |
| NE | | 1.2887 | [003] |
| NE | | 1.2912 | [003,029,023] |
| NE | ? | 1.4276 | [454] |
| NE | ? | 1.4304 | [454] |
| NE | ? | 1.4321 | [454] |
| NE | ? | 1.4330 | [454] |
| NE | ? | 1.4346 | [454] |
| NE | ? | 1.4368 | [454] |
| NE | | 1.484450 | [453] |
| NE | | 1.486926 | [453] |
| NE | | 1.487247 | [453] |
| NE | | 1.488759 | [453] |
| NE | | 1.489954 | [453] |
| NE | | 1.493623 | [453] |
| NE | | 1.5231 | [024,023,004] |
| NE | | 1.7162 | [003,023] |
| NE | | 1.8210 | [003,023] |

| | | WAVELENGTH IN AIR [MICROMETER] | |
|---|---|---|---|
| NE | | 1.8253 | [003,031] |
| NE | | 1.8276 | [003,031] |
| NE | | 1.8304 | [003,031] |
| NE | | 1.8403 | [003,031] |
| NE | | 1.8591 | [003,031] |
| NE | | 1.8597 | [003,029] |
| NE | | 1.9574 | [003,023] |
| NE | | 1.9577 | [003,023] |
| NE | | 2.0350 | [003,033,032] |
| NE | | 2.0353 | [003,023,033] |
| NE | | 2.1041 | [003,029,023] |
| NE | | 2.1708 | [003,029,023] |
| NE | | 2.3260 | [003,023] |
| NE | ? | 2.37 | [003,032] |
| NE | > | 2.3951 | [003,034,035] |
| NE | | 2.4219 | [003] |
| NE | | 2.4250 | [003,023] |
| NE | | 2.5393 | [003,033] |
| NE | | 2.5524 | [003,023] |
| NE | | 2.7574 | [003,033] |
| NE | | 2.7819 | [003,033] |
| NE | | 2.9448 | [003,033] |
| NE | | 2.9668 | [003,033] |
| NE | | 2.9805 | [003,033] |
| NE | | 3.0260 | [003,033] |
| NE | | 3.0268 | [003,033] |
| NE | | 3.3173 | [003,033] |
| NE | | 3.391 | [326] |
| NE | | 3.3333 | [003,033] |
| NE | | 3.3353 | [003,033] |
| NE | | 3.3500 | [006] |
| NE | | 3.3510 | [006] |
| NE | | 3.3804 | [003,033] |
| NE | | 3.3840 | [003,033] |
| NE | | 3.3903 | [003,033] |
| NE | > | 3.3913 | [003,026,033] |
| NE | | 3.4471 | [003,033] |
| NE | | 3.4489 | [003,033] |
| NE | | 3.475 | [326] |
| NE | | 3.4789 | [006] |
| NE | | 3.5835 | [003,033] |
| NE | | 3.6515 | [006] |
| NE | | 3.7736 | [003,033] |
| NE | | 3.9806 | [003,033] |
| NE | | 4.2171 | [006,313] |
| NE | | 5.1696 | [006] |
| NE | | 5.3243 | [006] |
| NE | | 5.3249 | [006] |
| NE | | 5.4033 | [003,033] |
| NE | | 5.405 | [326] |
| NE | | 5.6652 | [003,033] |
| NE | | 5.7053 | [006] |
| NE | | 5.7758 | [006] |
| NE | | 5.8844 | [006] |
| NE | | 5.9563 | [006] |
| NE | | 6.7769 | [006] |
| NE | | 6.8865 | [006] |

| | WAVELENGTH IN AIR [MICROMETER] | |
|---|---|---|
| NE | 6.9857 | [006] |
| NE | 7.3208 | [003,033] |
| NE | 7.4201 | [003,033] |
| NE | 7.4217 | [003,033] |
| NE | 7.4679 | [003,033] |
| NE | 7.4779 | [003,033] |
| NE | 7.4973 | [003,033] |
| NE | 7.5292 | [003] |
| NE | 7.5674 | [006] |
| NE | 7.5850 | [006] |
| NE | 7.6142 | [003,033] |
| NE | 7.6440 | [003] |
| NE | 7.6489 | [003,033] |
| NE | 7.6904 | [003] |
| NE | 7.6994 | [003,033] |
| NE | 7.7389 | [003,033] |
| NE | 7.7634 | [003,033] |
| NE | 7.7794 | [003,033] |
| NE | 7.8347 | [003,033] |
| NE | 7.8693 | [006] |
| NE | 7.9406 | [006] |
| NE | 7.9824 | [006] |
| NE | 8.0066 | [003,033] |
| NE | 8.0599 | [003,033] |
| NE | 8.1712 | [006] |
| NE | 8.3347 | [003,033] |
| NE | 8.3472 | [003,033] |
| NE | 8.8388 | [003,033] |
| NE | 8.8528 | [003,033] |
| NE | 9.0871 | [003,033] |
| NE | 10.060 | [003,033] |
| NE | 10.978 | [003,033] |
| NE | 11.857 | [003,033] |
| NE | 11.898 | [003,003] |
| NE | 12.831 | [003,033] |
| NE | 13.736 | [003] |
| NE | 13.756 | [003,033] |
| NE | 16.634 | [003,033] |
| NE | 16.664 | [003,033] |
| NE | 16.889 | [003,033] |

| | WAVELENGTH IN AIR [MICROMETER] | |
|---|---|---|
| NE | 16.943 | [003,033] |
| NE | 17.153 | [003,033] |
| NE | 17.184 | [003,033] |
| NE | 17.800 | [003,033] |
| NE | 17.837 | [003,033] |
| NE | 17.884 | [003,033] |
| NE | 18.392 | [003,033] |
| NE | 20.474 | [003,033] |
| NE | 21.746 | [003,033] |
| NE | 22.830 | [003,033] |
| NE | 25.416 | [003,033] |
| NE | 28.045 | [003,033] |
| NE | 31.544 | [003,033] |
| NE | 31.919 | [003,036] |
| NE | 32.007 | [003,033] |
| NE | 32.507 | [003,033] |
| NE | 33.815 | [003,033] |
| NE | 33.828 | [003,033] |
| NE | 34.543 | [003,033] |
| NE | 34.670 | [003,036] |
| NE | 35.592 | [003,036] |
| NE | 37.221 | [003,036] |
| NE | 41.730 | [003,036] |
| NE | 50.69 | [003,037] |
| NE | 52.40 | [003,037] |
| NE | 53.47 | [003,036] |
| NE | 54.00 | [003,036] |
| NE | 54.10 | [003,036] |
| NE | 55.51 | [003,037] |
| NE | 57.34 | [003,036] |
| NE | 68.31 | [003,038] |
| NE | 72.08 | [003,037] |
| NE | 85.01 | [003,038] |
| NE | 86.93 | [003,037] |
| NE | 88.47 | [003,037] |
| NE | 89.82 | [003,037] |
| NE | 106.0 | [003,037] |
| NE | 124.6 | [003,037] |
| NE | ? 126.1 | [003,037] |
| NE | ? 132.8 | [003,037] |

NUMBER OF LINES IN NEON 216

ACTIVE MEDIUM : ARGON
SYMBOL : AR

OPERATING CONDITIONS : NEARLY ALL LINES IN ARGON CAN BE GENERATED CONTINUOUSLY. TYPICAL CURRENT DENSITIES (THRESHOLD) WITH IONIZED AR 30-150 A/CM2. TYPICAL PRESSURE 0.01-0.6 TORR [106,107].
FOR THE EXCITATION OF UV LINES UP TO 0.33 MICROMETER, VERY HIGH CURRENT DENSITIES ARE NECESSARY (ONLY PULSED POSSIBLE).
THE STRONG LINES AT 351 AND 364 NM CAN BE GENERATED CONTINUOUSLY WITH 60 W OUTPUT POWER [457].
THE IR LINES (OVER 1.6 MICROMETER) ARE GENERATED IN A WEAK CONTINUOUS DISCHARGE AT 0.05 TORR. IN PULSED OPERATION, OUTPUT CAN BE INCREASED BY ADDITION OF SF6 [343]. SOME LINES CAN BE GENERATED IN TRANSVERSE PULSED DISCHARGE AT HIGH PRESSURE (30 TORR AR, 70 TORR HE [108], 17 ATM.[406]). RECENT REVIEW OF THE ARGON ION LASER [404].

| WAVELENGTH IN VACUUM [MICROMETER] | | | | WAVELENGTH IN VACUUM [MICROMETER] | | | |
|---|---|---|---|---|---|---|---|
| AR3+ | | 0.184340 | [427] | AR4+ | ? | 0.184343 | [427] |

| WAVELENGTH IN AIR [MICROMETER] | | | | WAVELENGTH IN AIR [MICROMETER] | | | |
|---|---|---|---|---|---|---|---|
| AR3+ | ? | 0.2113982 | [371] | AR+ | > | 0.476486 | [050,013,529] |
| AR3+ | ? | 0.2248840 | [371] | AR+ | > | 0.487986 | [043,050,049] |
| AR3+ | | 0.2513298 | [371] | AR+ | | 0.488903 | [013,040,048] |
| AR3+ | | 0.2621377 | [371] | AR+ | > | 0.496507 | [043,050,049] |
| AR3+ | | 0.2624882 | [012,371] | AR2+ | ? | 0.49928 | [040] |
| AR2+ | | 0.2753884 | [013,012,371] | AR+ | > | 0.501716 | [043,050,049] |
| AR2+ | | 0.2884216 | [012,014,371] | AR+ | | 0.506204 | [603] |
| AR2+ | | 0.2855374 | [371] | AR+ | | 0.514179 | [013,040,048] |
| AR3+ | | 0.2912924 | [013,012,371] | AR+ | > | 0.514532 | [043,050,049] |
| AR3+ | | 0.2926227 | [013,012,371] | AR+ | | 0.528690 | [050,013,049] |
| AR2+ | > | 0.3002642 | [013,371] | AR2+ | | 0.550220 | [013,040] |
| AR2+ | | 0.302405 | [013] | AR+ | | 0.64831 | [600] |
| AR2+ | | 0.305484 | [013] | AR+ | ? | 0.6730 | [168] |
| AR2+ | | 0.333613 | [013,012,014] | AR+ | | 0.68613 | [600] |
| AR2+ | | 0.334472 | [013,012,014] | AR | | 0.7503 | [013] |
| AR2+ | | 0.335849 | [013,012,014] | AR+ | | 0.877186 | [044,600] |
| AR2+ | > | 0.351112 | [013,039,047] | AR | | 0.912297 | [406,485] |
| AR2+ | | 0.351418 | [013,039,014] | AR | | 0.965778 | [406,485] |
| AR+ | | 0.357661 | [013] | AR | | 1.0470 | [406] |
| AR2+ | > | 0.363789 | [013,015,012] | AR+ | > | 1.092344 | [040,044,045] |
| AR2+ | ? | 0.37052 | [013] | AR | | 1.1448 | [406] |
| AR2+ | | 0.379532 | [013,014] | AR | | 1.21396 | [051] |
| AR2+ | | 0.385829 | [013] | AR | | 1.24028 | [051] |
| AR2+ | | 0.414671 | [013] | AR | | 1.27022 | [051,104] |
| AR2+ | | 0.418298 | [013,168] | AR | | 1.40948 | [112] |
| AR+ | | 0.437075 | [013,048,049] | AR | | 1.5046 | [570] |
| AR+ | | 0.448181 | [040,049] | AR | | 1.6180 | [003,004,005] |
| AR+ | | 0.454505 | [041,042,049] | AR | | 1.619395 | [003,004,112] |
| AR+ | > | 0.457935 | [041,042,049] | AR | | 1.652 | [343] |
| AR+ | | 0.460956 | [013] | AR | | 1.6941 | [003,004,005] |
| AR+ | | 0.465789 | [041,042,049] | AR | | 1.791437 | [003,004,485] |
| AR+ | | 0.472686 | [041,013,049] | AR | > | 2.0616 | [003,004,005] |

| | | WAVELENGTH IN AIR [MICROMETER] | | | | WAVELENGTH IN AIR [MICROMETER] | |
|---|---|---|---|---|---|---|---|
| AR | | 2.0986 | [003] | AR | | 3.708 | [324] |
| AR | | 2.1332 | [003] | AR | | 3.71439 | [323] |
| AR | | 2.1534 | [003,033] | AR | | 4.2033 | [046] |
| AR | | 2.2038 | [003,033] | AR | | 4.7138 | [046] |
| AR | | 2.2077 | [003,033,103] | AR | | 4.9146 | [003,033] |
| AR | | 2.31339 | [003,033,112] | AR | | 4.9199 | [046] |
| AR | | 2.3966 | [003,033,103] | AR | | 4.9496 | [003,033] |
| AR | | 2.5008 | [003,033] | AR | | 5.02338 | [323] |
| AR | | 2.5487 | [003,033] | AR | | 5.1203 | [003,033] |
| AR | | 2.5504 | [003,033] | AR | | 5.1205 | [003] |
| AR | > | 2.5627 | [003] | AR | | 5.3897 | [046] |
| AR | > | 2.5661 | [003,033,103] | AR | | 5.4666 | [003,033] |
| AR | | 2.6542 | [046] | AR | | 5.4680 | [003,033] |
| AR | | 2.6836 | [003,033] | AR | | 5.8022 | [046,103] |
| AR | | 2.71529 | [323] | AR | > | 5.8461 | [003,033] |
| AR | | 2.7357 | [003,033] | AR | | 6.0515 | [003,033] |
| AR | | 2.8195 | [003,033] | AR | > | 6.7443 | [046] |
| AR | | 2.8238 | [003,033] | AR | | 6.9410 | [003,033] |
| AR | | 2.862 | [103] | AR | | 6.9429 | [003,033] |
| AR | | 2.8776 | [003,033] | AR | > | 7.2147 | [003,033] |
| AR | | 2.8836 | [003,033] | AR | | 7.7982 | [003,033] |
| AR | | 2.9273 | [003,033] | AR | | 7.8002 | [003,033] |
| AR | | 2.9788 | [003,033] | AR | | 7.8042 | [003] |
| AR | | 3.0454 | [003,033] | AR | | 12.138 | [003,033] |
| AR | | 3.0988 | [003,033] | AR | | 12.188 | [003,033] |
| AR | > | 3.1325 | [003,033] | AR | | 15.032 | [003,033] |
| AR | | 3.1338 | [003,103] | AR | | 15.037 | [003,033] |
| AR | | 3.6312 | [323] | AR | | 26.937 | [003,033] |
| AR | | 3.70138 | [323] | AR | | 26.956 | [003,033] |

NUMBER OF LINES IN ARGON 124

---

ACTIVE MEDIUM : KRYPTON
SYMBOL : KR

OPERATING CONDITIONS : NEARLY ALL LINES IN KRYPTON CAN BE GENERATED CONTINUOUSLY. WITH IONIZED KRYPTON, CURRENT DENSITIES (THRESHOLD) OF 50-200 A/CM2 OR 7-10 KA/CM2 (VAC-UV) ARE NECESSARY. TYPICAL PRESSURE 0.01-0.25 TORR [109,049]. IN A TRANSVERSE PULSED DISCHARGE AT HIGH PRESSURE (50 TORR KR, 710 TORR HE), TWO INTENSE LINES IN THE NEAR INFRARED CAN BE GENERATED [108]. OPERATION UP TO 17 ATM. [406]. OPERATION IN HOLLOW CATHODE [541,600].

| | WAVELENGTH IN VACUUM [MICROMETER] | | | WAVELENGTH IN VACUUM [MICROMETER] | |
|---|---|---|---|---|---|
| KR3+ | 0.175641 | [427] | KR3+ | 0.195027 | [427] |
| KR4+ | 0.183243 | [427] | KR3+ | 0.196808 | [427] |

| | | WAVELENGTH IN AIR [MICROMETER] | |
|---|---|---|---|
| KR3+ | | 0.2051082 | [371] |
| KR3+ | > | 0.2191916 | [371] |
| KR3+ | | 0.2254638 | [371] |
| KR3+ | | 0.2338478 | [371] |
| KR3+ | ? | 0.2417843 | [371] |
| KR3+ | | 0.2649357 | [012,371] |
| KR3+ | | 0.2664398 | [012,371] |
| KR3+ | | 0.2741380 | [012,371] |
| KR2+ | | 0.3049704 | [013,371] |
| KR2+ | | 0.3124363 | [012,371] |
| KR2+ | | 0.3239512 | [013,012,371] |
| KR2+ | | 0.337496 | [013] |
| KR2+ | > | 0.3507420 | [012,047,371] |
| KR2+ | | 0.356423 | [012,371] |
| KR2+ | > | 0.406737 | [013,015] |
| KR2+ | > | 0.413133 | [013] |
| KR2+ | | 0.415444 | [013] |
| KR2+ | | 0.417179 | [013] |
| KR2+ | | 0.422658 | [013] |
| KR+ | | 0.431781 | [052,053,541] |
| KR+ | | 0.438654 | [052,053] |
| KR2+ | | 0.444329 | [013] |
| KR+ | | 0.457720 | [054,055,013] |
| KR+ | | 0.458285 | [052,053] |
| KR+ | | 0.461528 | [049] |
| KR+ | | 0.461915 | [055,013,049] |
| KR+ | | 0.463386 | [054,013,048] |
| KR+ | | 0.465016 | [013] |
| KR+ | > | 0.468041 | [055,013,049] |
| KR+ | | 0.469444 | [052,053,529] |
| KR+ | > | 0.476243 | [055,013,049] |
| KR+ | | 0.476573 | [055,013,049] |
| KR+ | > | 0.482517 | [055,013,049] |
| KR+ | | 0.484659 | [055,013,049] |
| KR2+ | | 0.501645 | [056,168] |
| KR+ | | 0.502240 | [050,048] |
| KR+ | | 0.512573 | [052,053] |
| KR+ | | 0.520831 | [040,055,049] |
| KR+ | | 0.530865 | [040,055,049] |
| KR+ | > | 0.568188 | [040,054,055] |
| KR+ | | 0.575298 | [050] |
| KR2+ | ? | 0.603717 | [057] |
| KR+ | ? | 0.60381 | [057] |
| KR+ | | 0.616880 | [048] |
| KR2+ | | 0.631022 | [057] |
| KR+ | ? | 0.631276 | [057,168] |
| KR+ | | 0.641661 | [058] |
| KR+ | | 0.647088 | [055,013,049] |
| KR+ | | 0.65100 | [600] |
| KR+ | | 0.657012 | [013,048,049] |
| KR+ | | 0.66029 | [057,168] |
| KR+ | | 0.676442 | [055,013,049] |
| KR+ | | 0.687084 | [013,048,049] |
| KR+ | > | 0.7435764 | [371] |
| KR | | 0.752546 | [105] |
| KR | | 0.7603 | [324] |
| KR+ | | 0.79314 | [049] |

| | | WAVELENGTH IN AIR [MICROMETER] | |
|---|---|---|---|
| KR+ | | 0.799322 | [013,044,049] |
| KR | > | 0.810433 | [059,112] |
| KR+ | ? | 0.828037 | [049,105] |
| KR+ | ? | 0.85878 | [044] |
| KR+ | | 0.86901 | [049] |
| KR | | 0.88058 | [600] |
| KR | | 0.8929 | [406] |
| KR | | 1.14582 | [112] |
| KR | | 1.31775 | [112] |
| KR | | 1.36225 | [112] |
| KR | | 1.44269 | [112] |
| KR | | 1.47648 | [112] |
| KR | | 1.4966 | [322] |
| KR | | 1.5330 | [322] |
| KR | | 1.68533 | [112] |
| KR | | 1.68965 | [003,004,112] |
| KR | | 1.6936 | [003,004,005] |
| KR | | 1.7843 | [003,004,005] |
| KR | | 1.8185 | [003,004,005] |
| KR | | 1.9211 | [003,004,005] |
| KR | | 2.04240 | [600] |
| KR | > | 2.1165 | [003,004,005] |
| KR | > | 2.19020 | [003,004,112] |
| KR | | 2.2475 | [323] |
| KR | | 2.4260 | [003,060] |
| KR | > | 2.52342 | [003,004,112] |
| KR | | 2.6260 | [003,033] |
| KR | | 2.6281 | [003,033] |
| KR | | 2.86134 | [003,033,112] |
| KR | | 2.8656 | [003,033] |
| KR | | 2.9836 | [003,033] |
| KR | | 2.9870 | [003,033] |
| KR | | 3.0664 | [003,033] |
| KR | | 3.0528 | [003,033] |
| KR | | 3.1508 | [003,033] |
| KR | | 3.3401 | [003,033] |
| KR | | 3.3411 | [003] |
| KR | | 3.4873 | [003,033] |
| KR | | 3.4885 | [003,033] |
| KR | | 3.774 | [324] |
| KR | | 3.956 | [324] |
| KR | | 3.9573 | [046] |
| KR | | 4.068 | [324] |
| KR | | 4.142 | [324] |
| KR | | 4.3736 | [003,033] |
| KR | | 4.3755 | [003] |
| KR | | 4.8760 | [003,033] |
| KR | | 4.8819 | [003,033] |
| KR | | 4.9963 | [046] |
| KR | | 4.9999 | [322] |
| KR | | 5.1298 | [046] |
| KR | | 5.2985 | [003,033] |
| KR | | 5.3004 | [003,033] |
| KR | ? | 5.5685 | [003,033] |
| KR | > | 5.5848 | [003,033] |
| KR | | 5.6290 | [003,033] |
| KR | | 7.0565 | [003,033] |

| WAVELENGTH IN AIR [MICROMETER] | | | WAVELENGTH IN AIR [MICROMETER] |
|---|---|---|---|
| KR | 7.3605 | [046] | |

NUMBER OF LINES IN KRYPTON 119

ACTIVE MEDIUM : XENON
SYMBOL : XE

OPERATING CONDITIONS : MANY LINES IN IONIZED XENON CAN BE EXCITED CONTINUOUSLY. TYPICAL CURRENT DENSITY 70-200 A/CM2.
WITH PULSED EXCITATION MANY LINES IN THE VISIBLE SPECTRAL REGION APPEAR SIMULTANOUSLY. TYPICAL PRESSURE 0.01-0.1 TORR.
IN ATOMIC XENON ALL LINES ABOVE 2.03 MICROMETER ARE EXCITED CONTINUOUSLY. PURE XE OR XE/HE MIXTURE. OTHERWISE TRANSVERSE PULSED DISCHARGE HIGH-PRESSURE XENON (UP TO 100 TORR) [116,126] OR 760 TORR AR-XE MIXTURE [389], OPERATION UP TO 17 ATM. [406].
OPERATION IN HOLLOW CATHODE [529,600].

| | | WAVELENGTH IN AIR [MICROMETER] | |
|---|---|---|---|
| XE3+ | ? | 0.2232442 | [371] |
| XE3+ | > | 0.2315357 | [371] |
| XE2+ | ? | 0.247718 | [012,167,318] |
| XE3+ | | 0.2526664 | [371] |
| XE2+ | | 0.2691939 | [167,318,371] |
| XE3+ | | 0.29837 | [013,318] |
| XE2+ | | 0.3079738 | [013,318,371] |
| XE3+ | > | 0.3246922 | [012,318,371] |
| XE3+ | | 0.3305957 | [318,371,522] |
| XE2+ | ? | 0.330599 | [013,012,318] |
| XE3+ | > | 0.3330869 | [013,318,371] |
| XE3+ | ? | 0.334974 | [012,167,318] |
| XE2+ | | 0.3454248 | [012,014,371] |
| XE3+ | | 0.348322 | [012,013,318] |
| XE2+ | | 0.359661 | [318,126] |
| XE3+ | > | 0.3645478 | [014,318,371] |
| XE2+ | | 0.366921 | [012,167,318] |
| XE2+ | | 0.374571 | [012] |
| XE3+ | | 0.375979 | [167,318] |
| XE2+ | | 0.3780990 | [013,014,371] |
| XE3+ | | 0.380322 | [012,167,318] |
| XE3+ | | 0.397302 | [012,167,318] |
| XE2+ | | 0.405005 | [057] |
| XE2+ | | 0.406048 | [013,014,371] |
| XE2+ | | 0.421401 | [013] |
| XE2+ | | 0.424024 | [013] |
| XE2+ | | 0.427259 | [013] |
| XE2+ | | 0.428588 | [013] |
| XE3+ | > | 0.430577 | [013,167,318] |
| XE2+ | | 0.441314 | [318,126] |
| XE2+ | | 0.443415 | [013,168] |
| XE3+ | | 0.455862 | [113,318] |
| XE+ | > | 0.460303 | [040,013,049] |
| XE3+ | | 0.464759 | [167,318] |
| XE3+ | | 0.465073 | [167,318] |
| XE2+ | | 0.467368 | [013,048,049] |
| XE2+ | | 0.468354 | [013] |
| XE2+ | | 0.472357 | [061,318] |
| XE2+ | | 0.474895 | [048,061,167] |
| XE+ | | 0.486249 | [062,053,529] |
| XE2+ | | 0.486946 | [013,048,049] |
| XE+ | | 0.488730 | [040] |
| XE3+ | > | 0.495414 | [013,167,318] |
| XE3+ | | 0.496508 | [040,013,318] |
| XE3+ | > | 0.500772 | [013,167,318] |
| XE+ | | 0.504492 | [040,013,049] |
| XE3+ | | 0.515703 | [113,318] |
| XE3+ | > | 0.515905 | [013,167,318] |
| XE2+ | | 0.523893 | [013,048] |
| XE+ | | 0.525992 | [040,013] |
| XE3+ | > | 0.526015 | [167,318] |
| XE3+ | | 0.525637 | [318,126] |
| XE+ | | 0.526042 | [040,013,048] |
| XE+ | | 0.526195 | [040,013,049] |
| XE+ | | 0.531387 | [062,053,529] |
| XE3+ | | 0.534131 | [318] |
| XE3+ | | 0.534331 | [113,318] |
| XE3+ | > | 0.535297 | [013,113,318] |
| XE3+ | > | 0.539461 | [013,113,318] |
| XE+ | ? | 0.539525 | [318] |
| XE2+ | | 0.540104 | [063] |
| XE+ | > | 0.541915 | [040,013,049] |
| XE3+ | | 0.549933 | [113,318] |
| XE2+ | ? | 0.552442 | [048] |

| | | WAVELENGTH IN AIR [MICROMETER] | | | | WAVELENGTH IN AIR [MICROMETER] | |
|---|---|---|---|---|---|---|---|
| XE3+ | | 0.55923 | [113,318] | XE | > | 2.65146 | [112,066,065] |
| XE+ | | 0.565938 | [040,318] | XE | | 2.6601 | [003] |
| XE+ | | 0.572691 | [040,053] | XE | | 2.6665 | [046] |
| XE+ | | 0.575103 | [040] | XE | | 2.8590 | [323] |
| XE+ | ? | 0.589330 | [167] | XE | | 3.1069 | [003,066,065] |
| XE3+ | | 0.595565 | [013,113,318] | XE | > | 3.2739 | [003,065] |
| XE+ | > | 0.597111 | [040,013,049] | XE | | 3.3085 | [046] |
| XE+ | | 0.609361 | [053] | XE | > | 3.3666 | [003,066,065] |
| XE2+ | | 0.61766 | [057,167,318] | XE | | 3.4014 | [046] |
| XE2+ | | 0.623824 | [057,168] | XE | | 3.4335 | [003,033] |
| XE+ | | 0.627081 | [040,013,049] | XE | > | 3.5070 | [116,067,065] |
| XE3+ | | 0.62865 | [318,126] | XE | | 3.6210 | [003] |
| XE2+ | | 0.63435 | [318] | XE | | 3.6509 | [003,033,116] |
| XE+ | | 0.652865 | [048,049] | XE | | 3.6788 | [003,004,066] |
| XE+ | | 0.66943 | [049] | XE | > | 3.6849 | [003,004,066] |
| XE3+ | | 0.669950 | [113,318] | XE | > | 3.8686 | [003] |
| XE+ | | 0.70723 | [049] | XE | > | 3.8940 | [003,004,066] |
| XE2+ | | 0.714894 | [061,049,318] | XE | > | 3.9955 | [003,004,066] |
| XE+ | | 0.76186 | [600] | XE | | 4.0196 | [432] |
| XE+ | > | 0.782763 | [044] | XE | | 4.1516 | [003] |
| XE+ | > | 0.798800 | [044] | XE | | 4.5381 | [046] |
| XE+ | | 0.823162 | [406,485] | XE | | 4.5665 | [046] |
| XE+ | | 0.833270 | [044] | XE | | 4.5694 | [046] |
| XE+ | | 0.840919 | [044,112] | XE | > | 4.6097 | [003] |
| XE2+ | | 0.85716 | [044,167,318] | XE | > | 5.0230 | [046] |
| XE+ | | 0.858251 | [044] | XE | | 5.02441 | [323] |
| XE+ | > | 0.871617 | [044,049] | XE | > | 5.3551 | [046] |
| XE | > | 0.904539 | [059,112] | XE | | 5.4735 | [046] |
| XE+ | > | 0.905930 | [044] | XE | > | 5.5739 | [003,004,066] |
| XE+ | | 0.926539 | [044] | XE | | 5.6019 | [046] |
| XE+ | | 0.928854 | [044] | XE | | 6.3103 | [046] |
| XE+ | > | 0.969859 | [044,114] | XE | | 6.3137 | [046] |
| XE | > | 0.979970 | [059] | XE | > | 7.3147 | [003,004,066] |
| XE+ | | 1.063385 | [044] | XE | | 7.4294 | [046] |
| XE+ | ? | 1.0950 | [044] | XE | | 9.0040 | [003,004,066] |
| XE | | 1.36562 | [112] | XE | | 9.7002 | [003,004,066] |
| XE | | 1.60519 | [112] | XE | | 11.289 | [003] |
| XE | | 1.73254 | [112] | XE | | 11.296 | [003,033] |
| XE | | 2.02623 | [112,064,065] | XE | | 12.263 | [003,004,066] |
| XE | | 2.3193 | [003,004,066] | XE | | 12.913 | [003,004,066] |
| XE | | 2.4825 | [003] | XE | | 18.500 | [003,033] |
| XE | | 2.51528 | [323] | XE | | 75.578 | [068] |
| XE | > | 2.6269 | [003,066,065] | | | | |

NUMBER OF LINES IN XENON 149

ACTIVE MEDIUM : IRON
SYMBOL : FE

OPERATING CONDITIONS : PULSED EXCITATION BY ELECTRICAL DISSOCIATION OF FE(CO)5.
PULSED EXCITATION IN IRON VAPOR WITH NE BUFFER GAS [560]. TEMPERATURE OF THE DISCHARGE TUBE 1680 C. ALSO PHOTO-DISSOCIATION OF FE(CO)5 BY KRF-LASER RADIATION [563].

| WAVELENGTH IN AIR [MICROMETER] | | | WAVELENGTH IN AIR [MICROMETER] | | |
|---|---|---|---|---|---|
| FE | 0.360 | [563] | FE | 0.556 | [563] |
| FE | 0.385 | [563] | FE | 0.563 | [563] |
| FE | 0.395 | [563] | FE | 6.8470 | [469] |
| FE | 0.4529 | [560] | FE | 8.4902 | [469] |
| FE | 0.540 | [563] | | | |

NUMBER OF LINES IN IRON 9

---

ACTIVE MEDIUM : NICKEL
SYMBOL : NI

OPERATING CONDITIONS : PULSED EXCITATION BY ELECTRICAL DISSOCIATION OF NI(CO)4 OR BY HOLLOW CATHODE DISCHARGES.

| WAVELENGTH IN AIR [MICROMETER] | | | WAVELENGTH IN AIR [MICROMETER] | | |
|---|---|---|---|---|---|
| NI+ | 0.79624 | [584] | NI | 1.3968 | [583] |
| NI+ | 0.79754 | [584] | NI | 1.4550 | [469] |

NUMBER OF LINES IN NICKEL 4

---

ACTIVE MEDIUM : HYDROGEN, ATOMIC
SYMBOL : H

OPERATING CONDITIONS : PULSED DISCHARGE IN 0.01 TORR H2 AND 3.5 TORR HE.

| WAVELENGTH IN AIR [MICROMETER] | | | WAVELENGTH IN AIR [MICROMETER] | | |
|---|---|---|---|---|---|
| H | 0.4340 | [010] | H | 1.8751 | [001] |
| H | 0.4861 | [010] | | | |

NUMBER OF LINES IN HYDROGEN, ATOMIC 3

---

ACTIVE MEDIUM : SODIUM
SYMBOL : NA

OPERATING CONDITIONS : PULSED DISCHARGE IN SODIUM VAPOR OF 0.001-0.003 TORR PRESSURE AND 1-10 TORR HE. EXCITATION BY PHOTODISSOCIATION OF NA-HALIDES WITH ARF-LASER [604].

| WAVELENGTH IN AIR [MICROMETER] | | | WAVELENGTH IN AIR [MICROMETER] | | |
|---|---|---|---|---|---|
| NA | 0.5866 | [604] | NA | 1.1382 | [283] |
| NA | 0.5890 | [604] | NA | 1.1404 | [283,604] |

NUMBER OF LINES IN SODIUM 4

ACTIVE MEDIUM : POTASSIUM
SYMBOL : K

OPERATING CONDITIONS : PULSED DISCHARGE IN POTASSIUM VAPOR OF 0.1 TORR PRESSURE AND 3-5 TORR H2. ALSO PHOTODISSOCIATION OF K2-VAPOR BY RUBY-LASER PULSES, AND OPTICAL PUMPING BY ND-YAG-SHG [410]. EXCITATION BY PHOTODISSOCIATION OF K-HALIDES WITH ARF-LASER [604].

| WAVELENGTH IN AIR [MICROMETER] | | | WAVELENGTH IN AIR [MICROMETER] | | |
|---|---|---|---|---|---|
| K | 0.4045 | [604] | K | 3.140 | [263] |
| K | 0.7665 | [604] | K | 3.15 | [604] |
| K | 0.7699 | [604] | K | 3.160 | [263] |
| K | 1.17 | [604] | K | 6.4 | [410] |
| K | 1.2434 | [283] | K | 7.9 | [410] |
| K | 1.2523 | [283,604] | K | 12.5 | [410] |
| K | 2.72 | [604] | K | 15.95 | [410] |

NUMBER OF LINES IN POTASSIUM 14

ACTIVE MEDIUM : RUBIDIUM
SYMBOL : RB

OPERATING CONDITIONS : PHOTODISSOCIATION OF RUBIDIUM VAPOR (RB2) BY RUBY-LASER PULSES. VAPOR GENERATED BY HEATING RUBIDIUM WITHIN A TUBULAR FURNACE TO 400 C [158]. ALSO OPTICAL PUMPING BY MEANS OF A DYE LASER [263] OR SHG-DYE LASER [568]. DYNAMIC STARK TUNING OF THE FIR-LINES. EXCITATION BY PHOTODISSOCIATION OF RB-HALIDES WITH ARF-LASER.

| WAVELENGTH IN AIR [MICROMETER] | | | WAVELENGTH IN AIR [MICROMETER] | | |
|---|---|---|---|---|---|
| RB | 0.4210 | [604] | RB | 1.53 | [604] |
| RB | 0.7619 | [604] | RB | 2.254 | [158,263] |
| RB | 0.7758 | [604] | RB | 2.293 | [158,263,604] |
| RB | 0.7800 | [604] | RB | 2.79 | [604] |
| RB | 0.7945 | [604] | RB | 49.68 | [568] |
| RB | 1.37 | [604] | RB | 50.93 | [568] |
| RB | 1.48 | [604] | | | |

NUMBER OF LINES IN RUBIDIUM 13

ACTIVE MEDIUM : CESIUM
SYMBOL : CS

OPERATING CONDITIONS : OPTICAL PUMPING OF CESIUM VAPOR ($CS_2$) BY MEANS OF THE HELIUM LINE AT 0.3880 MICROMETER. ALSO PHOTODISSOCIATION OF $CS_2$ VAPOR [158,263] AND CS-HALIDES [604] BY UV LASERS.

| | WAVELENGTH IN AIR [MICROMETER] | | | WAVELENGTH IN AIR [MICROMETER] | |
|---|---|---|---|---|---|
| CS | 0.4555 | [604] | CS | 2.95 | [604] |
| CS | 0.8521 | [604] | CS | 3.010 | [263,604] |
| CS | 0.8764 | [604] | CS | 3.095 | [158,263] |
| CS | 0.8943 | [604] | CS | 3.2040 | [004,200] |
| CS | 1.01 | [604] | CS | 3.489 | [263] |
| CS | 1.360 | [263,604] | CS | 3.613 | [263] |
| CS | 1.376 | [263] | CS | 4.22 | [604] |
| CS | 1.47 | [604] | CS | 7.1821 | [004,200] |

NUMBER OF LINES IN CESIUM 16

---

ACTIVE MEDIUM : COPPER
SYMBOL : CU

OPERATING CONDITIONS : PULSED DISCHARGE IN COPPER VAPOR AND HE. VAPOR GENERATED BY HEATING COPPER WITHIN THE DISCHARGE TUBE WITH A TUBULAR FURNACE TO 1420 C. EFFICIENCY 1.2% [157]. AT 18 KHZ REPETITION FREQUENCY A MEDIUM POWER OF 15 WATT HAS BEEN OBSERVED [311]. WHEN COPPER IODIDE IS USED, THE TEMPERATURE CAN BE REDUCED TO 600 C [328]. LOW TEMP. OP. IN COPPER ACETYLACETONATE AND COPPER NITRATE [486]. COPPER VAPOR CAN ALSO BE GENERATED BY DISCHARGE HEATING IN THE CARRIER GAS [345], ESPECIALLY IN COPPER HALIDES [374,496,562]. CW OPERATION IN A HOLLOW CATHODE MADE FROM COPPER. CARRIER-GAS MIXTURES: HE+AR, HE+NE OR HE+XE [362,363,503,601].

| | WAVELENGTH IN AIR [MICROMETER] | | | | WAVELENGTH IN AIR [MICROMETER] | |
|---|---|---|---|---|---|---|
| CU+ | 0.27032 | [497] | CU+ | | 0.506064 | [362] |
| CU+ | 0.24858 | [363] | CU | > | 0.510554 | [124,123] |
| CU+ | 0.25063 | [363] | CU | | 0.5700 | [328] |
| CU+ | 0.25905 | [363,497] | CU | > | 0.578213 | [124,123] |
| CU+ | 0.25988 | [363,497] | CU+ | | 0.72558 | [600] |
| CU+ | 0.450600 | [362] | CU+ | | 0.73999 | [600] |
| CU+ | 0.455592 | [362] | CU+ | | 0.740434 | [362,503] |
| CU+ | 0.467356 | [362] | CU+ | | 0.74382 | [600] |
| CU+ | 0.468199 | [362] | CU+ | | 0.766470 | [362,503] |
| CU+ | 0.485497 | [362] | CU+ | | 0.773868 | [362,503] |
| CU+ | 0.490973 | [362] | CU+ | | 0.777874 | [362] |
| CU+ | 0.493165 | [362] | CU+ | | 0.780519 | [362] |
| CU+ | 0.501261 | [362] | CU+ | | 0.780766 | [362,503,537] |
| CU+ | 0.502129 | [362] | CU+ | | 0.782566 | [362,503] |
| CU+ | 0.505178 | [362] | CU+ | | 0.784503 | [362] |

| WAVELENGTH IN AIR [MICROMETER] | | | | WAVELENGTH IN AIR [MICROMETER] | | | |
|---|---|---|---|---|---|---|---|
| CU+ | | 0.789583 | [362,503] | CU+ | | 1.7708 | [570] |
| CU+ | | 0.794442 | [362,569] | CU+ | ? | 1.8004 | [570] |
| CU+ | | 0.790257 | [362,569] | CU | | 1.8196 | [570] |
| CU+ | | 0.798817 | [362] | CU | | 1.8228 | [570] |
| CU+ | | 0.808858 | [569] | CU+ | ? | 1.8448 | [570] |
| CU+ | | 0.8096 | [581] | CU+ | | 1.9154 | [570] |
| CU+ | | 0.819228 | [569] | CU+ | ? | 1.9328 | [570] |
| CU+ | | 0.8277 | [570] | CU+ | | 1.9479 | [570] |
| CU+ | | 0.828321 | [569] | CU+ | | 1.9712 | [570] |
| CU+ | | 0.851104 | [569] | CU+ | | 2.0006 | [570] |
| CU+ | ? | 1.7438 | [570] | | | | |

NUMBER OF LINES IN COPPER 51

ACTIVE MEDIUM : SILVER
SYMBOL : AG

OPERATING CONDITIONS : CONTINUOUS OPERATION IN A SILVER HOLLOW CATHODE WITH HE OR NE AS BUFFER GAS. MULTILINE OUTPUT 1 W [537].

| WAVELENGTH IN AIR [MICROMETER] | | | | WAVELENGTH IN AIR [MICROMETER] | | | |
|---|---|---|---|---|---|---|---|
| AG+ | | 0.22434 | [409,507] | AG+ | | 1.6462 | [570] |
| AG+ | | 0.22774 | [409,507] | AG+ | ? | 1.6656 | [570] |
| AG+ | | 0.31807 | [409,507] | AG+ | | 1.7203 | [570] |
| AG+ | | 0.40859 | [424,507] | AG+ | | 1.7345 | [570] |
| AG+ | | 0.47884 | [424,507] | AG+ | | 1.7478 | [570] |
| AG+ | | 0.50273 | [424,507] | AG+ | ? | 1.7674 | [570] |
| AG+ | | 0.64027 | [442,507] | AG | | 1.8380 | [570] |
| AG+ | | 0.80054 | [442,507] | AG+ | | 1.8408 | [570] |
| AG+ | | 0.82547 | [442,507,569] | AG+ | | 1.8463 | [570] |
| AG+ | | 0.8263 | [570] | AG+ | | 1.8725 | [570] |
| AG+ | | 0.83244 | [442,507,569] | AG+ | | 1.8795 | [570] |
| AG+ | | 0.83795 | [569,570] | AG+ | | 1.8979 | [570] |
| AG+ | | 0.84032 | [442,507,569] | AG | | 1.9370 | [570] |
| AG+ | | 0.87476 | [569,570] | AG+ | | 1.9714 | [570] |
| AG+ | | 0.8772 | [570] | AG+ | | 1.9823 | [570] |
| AG+ | | 1.3759 | [570] | AG+ | | 2.0796 | [570] |
| AG+ | | 1.5982 | [570] | | | | |

NUMBER OF LINES IN SILVER 33

ACTIVE MEDIUM : GOLD
SYMBOL : AU

OPERATING CONDITIONS : PULSED DISCHARGE IN GOLD VAPOR WITH HE. VAPOR GENERATED BY HEATING GOLD WITHIN THE DISCHARGE TUBE WITH A TUBULAR FURNACE TO 1500 C. ALSO CONTINUOUS EXCITATION IN A GOLD-PLATED HOLLOW CATHODE WITH HELIUM AS BUFFER GAS [462].

| | | WAVELENGTH IN AIR [MICROMETER] | | | | WAVELENGTH IN AIR [MICROMETER] | |
|---|---|---|---|---|---|---|---|
| AU | | 0.2428 | [476] | AU+ | ? | 0.55221 | [462] |
| AU+ | | 0.25337 | [462] | AU+ | ? | 0.62123 | [462] |
| AU+ | | 0.26165 | [462] | AU | | 0.627818 | [284,285,476] |
| AU | | 0.2676 | [476] | AU+ | ? | 0.67014 | [462] |
| AU+ | | 0.28225 | [462] | AU+ | ? | 0.69029 | [462] |
| AU+ | | 0.28470 | [462] | AU+ | ? | 0.69403 | [462] |
| AU+ | ? | 0.28633 | [462] | AU+ | ? | 0.75558 | [462] |
| AU+ | ? | 0.28882 | [462] | AU+ | ? | 0.75929 | [462] |
| AU+ | | 0.28933 | [462] | AU+ | | 0.76005 | [462] |
| AU+ | | 0.29182 | [462] | AU+ | ? | 0.76067 | [462] |
| AU+ | ? | 0.29594 | [462] | AU+ | ? | 0.76351 | [462] |
| AU | | 0.3122 | [476,554] | AU+ | | 0.82729 | [569] |
| AU+ | ? | 0.55163 | [462] | AU+ | | 0.88676 | [569] |

NUMBER OF LINES IN GOLD 26

---

ACTIVE MEDIUM : BERYLLIUM
SYMBOL : BE

OPERATING CONDITIONS : PULSED EXCITATION IN BERYLLIUM VAPOR WITH HE OR NE AS BUFFER GAS. VAPOR GENERATED BY HEATING BERYLLIUM WITHIN THE DISCHARGE TUBE MADE OF BERYLLIUM OXIDE.

| | WAVELENGTH IN AIR [MICROMETER] | | | WAVELENGTH IN AIR [MICROMETER] | |
|---|---|---|---|---|---|
| BE | 0.4675 | [403] | BE | 1.2096 | [403] |
| BE | 0.5272 | [403] | | | |

NUMBER OF LINES IN BERYLLIUM 3

---

ACTIVE MEDIUM : MAGNESIUM
SYMBOL : MG

OPERATING CONDITIONS : PULSED AND CONTINUOUS DISCHARGE IN MAGNESIUM VAPOR WITH HE, NE OR AR. THE DISCHARGE TUBE CONTAINS MAGNESIUM AND IS HEATED TO 450 C BY A HEATER WINDING.

| | WAVELENGTH IN AIR [MICROMETER] | | | | WAVELENGTH IN AIR [MICROMETER] | |
|---|---|---|---|---|---|---|
| MG | 0.9218 | [299] | MG | | 3.67794 | [300] |
| MG | 0.9244 | [299] | MG | | 3.68154 | [300] |
| MG | 1.0952 | [299] | MG | ? | 3.86573 | [300] |
| MG | 1.0915 | [299] | MG | | 4.20018 | [300] |
| MG+ | 2.40415 | [300] | MG | | 4.36269 | [300] |
| MG+ | 2.41245 | [300] | | | | |

NUMBER OF LINES IN MAGNESIUM 11

ACTIVE MEDIUM : CALCIUM
SYMBOL : CA

OPERATING CONDITIONS : PULSED DISCHARGE IN CALCIUM VAPOR AND HE. CALCIUM VAPOR GENERATED BY HEATING CALCIUM WITHIN THE DISCHARGE TUBE WITH A TUBULAR FURNACE TO 500-700 C. SUPERRADIANCE. ALSO HOLLOW-CATHODE OPERATION [360]. CW OPERATION IN CA-H2 WITH ZEEMAN TUNABILITY [513].OPTICAL PUMPING OF CA VAPOR WITH INERT GASES BY KRF-LASER [561] OR SHG-DYE LASER [606].

| | WAVELENGTH IN AIR [MICROMETER] | | | | WAVELENGTH IN AIR [MICROMETER] | |
|---|---|---|---|---|---|---|
| CA | 0.535 | [561] | CA | | 0.6717 | [606] |
| CA | 0.586 | [561] | CA+ | > | 0.854209 | [123] |
| CA | 0.6102 | [606] | CA+ | > | 0.866214 | [123] |
| CA | 0.6122 | [606] | CA | | 1.9853 | [606] |
| CA | 0.6162 | [606] | CA | | 5.5457 | [152,513] |
| CA | 0.644981 | [360] | | | | |

NUMBER OF LINES IN CALCIUM 11

ACTIVE MEDIUM : STRONTIUM
SYMBOL : SR

OPERATING CONDITIONS : PULSED DISCHARGE IN STRONTIUM VAPOR AND HE. STRONTIUM VAPOR GENERATED BY HEATING STRONTIUM WITHIN THE DISCHARGE TUBE WITH A TUBULAR FURNACE TO 460 C. SUPERRADIANCE. ALSO HOLLOW-CATHODE OPERATION [360].

| | WAVELENGTH IN AIR [MICROMETER] | | | WAVELENGTH IN AIR [MICROMETER] | |
|---|---|---|---|---|---|
| SR | 0.638075 | [360] | SR | 3.0111 | [540] |
| SR+ | 1.033014 | [152,540] | SR | 6.4567 | [152,540] |
| SR+ | 1.091797 | [152,540] | | | |

NUMBER OF LINES IN STRONTIUM 5

ACTIVE MEDIUM : BARIUM
SYMBOL : BA

OPERATING CONDITIONS : PULSED DISCHARGE IN BARIUM VAPOR WITH HE, NE, AR OR H2. BARIUM VAPOR GENERATED BY HEATNG BARIUM WITHIN THE DISCHARGE TUBE WITH A TUBULAR FURNACE TO 500-850 C. SOME LINES SUPERRADIANT. ALSO HOLLOW-CATHODE OPERATION [360,361].

| WAVELENGTH IN AIR [MICROMETER] | | | | WAVELENGTH IN AIR [MICROMETER] | | | |
|---|---|---|---|---|---|---|---|
| BA+ | | 0.614172 | [360,361,386] | BA | | 2.9227 | [153,386] |
| BA+ | | 0.649690 | [360,361,386] | BA | > | 3.9578 | [153] |
| BA | | 0.712033 | [360,361] | BA | | 4.0069 | [153] |
| BA | > | 1.1303 | [153,386] | BA | | 4.33 | [386] |
| BA | > | 1.5000 | [153,386] | BA | ? | 4.6706 | [153,386] |
| BA | | 1.82 | [386] | BA | | 4.7156 | [153] |
| BA | | 1.9017 | [153] | BA | | 4.7171 | [153,386] |
| BA | | 2.1568 | [153,386] | BA | | 5.0309 | [153] |
| BA | | 2.3254 | [153,386] | BA | ? | 5.4798 | [153] |
| BA | | 2.4758 | [153] | BA | ? | 5.5636 | [153] |
| BA | | 2.5515 | [153,386] | BA | ? | 5.8899 | [153,386] |
| BA | | 2.5924 | [153] | BA | ? | 6.4546 | [153] |
| BA | | 2.9057 | [153] | | | | |

NUMBER OF LINES IN BARIUM 25

ACTIVE MEDIUM : ZINC
SYMBOL : ZN

OPERATING CONDITIONS : PULSED AND CONTINUOUS DISCHARGE IN ZINC VAPOR WITH HE, NE OR AR. ZINC IS HEATED IN THE DISCHARGE TUBE OR IN A SIDEARM TO 300-400 C BY A HEATER WINDING. ALSO CONTINUOUS DISCHARGE WITHIN A ZINC HOLLOW CATHODE [144,147,330] OR WITH HF-DISCHARGE [470]. PULSED EXCITATION BY ELECTRICAL DISSOCIATION OF ZN(CH3)2 [469].
CW OPERATION IN HE-ZN-HALIDE DISCHARGES [508].

| WAVELENGTH IN AIR [MICROMETER] | | | | WAVELENGTH IN AIR [MICROMETER] | | | |
|---|---|---|---|---|---|---|---|
| ZN+ | | 0.49116 | [141] | ZN+ | | 0.758848 | [092,142] |
| ZN+ | > | 0.492404 | [096,140,146] | ZN+ | | 0.761290 | [092] |
| ZN+ | > | 0.58944 | [140,141,143] | ZN+ | | 0.77325 | [330] |
| ZN+ | | 0.6021 | [144] | ZN+ | | 0.775786 | [096] |
| ZN+ | > | 0.610253 | [092,143] | ZN+ | | 1.8308 | [469] |
| ZN+ | | 0.747879 | [092,140,143] | ZN+ | | 5.0848 | [469] |

NUMBER OF LINES IN ZINC 12

ACTIVE MEDIUM : CADMIUM
SYMBOL : CD

OPERATING CONDITIONS : PULSED OR CONTINUOUS DISCHARGE IN CADMIUM VAPOR WITH HE OR NE. CADMIUM IS HEATED IN THE DISCHARGE TUBE OR IN A SIDEARM TO 200-320 C BY A HEATER WINDING. IF THE CADMIUM IS PLACED IN THE TUBE, THE ENERGY SUPPLIED BY THE DISCHARGE CURRENT IS SUFFICIENT TO GENERATE THE APPROPRIATE VAPOR PRESSURE [148].
CONTINUOUS DISCHARGE WITHIN A CADMIUM HOLLOW CATHODE [144,147].
CW OPERATION IN HE-CD-HALIDE DISCHARGES [508].
OTHER LITERATURE ON THE OPERATION OF CADMIUM LASERS [294,295,296,297,438]. PULSED EXCITATION BY ELECTRICAL DISSOCIATION OF CD(CH3)2 [469]. RECOMBINATION EXCITATION IN LASER-PRODUCED PLASMAS [594].

| | | WAVELENGTH IN AIR [MICROMETER] | | | | WAVELENGTH IN AIR [MICROMETER] | |
|---|---|---|---|---|---|---|---|
| CD+ | > | 0.3250 | [148,149] | CD+ | | 0.80669 | [141,142] |
| CD+ | > | 0.441563 | [092,145,148] | CD+ | | 0.85300 | [141] |
| CD+ | | 0.48820 | [146] | CD+ | | 0.88778 | [141] |
| CD+ | | 0.50259 | [146] | CD | | 1.40 | [594] |
| CD+ | | 0.533749 | [096,142,147] | CD | | 1.43 | [594] |
| CD+ | | 0.537804 | [096,142,147] | CD | | 1.45 | [594] |
| CD+ | | 0.63548 | [141,147] | CD | | 1.64 | [594] |
| CD+ | | 0.63601 | [141,147] | CD+ | | 3.2862 | [469] |
| CD+ | | 0.72369 | [141,147] | CD | | 13.185 | [469] |
| CD+ | ? | 0.78443 | [141,147] | CD | | 14.578 | [469] |

NUMBER OF LINES IN CADMIUM 20

---

ACTIVE MEDIUM : MERCURY
SYMBOL : HG

OPERATING CONDITIONS : PULSED DISCHARGE IN 0.001 TORR MERCURY VAPOR AND 1 TORR HE.
ALSO HOLLOW-CATHODE DISCHARGE [150,366]. ATOMIC LINES IN PULSED, PARTLY IN CONTINUOUS [122] OPERATION: HG/HE OR HG/AR MIXTURES.
OPTICAL PUMPING WITH $C_W$ MERCURY LAMP [364,398]. ALSO PULSED EXCITATION BY ELECTRICAL DISSOCIATION OF HG(CH3)2 [394].
CONTINUOUS EXCITATION WITH HF-DISCHARGE [470].

| | | WAVELENGTH IN AIR [MICROMETER] | | | | WAVELENGTH IN AIR [MICROMETER] | |
|---|---|---|---|---|---|---|---|
| HG | | 0.365 | [528] | HG+ | | 0.74181 | [151] |
| HG2+ | | 0.479701 | [040,097,431] | HG+ | | 0.79447 | [150,151,366] |
| HG2+ | | 0.5210 | [431] | HG+ | | 0.85498 | [100] |
| HG | | 0.5461 | [364,528,545] | HG+ | | 0.8622 | [100] |
| HG+ | > | 0.56773 | [098,099,100] | HG+ | ? | 0.8677 | [100] |
| HG+ | > | 0.61499 | [098,099,100] | HG+ | | 0.93968 | [100] |
| HG2+ | | 0.65015 | [431] | HG+ | | 1.0583 | [098,100] |
| HG+ | ? | 0.7065 | [003] | HG | | 1.11768 | [120] |
| HG+ | | 0.73466 | [098,100] | HG | ? | 1.2222 | [003] |

| WAVELENGTH IN AIR [MICROMETER] | | | | WAVELENGTH IN AIR [MICROMETER] | | |
|---|---|---|---|---|---|---|
| HG | ? | 1.2246 | [003] | HG | 1.6942 | [100,120] |
| HG | ? | 1.2545 | [100] | HG | 1.7073 | [120] |
| HG | ? | 1.2760 | [003] | HG | 1.71099 | [100,120] |
| HG | ? | 1.2981 | [100] | HG | 1.7329 | [120] |
| HG | ? | 1.3655 | [100] | HG | 1.8130 | [122,100] |
| HG | | 1.3675 | [120] | HG | 3.93 | [120] |
| HG | | 1.5295 | [121,100,120] | HG | 5.88 | [120,394] |
| HG+ | | 1.5555 | [100] | HG | 5.9817 | [394] |
| HG | | 1.6920 | [100,120] | HG | 6.49 | [120] |

NUMBER OF LINES IN MERCURY 36

ACTIVE MEDIUM : BORON
SYMBOL : B

OPERATING CONDITIONS : PULSED DISCHARGE IN BCL3 OF 0,5 TORR PRESSURE.

| WAVELENGTH IN AIR [MICROMETER] | | | WAVELENGTH IN AIR [MICROMETER] |
|---|---|---|---|
| B+ | 0.345134 | [093] | |

NUMBER OF LINES IN BORON 1

ACTIVE MEDIUM : ALUMINUM
SYMBOL : AL

OPERATING CONDITIONS : EXCITATION BY ELECTRICAL DISCHARGE WITHIN A ALUMINUM-HOLLOW CATHODE.
BUFFER GAS IS HELIUM OR NEON.

| WAVELENGTH IN AIR [MICROMETER] | | | WAVELENGTH IN AIR [MICROMETER] | | |
|---|---|---|---|---|---|
| AL+ | 0.358744 | [447,471] | AL+ | 0.705656 | [447,471] |
| AL+ | 0.691996 | [447,471] | AL+ | 0.747137 | [447] |
| AL+ | 0.704206 | [447,471] | | | |

NUMBER OF LINES IN ALUMINUM 5

ACTIVE MEDIUM : GALLIUM
SYMBOL : GA

OPERATING CONDITIONS : PULSED EXCITATION BY ELECTRICAL DISSOCIATION OF GA(CH3)3 OR PHOTODISSOCIATION OF GA-IODIDE WITH ARF-LASER.
WITH ARF-LASER.

| | WAVELENGTH IN AIR [MICROMETER] | | | WAVELENGTH IN AIR [MICROMETER] | |
|---|---|---|---|---|---|
| GA | 0.4172 | [585] | GA | 5.7534 | [394] |
| GA | 1.7363 | [394] | GA | 6.1460 | [394] |

NUMBER OF LINES IN GALLIUM 4

ACTIVE MEDIUM : INDIUM
SYMBOL : IN

OPERATING CONDITIONS : PULSED DISCHARGE IN INDIUM VAPOR WITH HE OR NE. THE DISCHARGE TUBE CONTAINS INDIUM AND IS HEATED TO 200-320 C BY MEANS OF A HEATER WINDING. ALSO PULSED EXCITATION BY ELECTRICAL DISSOCIATION OF IN(CH3)3 [394] OR PHOTODISSOCIATION OF INI VAPOR BY ARF-LASER RADIATION [467].

| | WAVELENGTH IN AIR [MICROMETER] | | | WAVELENGTH IN AIR [MICROMETER] | |
|---|---|---|---|---|---|
| IN | 0.4511 | [467] | IN | 1.8732 | [394] |
| IN+ > | 0.468082 | [092] | IN | 2.3779 | [394] |

NUMBER OF LINES IN INDIUM 4

ACTIVE MEDIUM : THALLIUM
SYMBOL : TL

OPERATING CONDITIONS : EXCITATION WITH SHORT RISE-TIME PULSES IN 0.01 TORR OF THALLIUM WITH NE OR HE AT SEVERAL TORR. SUPERRADIANCE. ALSO PULSED EXCITATION BY ELECTRICAL DISSOCIATION OF TL(CH3)3 [394]. CONTINUOUS EXCITATION BY HF-DISCHARGE [470] OR HOLLOW CATHODE DISCHARGE [582].

| | WAVELENGTH IN AIR [MICROMETER] | | | WAVELENGTH IN AIR [MICROMETER] | |
|---|---|---|---|---|---|
| TL | 0.5152 | [470] | TL | 3.8125 | [394] |
| TL | 0.53503 | [005,156] | TL | 5.1059 | [394] |
| TL | 0.5949 | [470,582] | TL | 10.449 | [394] |
| TL | 0.6950 | [470] | | | |

NUMBER OF LINES IN THALLIUM 7

ACTIVE MEDIUM : CARBON
SYMBOL : C

OPERATING CONDITIONS : LASER EMISSION FROM IONIZED CARBON IN A PULSED DISCHARGE IN CO2 OR AIR. THE ATOMIC LINES ARE GENERATED IN CONTINUOUS RF DISCHARGE IN 0.01 TORR CO OR CO2 AND 2 TORR HE OR NE. DIRECT NUCLEAR PUMPING [501].

| | WAVELENGTH IN AIR [MICROMETER] | | | WAVELENGTH IN AIR [MICROMETER] | |
|---|---|---|---|---|---|
| C3+ | 0.15482 | [293] | C+ | 0.67838 | [329] |
| C3+ | 0.15508 | [293] | C | 1.0691 | [003,088] |
| C2+ | 0.464745 | [085,013] | C | 1.4540 | [003,088,501] |
| C2+ | 0.465016 | [085] | C | 2.0645 | [003] |
| C+ ? | 0.49541 | [013] | C | 3.4046 | [003] |
| C+ | 0.51457 | [329] | C | 3.5155 | [003] |
| C+ | 0.65780 | [329] | C | 5.5956 | [003] |

NUMBER OF LINES IN CARBON 14

---

ACTIVE MEDIUM : SILICON
SYMBOL : SI

OPERATING CONDITIONS : LASER EMISSION FROM IONIZED SILICON IN A PULSED DISCHARGE IN PF6. SILICON VAPOR IS GENERATED BY THE REACTION OF FLUORINE WITH THE WALLS OF THE DISCHARGE VESSEL.
ATOMIC LINES: PULSED DISCHARGE IN SICL4, IN SOME CASES WITH NEON.

| | WAVELENGTH IN AIR [MICROMETER] | | | WAVELENGTH IN AIR [MICROMETER] | |
|---|---|---|---|---|---|
| SI2+ | 0.455259 | [069,070] | SI+ | 0.667193 | [069,070] |
| SI2+ | 0.456784 | [069,070] | SI | 1.1984 | [091] |
| SI+ | 0.634724 | [069,070] | SI | 1.2034 | [091] |
| SI+ | 0.637148 | [069,070] | SI | 1.5883 | [091] |

NUMBER OF LINES IN SILICON 8

---

ACTIVE MEDIUM : GERMANIUM
SYMBOL : GE

OPERATING CONDITIONS : PULSED DISCHARGE IN GERMANIUM VAPOR WITH HE OR NE. GERMANIUM VAPOR IS GENERATED BY PLACING GE IN THE DISCHARGE TUBE HEATED WITH A TUBULAR FURNACE TO 900-1000 C. ALSO PULSED EXCITATION BY ELECTRICAL DISSOCIATION OF GECL4 [394].

| WAVELENGTH IN AIR [MICROMETER] | | | | WAVELENGTH IN AIR [MICROMETER] | |
|---|---|---|---|---|---|
| GE+ | 0.513175 | [092] | GE | 1.9809 | [394] |
| GE+ > | 0.517865 | [092] | GE | 2.0200 | [394] |

NUMBER OF LINES IN GERMANIUM 4

---

ACTIVE MEDIUM : TIN
SYMBOL : SN

OPERATING CONDITIONS : PULSED DISCHARGE IN TIN VAPOR WITH NE OR HE. THE DISCHARGE TUBE CONTAINS TIN AND IS HEATED TO 1000 C BY MEANS OF A TUBULAR FURNACE. ALSO PULSED DISCHARGE IN SNCL4 VAPOR.
CONTINUOUS WAVE OPERATION IN HE/NE MIXTURE AT SN PRESSURE OF 1.5 MILLITORR [139].

| WAVELENGTH IN AIR [MICROMETER] | | | | WAVELENGTH IN AIR [MICROMETER] | |
|---|---|---|---|---|---|
| SN+ | 0.5589 | [244] | SN | 1.061 | [407] |
| SN+ > | 0.579918 | [092,244] | SN+ | 1.062 | [244] |
| SN+ > | 0.645350 | [092,139] | SN+ | 1.074 | [244] |
| SN+ | 0.684405 | [092,139] | SN | 4.6146 | [469] |
| SN | 0.6579 | [093,094,244] | | | |

NUMBER OF LINES IN TIN 9

---

ACTIVE MEDIUM : LEAD
SYMBOL : PB

OPERATING CONDITIONS : PULSED DISCHARGE IN LEAD VAPOR WITH HE, NE OR AR. THE DISCHARGE TUBE CONTAINS LEAD AND IS HEATED TO 900 C BY MEANS OF A TUBULAR FURNACE. DISCHARGE HEATING OF LEAD [433].
AT THE 0.72 MICROMETER LINE A GAIN OF 600 DB/M HAS BEEN MEASURED [137]. ALL LINES IN ATOMIC LEAD ARE SUPERRADIANT.
ALSO PULSED EXCITATION BY ELECTRICAL DISSOCIATION OF PB(CH3)4 [394].

| WAVELENGTH IN AIR [MICROMETER] | | | | WAVELENGTH IN AIR [MICROMETER] | |
|---|---|---|---|---|---|
| PB | 0.363954 | [138] | PB | 0.72291 | [095,546] |
| PB | 0.405779 | [138,433] | PB | 3.1738 | [394] |
| PB | 0.4062 | [138,433] | PB | 7.1740 | [394] |
| PB+ > | 0.53721 | [092] | PB | 7.9399 | [394] |

NUMBER OF LINES IN LEAD 8

---

ACTIVE MEDIUM : NITROGEN, ATOMIC
SYMBOL : N

OPERATING CONDITIONS : PULSED DISCHARGE IN AIR, N2, OR NH3 AT PRESSURES OF 0.001-0.1 TORR FOR EXCITATION OF IONIZED NITROGEN.
GENERATION OF THE ATOMIC LINES IN A PULSED DISCHARGE IN PURE N2 OR N2/HE. ALSO PULSED OR CONTINUOUS DISCHARGE IN MIXTURES OF NO, N2O/HE, NE.

| | | WAVELENGTH IN AIR [MICROMETER] | | | | WAVELENGTH IN AIR [MICROMETER] | |
|---|---|---|---|---|---|---|---|
| N2+ | | 0.336734 | [012] | N | | 0.86284 | [086,536] |
| N3+ | | 0.347867 | [085,012] | N | | 0.90455 | [536] |
| N3+ | | 0.348296 | [012] | N | | 0.93662 | [086] |
| N+ | | 0.399501 | [090] | N | | 0.93921 | [086] |
| N2+ | | 0.409732 | [085] | N | | 1.0568 | [430] |
| N2+ | | 0.410338 | [085] | N | | 1.0611 | [430] |
| N2+ | | 0.451088 | [085] | N | | 1.0623 | [430] |
| N2+ | | 0.451487 | [085] | N | | 1.34295 | [086] |
| N+ | | 0.463055 | [085] | N | | 1.35818 | [003,088,086] |
| N+ | | 0.566663 | [040] | N | ? | 1.45423 | [003,088,086] |
| N+ | | 0.567601 | [040] | N | ? | 3.7942 | [086] |
| N+ | > | 0.567956 | [040,013] | N | ? | 3.8154 | [086] |
| N | | 0.8594 | [536] | | | | |

NUMBER OF LINES IN NITROGEN, ATOMIC 25

ACTIVE MEDIUM : PHOSPHORUS
SYMBOL : P

OPERATING CONDITIONS : PULSED DISCHARGE IN PF5 OF 0.04 TORR. SOME LINES ALSO CONTINUOUS WAVE [286].
THE IR LINES WITH PULSED OPERATION IN PHOSPHORUS VAPOR OF 0.002-0.2 TORR AND HE OR NE AS CARRIER GAS.

| | | WAVELENGTH IN AIR [MICROMETER] | | | | WAVELENGTH IN AIR [MICROMETER] | |
|---|---|---|---|---|---|---|---|
| P3+ | | 0.334769 | [069] | P | | 1.008 | [320] |
| P2+ | | 0.442208 | [069] | P | | 1.116 | [320] |
| P+ | > | 0.602421 | [069,286] | P | | 1.119 | [320] |
| P+ | > | 0.603421 | [069] | P | | 1.154 | [320] |
| P+ | > | 0.604325 | [069,286] | P | | 1.178 | [320] |
| P+ | | 0.608786 | [069] | P | | 1.571 | [320] |
| P+ | | 0.616577 | [069] | P | | 1.648 | [320] |
| P | ? | 0.667193 | [069] | P | | 1.894 | [320] |
| P | | 0.784563 | [286] | P | | 2.060 | [320] |

NUMBER OF LINES IN PHOSPHORUS 18

ACTIVE MEDIUM : ARSENIC
SYMBOL : AS

OPERATING CONDITIONS : PULSED TOROIDAL DISCHARGE IN ARSENIC VAPOR AND 0.1 TORR NE [078]. CONTINUOUS DISCHARGE IN A MIXTURE OF AS/HE WITH A MULTIPLE-ANODE HOLLOW-CATHODE DEVICE [307]. ALSO EXCITATION BY PULSED ELECTRICAL DISSOCIATION OF ASCL3 [394]

| | WAVELENGTH IN AIR [MICROMETER] | | | | WAVELENGTH IN AIR [MICROMETER] | |
|---|---|---|---|---|---|---|
| AS+ | | 0.538520 | [307] | AS | | 1.1519 | [394] |
| AS+ | | 0.549695 | [307] | AS | | 1.152 | [072] |
| AS+ | | 0.549773 | [078,307] | AS | | 1.1521 | [394] |
| AS+ | | 0.555809 | [078,307] | AS | | 1.294 | [072] |
| AS+ | | 0.56516 | [078] | AS | | 1.412 | [072] |
| AS+ | | 0.583790 | [307] | AS | | 1.463 | [072] |
| AS+ | | 0.617027 | [078,307] | AS | | 1.8049 | [394] |
| AS+ | | 0.651174 | [307] | AS | | 1.807 | [072] |
| AS+ | | 0.710272 | [307] | AS | ? | 1.9750 | [394] |
| AS | | 1.045 | [072] | AS | | 2.0277 | [394] |
| AS | | 1.061 | [072] | AS | | 2.4460 | [394] |
| AS | | 1.124 | [072] | AS | | 2.9805 | [394] |
| AS | ? | 1.4255 | [394] | AS | | 5.2865 | [394] |

NUMBER OF LINES IN ARSENIC 26

---

ACTIVE MEDIUM : ANTIMONY
SYMBOL : SB

OPERATING CONDITIONS : PULSED TOROIDAL DISCHARGE IN 0.002 TORR ANTIMONY VAPOR AND 0.2 TORR NE. ALSO PULSED EXCITATION BY ELECTRICAL DISSOCIATION OF SB(CH3)3 [394].

| | | WAVELENGTH IN AIR [MICROMETER] | | | | WAVELENGTH IN AIR [MICROMETER] | |
|---|---|---|---|---|---|---|---|
| SB+ | | 0.61299 | [089] | SB | ? | 12.033 | [394] |

NUMBER OF LINES IN ANTIMONY 2

---

ACTIVE MEDIUM : BISMUTH
SYMBOL : BI

OPERATING CONDITIONS : PULSED DISCHARGE IN BISMUTH VAPOR WITH HE OR NE [154] OR IN BI(CH3)3 WITH HE [394].

| WAVELENGTH IN AIR [MICROMETER] | | | WAVELENGTH IN AIR [MICROMETER] | | |
|---|---|---|---|---|---|
| BI2+ | 0.456084 | [154] | BI2+ | 0.75990 | [154] |
| BI | 0.4722 | [512] | BI2+ | 0.80689 | [154] |
| BI+ | 0.571921 | [154] | BI | 5.3264 | [394] |

NUMBER OF LINES IN BISMUTH 6

ACTIVE MEDIUM : VANADIUM
SYMBOL : V

OPERATING CONDITIONS : PULSED EXCITATION BY ELECTRICAL DISSOCIATION OF VCL4.

| WAVELENGTH IN AIR [MICROMETER] | | | WAVELENGTH IN AIR [MICROMETER] | | |
|---|---|---|---|---|---|
| V | 2.0195 | [469] | V | 2.4473 | [469] |

NUMBER OF LINES IN VANADIUM 2

ACTIVE MEDIUM : OXYGEN, ATOMIC
SYMBOL : O

OPERATING CONDITIONS : ALL LINES OF IONIZED OXYGEN IN A PULSED DISCHARGE OF 0.001-0.1 TORR O2. TYPICAL CURRENT DENSITIES 500-2000 A/CM2.
IN ATOMIC OXYGEN, NEARLY ALL LINES HAVE BEEN OBSERVED CONTINUOUSLY. IN SOME CASES COOLING WITH LIQUID N2 IS NECESSARY.
GAS MIXTURE O2/AR OR O2/NE. WITH ADDITION OF HE LASING UP TO A PRESSURE OF 200 TORR IS POSSIBLE [429].
EXTENSIVE DATA ON THE EXCITATION MECHANISM OF THE 844 NM LINES [445].

| WAVELENGTH IN AIR [MICROMETER] | | | | WAVELENGTH IN AIR [MICROMETER] | | | |
|---|---|---|---|---|---|---|---|
| O4+ | | 0.2640 | [314] | O+ | > | 0.441488 | [085,013] |
| O4+ | | 0.278139 | [314] | O+ | > | 0.441697 | [085,013] |
| O2+ | > | 0.298378 | [013,012] | O+ | ? | 0.460552 | [085] |
| O2+ | | 0.304713 | [013,012] | O+ | | 0.464914 | [013] |
| O3+ | | 0.306345 | [012] | O2+ | > | 0.559237 | [040,085,013] |
| O3+ | ? | 0.338128 | [012] | O+ | | 0.66402 | [127] |
| O3+ | ? | 0.338133 | [012] | O | | 0.672136 | [085] |
| O3+ | | 0.338554 | [012] | O | | 0.844628 | [130,131,132] |
| O+ | > | 0.374949 | [085,013,012] | O | | 0.844638 | [130,131,132] |
| O2+ | | 0.375426 | [566] | O | | 0.844672 | [130,131,132] |
| O2+ | > | 0.375467 | [085,013,012] | O | | 0.844680 | [130,131,132] |
| O2+ | > | 0.375988 | [085,013,012] | O | | 2.652 | [342] |
| O+ | > | 0.434738 | [085,013] | O | | 2.89 | [129] |
| O+ | > | 0.435128 | [085,013] | O | | 4.5607 | [133] |

| | WAVELENGTH IN AIR [MICROMETER] | | | WAVELENGTH IN AIR [MICROMETER] | |
|---|---|---|---|---|---|
| O | 4.5632 | [128,129] | O | 6.858 | [129] |
| O | 5.981 | [129] | O | 6.8731 | [086] |
| O | 6.8161 | [086] | O | 10.400 | [129] |

NUMBER OF LINES IN OXYGEN, ATOMIC 34

---

ACTIVE MEDIUM : SULFUR
SYMBOL : S

OPERATING CONDITIONS : ALL LINES OF IONIZED SULFUR WITH A PULSED HIGH-CURRENT DISCHARGE IN SO2, SF6 OR H2S. PRESSURE 0.01-0.05 TORR. ATOMIC LINES IN PULSED OR CONTINUOUS OPERATION. CW OR LONG-PULSE EMISSION IN POSITIVE COLUMN DISCHARGE IN SULFUR VAPOR WITH NE AS BUFFER GAS [405]. EXCITATION BY OCS-PHOTODISSOCIATION [578].

| | | WAVELENGTH IN AIR [MICROMETER] | | | | WAVELENGTH IN AIR [MICROMETER] | |
|---|---|---|---|---|---|---|---|
| S2+ | ?> | 0.2638964 | [087,505] | S+ | | 0.547374 | [087,286] |
| S2+ | | 0.3324859 | [087,505] | S+ | | 0.556511 | [087] |
| S2+ | | 0.3497332 | [087,505] | S+ | | 0.550990 | [087] |
| S2+ | > | 0.3709354 | [087,505] | S+ | > | 0.564012 | [087,286] |
| S+ | | 0.492560 | [087] | S+ | > | 0.564716 | [087,286] |
| S+ | | 0.50116 | [405] | S+ | | 0.581935 | [087] |
| S+ | | 0.501424 | [087] | S | | 0.7725 | [578] |
| S+ | | 0.503262 | [087] | S | | 1.0455 | [003,088] |
| S+ | > | 0.516032 | [087] | S | | 1.0636 | [003,088] |
| S+ | > | 0.521962 | [087] | S | | 1.402 | [308] |
| S+ | > | 0.532088 | [087,286] | S | | 1.5422 | [484] |
| S+ | > | 0.534583 | [087,286] | S | | 1.6543 | [484] |
| S+ | | 0.542874 | [087,286] | S | | 2.4363 | [484] |
| S+ | > | 0.543287 | [087,286] | S | | 2.7799 | [133] |
| S+ | > | 0.545388 | [087,286] | S | | 3.3892 | [133] |

NUMBER OF LINES IN SULFUR 30

---

ACTIVE MEDIUM : SELENIUM
SYMBOL : SE

OPERATING CONDITIONS : ALL LINES CONTINUOUS WAVE. TYPICAL CURRENT DENSITY 1.5-15 A/CM2. TYPICAL GAS MIXTURE 0.005 TORR SE, 6-8 TORR HE. SELENIUM VAPOR IS GENERATED AT AN APPROPRIATE PRESSURE BY PLACING A PIECE EXCITATION BY HF-DISCHARGE [470]. PULSED EXCITATION BY ELECTRICAL [469] AND PHOTODISSOCIATION [564] OF OCSE. WHITE-LIGHT LASER [602].

| | | WAVELENGTH IN AIR [MICROMETER] | | | | WAVELENGTH IN AIR [MICROMETER] | |
|---|---|---|---|---|---|---|---|
| SE+ | | 0.446760 | [135] | SE+ | | 0.562313 | [135] |
| SE+ | | 0.460434 | [134] | SE+ | | 0.569788 | [134] |
| SE+ | | 0.461877 | [135] | SE+ | | 0.574762 | [134] |
| SE+ | | 0.464844 | [134] | SE+ | | 0.584268 | [135] |
| SE+ | | 0.471823 | [135] | SE+ | | 0.586627 | [135] |
| SE+ | | 0.474097 | [135] | SE+ | | 0.605596 | [134] |
| SE+ | | 0.476365 | [134] | SE+ | | 0.606583 | [135] |
| SE+ | | 0.476552 | [135] | SE+ | | 0.610196 | [135] |
| SE+ | | 0.484063 | [134] | SE+ | | 0.644425 | [134] |
| SE+ | | 0.484496 | [134] | SE+ | | 0.649048 | [134] |
| SE | | 0.489 | [564] | SE+ | | 0.653495 | [134] |
| SE+ | > | 0.497566 | [134] | SE+ | | 0.706389 | [135] |
| SE+ | > | 0.499275 | [134] | SE+ | | 0.739199 | [135] |
| SE+ | > | 0.506865 | [134,470] | SE+ | | 0.767482 | [135] |
| SE+ | | 0.509650 | [078,134] | SE+ | | 0.772404 | [135] |
| SE+ | | 0.514214 | [134] | SE | | 0.777 | [564] |
| SE+ | > | 0.517598 | [134,470] | SE+ | | 0.779615 | [135] |
| SE+ | > | 0.522751 | [078,134,470] | SE+ | | 0.783881 | [135] |
| SE+ | | 0.525307 | [134] | SE+ | | 0.830952 | [135] |
| SE+ | | 0.525363 | [134] | SE+ | > | 0.92493 | [134] |
| SE+ | | 0.527111 | [134] | SE+ | | 0.995515 | [134] |
| SE+ | > | 0.530535 | [134,470] | SE+ | > | 1.040881 | [134] |
| SE+ | | 0.552242 | [134] | SE+ | > | 1.256678 | [134] |
| SE+ | | 0.556693 | [135] | SE | | 6.3672 | [469] |
| SE+ | | 0.559116 | [134] | | | | |

NUMBER OF LINES IN SELENIUM 49

ACTIVE MEDIUM : TELLURIUM
SYMBOL : TE

OPERATING CONDITIONS : PULSED OR CONTINUOUS DISCHARGE. TELLURIUM VAPOR IS GENERATED AT AN APPROPRIATE PRESSURE BY PLACING A PIECE OF TELLURIUM IN A HEATED DISCHARGE TUBE OR IN A HEATED SIDEARM. TEMPERATURE 420 C. PULSED EXCITATION BY ELECTRICAL DISSOCIATION OF TE(CH3)2 [469].

| | | WAVELENGTH IN AIR [MICROMETER] | | | | WAVELENGTH IN AIR [MICROMETER] | |
|---|---|---|---|---|---|---|---|
| TE+ | | 0.48429 | [305] | TE+ | | 0.57563 | [305] |
| TE+ | ? | 0.50204 | [305] | TE+ | | 0.58511 | [305] |
| TE+ | ? | 0.52564 | [305] | TE+ | > | 0.59361 | [305,136] |
| TE+ | > | 0.54498 | [305] | TE+ | | 0.59726 | [305] |
| TE+ | ? | 0.54540 | [136] | TE+ | > | 0.59747 | [305] |
| TE+ | > | 0.54791 | [305] | TE+ | | 0.60145 | [305] |
| TE+ | | 0.55762 | [089] | TE+ | ? | 0.60823 | [305] |
| TE+ | | 0.55764 | [136] | TE+ | > | 0.62307 | [305] |
| TE+ | | 0.56405 | [136] | TE+ | | 0.62454 | [305,136] |
| TE+ | > | 0.56662 | [305] | TE+ | ? | 0.63497 | [089] |
| TE+ | > | 0.57081 | [305,089,136] | TE+ | ? | 0.65851 | [305] |
| TE+ | > | 0.57416 | [305] | TE+ | | 0.66486 | [305] |
| TE+ | | 0.57559 | [305] | TE+ | | 0.66761 | [305] |

| | WAVELENGTH IN AIR [MICROMETER] | | | | WAVELENGTH IN AIR [MICROMETER] | |
|---|---|---|---|---|---|---|
| TE+ | | 0.68851 | [305] | TE+ | | 0.89721 | [305] |
| TE+ | > | 0.70391 | [305,136] | TE+ | ? | 0.89982 | [305] |
| TE+ | | 0.78017 | [305] | TE+ | | 0.93779 | [305] |
| TE+ | | 0.79217 | [305] | TE | | 3.1720 | [469] |
| TE+ | ? | 0.86046 | [305] | TE | | 6.7595 | [469] |
| TE+ | ? | 0.87338 | [305] | TE | | 7.7856 | [469] |

NUMBER OF LINES IN TELLURIUM 38

---

ACTIVE MEDIUM : FLUORINE, ATOMIC
SYMBOL : F

OPERATING CONDITIONS : PULSED DISCHARGE IN 0.002 TORR F2. ALSO PULSED DISCHARGE IN A MIXTURE OF CF4, SF6, C2F6 OR NF3 AND HELIUM.

| | | WAVELENGTH IN AIR [MICROMETER] | | | | WAVELENGTH IN AIR [MICROMETER] | |
|---|---|---|---|---|---|---|---|
| F2+ | | 0.275958 | [069] | F | | 0.720237 | [110,460,474] |
| F3+ | | 0.282612 | [069] | F | | 0.72043 | [110,396] |
| F2+ | | 0.3121501 | [069,505] | F | | 0.7310102 | [211,396,474] |
| F2+ | | 0.317413 | [069] | F | | 0.7398688 | [474,485] |
| F+ | | 0.320276 | [069] | F | | 0.74257 | [586] |
| F+ | | 0.402472 | [069] | F | | 0.74827 | [586] |
| F | | 0.6239651 | [485] | F | | 0.748914 | [460,474] |
| F | | 0.6348508 | [474,485] | F | | 0.75150 | [586] |
| F | | 0.6413651 | [485] | F | | 0.7552235 | [474,485] |
| F | | 0.696635 | [396,460,474] | F | | 0.775470 | [460,474] |
| F | | 0.70394 | [110,396] | F | | 0.780022 | [118,460,474] |
| F | | 0.703745 | [110,118,460] | F | ? | 1.5900 TO | [118] |
| F | | 0.712788 | [110,460,474] | F | ? | 9.3462 | [118] |
| F | | 0.71298 | [110,396] | | | | |

NUMBER OF LINES IN FLUORINE, ATOMIC 27

---

ACTIVE MEDIUM : CHLORINE, ATOMIC
SYMBOL : CL

OPERATING CONDITIONS : IN IONIZED CHLORINE THE UV LINES UP TO 0.37 MICROMETER CAN BE GENERATED ONLY IN PULSED OPERATION. TYPICAL PRESSURE 0.002 TORR. THE VISIBLE LINES APPEAR CONTINUOUSLY IN A TOROIDAL DISCHARGE OF 0.05 TORR. THE LINES IN ATOMIC CHLORINE ARE EXCITED (PULSED OR CONT.) IN HCL OR CL2/HE.

| WAVELENGTH IN AIR [MICROMETER] | | | WAVELENGTH IN AIR [MICROMETER] | | |
|---|---|---|---|---|---|
| CL2+ | 0.2632686 | [069,070,505] | CL+ | 0.490483 | [039,072] |
| CL2+ | 0.3191424 | [069,505] | CL+ | 0.491781 | [039,072] |
| CL2+ | 0.3392861 | [069,505] | CL+ | 0.507829 | [039,072] |
| CL2+ > | 0.3393444 | [069,505] | CL+ | 0.510310 | [071] |
| CL2+ | 0.3530016 | [069,505] | CL+ > | 0.521776 | [039,072] |
| CL2+ | 0.3560632 | [069,505] | CL+ | 0.522136 | [039,072] |
| CL2+ | 0.360210 | [069] | CL+ | 0.539216 | [039,072] |
| CL2+ | 0.361283 | [069] | CL+ | 0.609473 | [039,072] |
| CL2+ | 0.362268 | [069] | CL | 0.9451 | [073] |
| CL2+ | 0.3720436 | [069,505] | CL | 1.3859 | [074] |
| CL2+ | 0.3748770 | [069,505] | CL | 1.3891 | [074] |
| CL+ | 0.413250 | [071] | CL | 1.9755 | [075,076] |
| CL+ | 0.474042 | [071] | CL | 2.0199 | [075,076] |
| CL+ | 0.476871 | [071,069] | CL | 2.4466 | [074,303] |
| CL+ | 0.478134 | [039,072] | CL | 3.0664 | [303] |
| CL+ | 0.489685 | [039,072] | | | |

NUMBER OF LINES IN CHLORINE, ATOMIC 31

ACTIVE MEDIUM : BROMINE, ATOMIC
SYMBOL : BR

OPERATING CONDITIONS : CONTINUOUS DISCHARGE IN HBR. PULSED DISCHARGE IN 0.04 TORR BR2. ALSO CHEMICAL PUMPING BY FLASH PHOTOLYSIS [592].

| WAVELENGTH IN AIR [MICROMETER] | | | WAVELENGTH IN AIR [MICROMETER] | | |
|---|---|---|---|---|---|
| BR3+?> | 0.2362465 | [505] | BR+ > | 0.533203 | [077,078] |
| BR3+ ? | 0.2581246 | [505] | BR+ | 0.611756 | [079] |
| BR2+ ? | 0.2787619 | [505] | BR+ | 0.616878 | [079] |
| BR+ | 0.474266 | [077] | BR | 2.2854 | [080] |
| BR+ | 0.505463 | [077] | BR | 2.3511 | [080] |
| BR+ > | 0.518238 | [077,078] | BR | 2.714 | [592] |
| BR+ | 0.523826 | [077] | BR | 2.8375 | [080] |

NUMBER OF LINES IN BROMINE, ATOMIC 14

ACTIVE MEDIUM : IODINE, ATOMIC
SYMBOL : I

OPERATING CONDITIONS : PULSED DISCHARGE IN 0.1 TORR IX AND 2-4 TORR HE. CW OPERATION IN POSITIVE-COLUMN [165,368] AND HOLLOW-CATHODE [367,436] DISCHARGES. THE IR LINES ARE GENERATED BY A PULSED OR CONTINUOUS DISCHARGE IN HI OR I/HE, HIGH-POWER EMISSION AT 1.315 MICROMETER BY PHOTODISSOCIATION OF C3F7I [166]. CLOSED CYCLE OPERATION [440]. PURELY CHEMICAL EXCITATION [489].

| WAVELENGTH IN AIR [MICROMETER] | | | | WAVELENGTH IN AIR [MICROMETER] | | | |
|---|---|---|---|---|---|---|---|
| I+ | | 0.448855 | [367] | I+ | | 0.761850 | [367] |
| I+ | ? | 0.453379 | [083] | I+ | | 0.773578 | [367] |
| I+ | ? | 0.467440 | [083] | I+ | > | 0.817007 | [367] |
| I+ | | 0.467553 | [367] | I+ | | 0.825381 | [040,081,165] |
| I+ | ? | 0.493467 | [083] | I+ | > | 0.880428 | [040,081,165] |
| I+ | | 0.498692 | [040,081,367] | I+ | | 0.887761 | [367] |
| I+ | | 0.521408 | [367] | I | ? | 0.98 | [081] |
| I+ | | 0.521627 | [040,081,367] | I | ? | 1.01 | [081] |
| I+ | > | 0.540736 | [040,165,081] | I | ? | 1.03 | [081] |
| I+ | ? | 0.5419 | [081] | I | ? | 1.06 | [081] |
| I+ | | 0.559312 | [367] | I | | 1.315 | [084,117,166] |
| I+ | | 0.562569 | [040,081] | I | | 1.4542 | [074] |
| I+ | > | 0.567808 | [040,165,081] | I | | 1.553 | [262] |
| I+ | > | 0.576072 | [040,165,047] | I | | 2.5986 | [074] |
| I+ | > | 0.612749 | [040,165,081] | I | | 2.7572 | [080] |
| I+ | | 0.633997 | [367] | I | | 3.0360 | [119] |
| I+ | | 0.651618 | [083,165,367] | I | | 3.2363 | [080] |
| I+ | | 0.658521 | [040,165,081] | I | | 3.4296 | [080,119] |
| I+ | | 0.606893 | [367] | I | | 4.331 | [262] |
| I+ | | 0.620486 | [367] | I | | 4.858 | [119] |
| I+ | | 0.662235 | [367] | I | > | 4.8619 | [119] |
| I+ | | 0.682523 | [261,367] | I | > | 5.4972 | [119] |
| I+ | | 0.690477 | [040,081] | I | | 6.7198 | [119] |
| I+ | > | 0.703299 | [040,165,081] | I | | 6.902 | [119] |
| I+ | | 0.713897 | [165,261] | I | | 9.326 | [119] |

NUMBER OF LINES IN IODINE, ATOMIC 50

ACTIVE MEDIUM : MANGANESE
SYMBOL : MN

OPERATING CONDITIONS : PULSED DISCHARGE IN MANGANESE VAPOR AND HE. VAPOR GENERATED BY HEATING MANGANESE WITHIN THE DISCHARGE TUBE WITH A TUBULAR FURNACE TO 1060 C. WHEN MANGANESE CHLORIDE IS USED, THE TEMP. CAN BE REDUCED TO 680 C [339].

| WAVELENGTH IN AIR [MICROMETER] | | | | WAVELENGTH IN AIR [MICROMETER] | | | |
|---|---|---|---|---|---|---|---|
| MN | > | 0.534106 | [125,123] | MN | | 1.32938 | [125,123] |
| MN | | 0.542036 | [125,123] | MN | | 1.33190 | [125,123] |
| MN | | 0.547064 | [125,123] | MN | | 1.36267 | [125,123] |
| MN | | 0.551677 | [125,123] | MN | | 1.38642 | [125,123] |
| MN | | 0.553776 | [125,123] | MN | | 1.39975 | [125,123] |
| MN | > | 1.28998 | [125,123] | | | | |

NUMBER OF LINES IN MANGANESE 11

ACTIVE MEDIUM : SAMARIUM
SYMBOL : SM

OPERATING CONDITIONS : PULSED DISCHARGE IN SAMARIUM VAPOR WITH HE, NE OR AR. VAPOR GENERATED BY HEATING SAMARIUM WITHIN THE DISCHARGE TUBE.

| WAVELENGTH IN AIR [MICROMETER] | | | | WAVELENGTH IN AIR [MICROMETER] | | | |
|---|---|---|---|---|---|---|---|
| SM | ? | 1.912 | | SM | > | 3.4654 | [159] |
| SM | ? | 2.0482 | [159] | SM | > | 3.5361 | [159] |
| SM | > | 2.6998 | [159] | SM | > | 4.1368 | [159] |
| SM | > | 2.9663 | [159] | SM | > | 4.8656 | [159] |

NUMBER OF LINES IN SAMARIUM 8

---

ACTIVE MEDIUM : EUROPIUM
SYMBOL : EU

OPERATING CONDITIONS : PULSED DISCHARGE IN EUROPIUM VAPOR WITH HE, NE OR AR. VAPOR GENERATED BY HEATING EUROPIUM WITHIN THE DISCHARGE TUBE.

| WAVELENGTH IN AIR [MICROMETER] | | | | WAVELENGTH IN AIR [MICROMETER] | | | |
|---|---|---|---|---|---|---|---|
| EU+ | | 0.6645 | [558] | EU | | 2.7174 | [159] |
| EU+ | | 0.9898 | [558] | EU | | 4.3202 | [159] |
| EU+ | | 1.002 | [558] | EU | | 4.6935 | [159] |
| EU+ | | 1.016 | [558] | EU | > | 5.0647 | [159] |
| EU+ | | 1.361 | [558] | EU | | 5.2811 | [159] |
| EU+ | | 1.477 | [558] | EU | > | 5.4292 | [159] |
| EU | | 1.66 | [557] | EU | | 5.7706 | [159] |
| EU | > | 1.7596 | [159] | EU | | 5.9479 | [159] |
| EU | | 2.5811 | [159] | EU | | 6.0576 | [159] |

NUMBER OF LINES IN EUROPIUM 18

---

ACTIVE MEDIUM : THULIUM
SYMBOL : TM

OPERATING CONDITIONS : PULSED DISCHARGE IN THULIUM VAPOR WITH HE,NE OR AR. VAPOR GENERATED BY HEATING THULIUM WITHIN THE DISCHARGE TUBE.

| WAVELENGTH IN AIR [MICROMETER] | | | | WAVELENGTH IN AIR [MICROMETER] | | | |
|---|---|---|---|---|---|---|---|
| TM | ? | 1.304 | [160] | TM | > | 1.675404 | [160] |
| TM | | 1.310058 | [160] | TM | | 1.7319 | [159] |
| TM | | 1.338008 | [160] | TM | > | 1.958443 | [160] |
| TM | | 1.433973 | [160] | TM | ? | 1.973 | [160] |
| TM | | 1.448509 | [160] | TM | | 1.994160 | [160] |
| TM | ? | 1.500 | [160] | TM | ? | 2.107 | [160] |
| TM | ? | 1.637914 | [160] | TM | | 2.384515 | [160] |

NUMBER OF LINES IN THULIUM 14

ACTIVE MEDIUM : YTTERBIUM
SYMBOL : YB

OPERATING CONDITIONS : PULSED DISCHARGE IN YTTERBIUM VAPOR WITH HE, NE OR AR. VAPOR GENERATED BY HEATING YTTERBIUM WITHIN THE DISCHARGE TUBE.

| WAVELENGTH IN AIR [MICROMETER] | | | | WAVELENGTH IN AIR [MICROMETER] | | | |
|---|---|---|---|---|---|---|---|
| YB | | 1.0322 | [159] | YB | | 1.7977 | [387] |
| YB | | 1.2548 | [159] | YB+ | | 1.8057 | [387] |
| YB+ | | 1.2714 | [387] | YB | | 1.9830 | [159] |
| YB+ | | 1.3453 | [387] | YB | | 2.0036 | [159] |
| YB | | 1.4280 | [159] | YB | | 2.1161 | [159] |
| YB | | 1.4787 | [387] | YB+ | ? | 2.1480 | [387] |
| YB+ | > | 1.6498 | [159] | YB+ | > | 2.4377 | [159] |
| YB | | 1.7454 | [159] | YB | | 4.8009 | [159] |

NUMBER OF LINES IN YTTERBIUM 16

ACTIVE MEDIUM : NOBLE GASES, MOLECULAR
SYMBOL : XE2,AR2,KR2

OPERATING CONDITIONS : EXCITATION WITH HIGH-ENERGY ELECTRON BEAMS IN NOBLE GASES OF HIGH PRESSURE. IN THE DISPERSIVE-RESONATOR TUNING RANGE 5-6 NM [192]. HIGH-EFFICIENCY XENON LASER [356]. LONG TIME OPERATION WITH GAS PURIFICATION [547].

| WAVELENGTH IN AIR [MICROMETER] | | | WAVELENGTH IN AIR [MICROMETER] | | |
|---|---|---|---|---|---|
| XE2 | 0.1722 | [355,357,358] | KR2 | 0.1457 | [359] |
| AR2 | 0.1261 | [115] | | | |

NUMBER OF LINES IN NOBLE GASES, MOLECULAR 3

ACTIVE MEDIUM : NOBLE-GAS HALIDES
SYMBOL : ARF,ARCL,KRF,ETC.

OPERATING CONDITIONS : EXCITATION WITH HIGH-ENERGY ELECTRON BEAMS WITHIN A HIGH-PRESSURE MIXTURE OF NOBLE GASES AND HALIDES OR NF3. PEAK POWER IN KRF 1.9 GW [393]. ALSO OPERATION WITH TRANSVERSE DISCHARGE AT ATMOSPHERIC PRESSURE WITH FAST GAS RECIRCULATION, MEAN OUTPUT 10 W AT 1 KHZ [532]. LONG PULSE OPERATION IN SUPERSONIC FLOW AFTER ELECTRON-BEAM EXCITATION [516]. REVIEW PAPER [491].

| | WAVELENGTH IN AIR [MICROMETER] | | | WAVELENGTH IN AIR [MICROMETER] | |
|---|---|---|---|---|---|
| ARF | 0.1933 | [393,451] | XECL | 0.30792 | [351,450] |
| ARCL | 0.1750 | [448] | XECL | 0.30816 | [351,450] |
| KRBR | 0.2065 | [449] | XEF | 0.34875 | [450] |
| KRCL | 0.2229 | [415,426] | XEF | 0.35091 | [450] |
| KRF | 0.2481 | [352,393] | XEF | 0.35097 | [450] |
| KRF | 0.2484 | [351,352] | XEF | 0.35114 | [373,450] |
| KRF | 0.2485 | [354] | XEF | 0.35305 | [350,373,450] |
| KRF | 0.2495 | [354] | XEF | 0.35354 | [450] |
| XEBR | 0.2818 | [353,490] | XEF | 0.483 | [571] |

NUMBER OF LINES IN NOBLE-GAS HALIDES 18

---

ACTIVE MEDIUM : NOBLE-GAS OXIDES
SYMBOL : XEO,KRO

OPERATING CONDITIONS : EXCITATION WITH HIGH-ENERGY ELECTRON BEAMS WITHIN A HIGH-PRESSURE MIXTURE OF NOBLE GASES AND O2.

| | WAVELENGTH IN AIR [MICROMETER] | | | WAVELENGTH IN AIR [MICROMETER] | |
|---|---|---|---|---|---|
| XEO | 0.5300 | [384] | ARO | 0.5577 | [579] |
| XEO | 0.5550 | [384] | KRO | 0.5578 | [384] |

NUMBER OF LINES IN NOBLE-GAS OXIDES 4

---

ACTIVE MEDIUM : MERCURY HALIDES
SYMBOL : HGBR,HGCL

OPERATING CONDITIONS : HIGH-INTENSITY ELECTRON-BEAM EXCITATION OF HIGH-PRESSURE RARE-GAS MIXTURES CONTAINING SMALL AMOUNTS OF HG AND HALIDE COMPOUNDS. PHOTODISSOCIATION OF HGBR2 BY OPTICAL PUMPING WITH ARF-LASER [504] OR BY DISSOCIATION IN ELECTRICAL DISCHARGE [553].

| | WAVELENGTH IN AIR [MICROMETER] | | | WAVELENGTH IN AIR [MICROMETER] | |
|---|---|---|---|---|---|
| HGBR | 0.5018 | [480] | HGBR | 0.5042 | [504] |
| HGBR | 0.5020 | [504] | HGBR | 0.5046 | [504] |
| HGBR | 0.5023 | [504] | HGCL | 0.55762 | [481,502] |
| HGBR | 0.5026 | [504] | HGCL | 0.55835 | [502,534] |
| HGBR | 0.5039 | [504] | | | |

NUMBER OF LINES IN MERCURY HALIDES 9

ACTIVE MEDIUM : HYDROGEN, MOLECULAR
SYMBOL : H2

OPERATING CONDITIONS : PULSED DISCHARGE WITH HIGH CURRENT DENSITY IN H2 OF 0.1-3 TORR PRESSURE. TO GENERATE THE UV LINES, VERY SHORT (2.5 NS) AND INTENSE (SOME 100 KA) CURRENT PULSES ARE NECESSARY [161]. ALSO EXCITATION BY PULSED ELECTRON BEAMS (400 KEV) IN 10-100 TORR H2. [162,290] THE LINES IN PARA H2 ARE DESIGNATED H2 P.

| | WAVELENGTH IN VACUUM [MICROMETER] | | TRANSITION |
|---|---|---|---|
| H2 | 0.115976 | [290] | 1- 4 R(1) |
| H2 | 0.116136 | [325,290,291] | 1- 4 Q(1) |
| H2 | 0.116617 | [325,290] | 1- 4 P(3) |
| H2 | 0.117436 | [290] | 2- 5 R(1) |
| H2 | 0.117586 | [325,290] | 2- 5 Q(1) |
| H2 | 0.118050 | [325,290] | 2- 5 P(3) |
| H2 | 0.118936 | [325,290] | 3- 6 Q(1) |
| H2 | 0.120497 | [290] | 1- 5 R(1) |
| H2 | 0.120668 | [325,290] | 1- 5 Q(1) |
| H2 | 0.121734 | [325,290] | 2- 6 R(1) |
| H2 | 0.121900 | [325,290] | 2- 6 Q(1) |
| H2 | 0.122358 | [325,290] | 2- 6 P(3) |
| H2 | 0.123004 | [325,290,291] | 3- 7 Q(1) |
| H2 | 0.123956 | [325,290] | 4- 8 Q(1) |
| H2 P | 0.109816 | [325] | 0- 2 R(0) |
| H2 P | 0.110205 | [325] | 0- 2 P(2) |
| H2 P | 0.111515 | [325] | 1- 3 R(0) |
| H2 P | 0.111894 | [325] | 1- 3 P(2) |
| H2 P | 0.114462 | [325] | 0- 3 R(0) |
| H2 P | 0.114862 | [325] | 0- 3 P(2) |
| H2 P | 0.116003 | [325] | 1- 4 R(0) |
| H2 P | 0.116390 | [325] | 1- 4 P(2) |
| H2 P | 0.117456 | [325] | 2- 5 R(0) |
| H2 P | 0.117830 | [325] | 2- 5 P(2) |
| H2 P | 0.120536 | [325] | 1- 5 R(0) |
| H2 P | 0.120929 | [325] | 1- 5 P(2) |
| H2 P | 0.121767 | [325] | 2- 6 R(0) |
| H2 P | 0.122143 | [325] | 2- 6 P(2) |
| H2 P | 0.122874 | [325] | 3- 7 R(0) |
| H2 P | 0.123230 | [325] | 3- 7 P(2) |
| H2 P | 0.123833 | [325] | 4- 8 R(0) |

| | WAVELENGTH IN VACUUM [MICROMETER] | | TRANSITION |
|---|---|---|---|
| H2 P | 0.124167 | [325] | 4- 6 P(2) |
| H2 P | 0.124620 | [325] | 5- 9 R(0) |
| H2 P | 0.125202 | [325] | 6-10 R(0) |
| H2 P | 0.126839 | [325] | 3- 8 R(0) |
| | | | |
| H2 | 0.134226 | [325] | 0- 4 P(3) |
| H2 | 0.140264 | [325] | 0- 5 P(3) |
| H2 | 0.143622 | [325] | 1- 6 P(3) |
| H2 | 0.144049 | [325] | 3- 7 P(3) |
| H2 | 0.146383 | [325] | 0- 6 P(3) |
| H2 | 0.14942 | [327] | 11-14 P(3) |
| H2 | 0.149522 | [325] | 1- 7 P(3) |
| H2 | 0.151867 | [325] | 4- 9 P(3) |
| H2 | 0.152325 | [161] | 2- 8 P(3) |
| H2 | 0.155345 | [325] | 1- 8 P(3) |
| H2 | 0.15655 | [327] | 8-14 P(1) |
| H2 | 0.156725 | [325,161,292] | 8-14 P(3) |
| H2 | 0.157199 | [161] | 2- 9 P(1) |
| H2 | 0.15743 | [327] | 2- 9 P(2) |
| H2 | 0.157739 | [325,161] | 2- 9 P(3) |
| H2 | 0.157919 | [325,161] | 7-13 P(1) |
| H2 | 0.157998 | [325,161] | 7-13 P(2) |
| H2 | 0.158077 | [325,161,292] | 7-13 P(3) |
| H2 | 0.159131 | [325,161] | 3-10 P(1) |
| H2 | 0.159340 | [325,161] | 3-10 P(2) |
| H2 | 0.159606 | [325,161,292] | 3-10 P(3) |
| H2 | 0.160448 | [325,161,292] | 4-11 P(1) |
| H2 | 0.160623 | [325,161] | 4-11 P(2) |
| H2 | 0.160751 | [325,161,292] | 6-13 P(1) |
| H2 | 0.160839 | [325,161,292] | 4-11 P(3) |
| H2 | 0.160902 | [325,161,292] | 6-13 P(3) |
| H2 | 0.160829 | [325] | 6-13 P(2) |
| H2 | 0.161033 | [325,161,292] | 5-12 P(1) |
| H2 | 0.161166 | [325] | 5-12 P(2) |
| H2 | 0.161319 | [325] | 5-12 P(3) |
| H2 | 0.16148 | [327] | 5-12 P(4) |
| H2 | 0.16165 | [327] | 5-12 P(5) |
| | | | |
| H2 P | 0.121946 | [325] | 0- 3 P(2) |
| H2 P | 0.133856 | [325] | 0- 4 P(2) |
| H2 P | 0.135984 | [325] | 4- 6 P(2) |
| H2 P | 0.136799 | [325] | 6- 7 P(2) |
| H2 P | 0.139895 | [325] | 0- 5 P(2) |
| H2 P | 0.140728 | [625] | 0- 5 P(4) |
| H2 P | 0.143262 | [325] | 1- 6 P(2) |
| H2 P | 0.143757 | [325] | 3- 7 P(2) |
| H2 P | 0.144061 | [325] | 7- 9 P(2) |
| H2 P | 0.146017 | [325] | 0- 6 P(2) |
| H2 P | 0.146411 | [325] | 6- 9 P(2) |
| H2 P | 0.146841 | [625] | 0- 6 P(4) |
| H2 P | 0.148652 | [325] | 1- 7 R(0) |
| H2 P | 0.149171 | [325] | 1- 7 P(2) |
| H2 P | 0.151570 | [325] | 4- 9 P(2) |
| H2 P | 0.151994 | [325,289] | 2- 8 P(2) |
| H2 P | 0.153494 | [325] | 5-10 P(2) |
| H2 P | 0.154493 | [325] | 1- 8 R(0) |
| H2 P | 0.155010 | [325] | 1- 8 P(2) |

| | WAVELENGTH IN VACUUM [MICROMETER] | | TRANSITION |
|---|---|---|---|
| H2 P | 0.156629 | [325] | 8-14 R(0) |
| H2 P | 0.156644 | [325] | 8-14 P(4) |
| H2 P | 0.156753 | [325,289] | 8-14 P(2) |
| H2 P | 0.157434 | [325,289] | 2- 9 P(2) |
| H2 P | 0.157771 | [325,289] | 7-13 R(0) |
| H2 P | 0.157998 | [325,289] | 7-13 P(2) |
| H2 P | 0.158110 | [289] | 2- 9 P(4) |
| H2 P | 0.158140 | [289] | 7-13 P(4) |
| H2 P | 0.158899 | [325,289] | 3-10 R(0) |
| H2 P | 0.159340 | [325,289] | 3-10 P(2) |
| H2 P | 0.159926 | [325,289] | 3-10 P(4) |
| H2 P | 0.160236 | [325] | 4-11 R(0) |
| H2 P | 0.160594 | [325,289] | 6-13 R(0) |
| H2 P | 0.160623 | [325] | 4-11 P(2) |
| H2 P | 0.160829 | [325] | 6-13 P(2) |
| H2 P | 0.160844 | [325] | 5-12 R(0) |
| H2 P | 0.160961 | [325,289] | 6-13 P(4) |
| H2 P | 0.161019 | [325,289] | 4-11 P(4) |
| H2 P | 0.161033 | [289] | 5-12 P(1) |
| H2 P | 0.161165 | [325,289] | 5-12 P(2) |
| H2 P | 0.161318 | [289] | 5-12 P(3) |
| H2 P | 0.161485 | [325,289] | 5-12 P(4) |

| | WAVELENGTH IN AIR [MICROMETER] | | | WAVELENGTH IN AIR [MICROMETER] | |
|---|---|---|---|---|---|
| H2 | 0.752464 | [163] | H2 | 1.305662 | [086] |
| H2 | 0.834950 | [086] | H2 | 1.316109 | [086] |
| H2 | 0.887613 | [086] | H2 | 1.50 | [401] |
| H2 | 0.889882 | [086] | H2 | 1.564193 | [086] |
| H2 | 1.116220 | [086] | H2 | 1.581950 | [086] |
| H2 | 1.122205 | [086] | H2 | 1.63 | [401] |

NUMBER OF LINES IN HYDROGEN, MOLECULAR 120

ACTIVE MEDIUM : DEUTERIUM, MOLECULAR
SYMBOL : D2

OPERATING CONDITIONS : PULSED DISCHARGE WITH HIGH CURRENT DENSITY IN D2 OF 0.1-3 TORR PRESSURE. TO EXCITE THE UV LINES, VERY SHORT (2.5 NS) AND INTENSE (SOME 100 KA) CURRENT PULSES ARE NECESSARY.

| | WAVELENGTH IN VACUUM [MICROMETER] | | TRANSITION |
|---|---|---|---|
| D2 | 0.111336 | [325] | 1- 4 R(0) |
| D2 | 0.113770 | [325] | 1- 5 R(0) |
| D2 | 0.114757 | [325] | 1- 5 P(2) |
| D2 | 0.115650 | [325] | 2- 6 R(0) |
| D2 | 0.118811 | [325] | 2- 7 R(0) |
| D2 | 0.119753 | [325] | 3- 8 R |

| | | WAVELENGTH IN VACUUM [MICROMETER] | | TRANSITION |
|---|---|---|---|---|
| D2 | | 0.120640 | [325] | 4- 9 R(0) |
| D2 | | 0.122800 | [325] | 3- 9 R(0) |
| D2 | | 0.123556 | [325] | 4-10 R(0) |
| D2 | | 0.124239 | [325] | 5-11 R(0) |
| D2 | | 0.124831 | [325] | 6-12 R(0) |
| D2 | | 0.125329 | [325] | 7-13 R(0) |
| D2 | | 0.115840 | [325] | 2- 6 P(2) |
| D2 | | 0.119005 | [325] | 2- 7 P(2) |
| D2 | | 0.119940 | [325] | 3- 8 P(2) |
| D2 | | 0.120821 | [325] | 4- 9 P(2) |
| D2 | | 0.124412 | [325] | 5-11 P(2) |
| D2 | | 0.124997 | [325] | 6-12 P(2) |
| D2 | | 0.130363 | [325] | 0- 5 P(2) |
| D2 | | 0.134590 | [325] | 0- 6 P(2) |
| D2 | | 0.138879 | [325] | 0- 7 P(2) |
| D2 | | 0.143217 | [325] | 0- 8 P(2) |
| D2 | | 0.157585 | [325] | 3-13 P(2) |
| D2 | | 0.158634 | [325] | 7-16 P(1) |
| D2 | | 0.158642 | [325] | 10-19 P(2) |
| D2 | | 0.158675 | [325,164,289] | 10-19 P(3) |
| D2 | | 0.158714 | [164,289] | 10-19 P(4) |
| D2 | | 0.158720 | [325] | 7-16 P(2) |
| D2 | ? | 0.15890 | [327] | |
| D2 | ? | 0.15923 | [327] | |
| D2 | | 0.158983 | [325] | 4-14 P(2) |
| D2 | | 0.159130 | [164,289] | 9-18 P(2) |
| D2 | | 0.159137 | [325] | 8-17 P(2) |
| D2 | | 0.159226 | [325] | 8-17 P(3) |
| D2 | | 0.159257 | [164,289] | 9-18 P(4) |
| D2 | ? | 0.160044 | [292] | |
| D2 | | 0.160066 | [325,164,289] | 5-15 P(2) |
| D2 | | 0.160210 | [325,327] | 5-15 P(3) |
| D2 | | 0.160354 | [164,289] | 5-15 P(4) |
| D2 | | 0.160578 | [325] | 9-19 R(0) |
| D2 | ? | 0.16063 | [327] | |
| D2 | | 0.160650 | [325] | 9-19 P(1) |
| D2 | | 0.160769 | [325] | 6-16 P(1) |
| D2 | | 0.160681 | [325] | 9-19 P(2) |
| D2 | | 0.160848 | [325,164,289] | 6-16 P(2) |
| D2 | | 0.160955 | [325,327] | 6-16 P(3) |
| D2 | | 0.161075 | [325] | 7-17 R(0) |
| D2 | | 0.161080 | [164,289] | 6-16 P(4) |
| D2 | | 0.161147 | [325] | 8-18 P(1) |
| D2 | | 0.161165 | [161,292] | 5-12 P(2) |
| D2 | | 0.161171 | [325] | 7-17 P(1) |
| D2 | | 0.161198 | [164,289] | 8-18 P(2) |
| D2 | | 0.161236 | [325,164,289] | 7-17 P(2) |
| D2 | | 0.161251 | [325,164,289] | 8-18 P(2) |
| D2 | | 0.161257 | [325] | 8-18 P(3) |
| D2 | | 0.161318 | [161] | 5-12 P(3) |
| D2 | | 0.161320 | [164,289] | 8-18 P(4) |
| D2 | | 0.161324 | [325] | 7-17 P(3) |
| D2 | | 0.161412 | [164,289] | 7-17 P(4) |
| D2 | | 0.161658 | [325] | 9-20 P(2) |

| | WAVELENGTH IN AIR [MICROMETER] | | | WAVELENGTH IN AIR [MICROMETER] | |
|---|---|---|---|---|---|
| D2 | 0.827752 | [086] | D2 | 0.953005 | [086] |
| D2 | 0.944156 | [086] | D2 | 1.477548 | [086] |
| D2 | 0.952367 | [086] | | | |

NUMBER OF LINES IN DEUTERIUM, MOLECULAR 65

ACTIVE MEDIUM : HYDROGEN DEUTERIUM
SYMBOL : HD

OPERATING CONDITIONS : PULSED DISCHARGE WITH HIGH CURRENT DENSITY IN HD OF 0.1-3 TORR PRESSURE.

| | | WAVELENGTH IN VACUUM [MICROMETER] | | TRANSITION |
|---|---|---|---|---|
| HD | | 0.113864 | [325] | 1- 4 R(0) |
| HD | | 0.114154 | [325] | 1- 4 P(2) |
| HD | | 0.115198 | [325] | 2- 5 R(0) |
| HD | | 0.117806 | [325] | 1- 5 R(0) |
| HD | | 0.118995 | [325] | 2- 6 R(0) |
| HD | | 0.119281 | [325] | 2- 6 P(2) |
| HD | | 0.120103 | [325] | 3- 7 R(0) |
| HD | | 0.121125 | [325] | 4- 8 R(0) |
| HD | | 0.122837 | [325] | 6-10 R(0) |
| HD | ? | 0.124567 | [325] | |
| HD | | 0.125276 | [325] | 0- 3 P(2) |
| HD | | 0.130334 | [325] | 0- 4 P(2) |
| HD | | 0.135507 | [325] | 0- 5 P(2) |
| HD | | 0.140770 | [325] | 0- 6 P(2) |
| HD | | 0.148843 | [325] | 1- 8 P(2) |
| HD | | 0.151359 | [325] | 2- 9 P(2) |
| HD | | 0.152989 | [325] | 5-11 P(2) |
| HD | | 0.156201 | [325] | 2-10 P(2) |
| HD | | 0.157136 | [325] | 9-16 R(0) |
| HD | | 0.157242 | [325,289] | 9-16 P(2) |
| HD | | 0.157267 | [325] | 9-16 P(3) |
| HD | | 0.158008 | [325] | 6-15 R(0) |
| HD | | 0.158085 | [325] | 3-11 P(2) |
| HD | | 0.158185 | [325] | 8-15 P(2) |
| HD | | 0.158253 | [325] | 8-15 P(3) |
| HD | | 0.158305 | [325] | 3-11 P(3) |
| HD | | 0.159378 | [325] | 4-12 P(1) |
| HD | | 0.159524 | [325,289] | 4-12 P(2) |
| HD | | 0.159713 | [325,289] | 4-12 P(3) |
| HD | | 0.160233 | [325] | 5-13 R(0) |
| HD | | 0.160365 | [325,289] | 5-13 P(1) |
| HD | | 0.160496 | [325,289] | 5-13 P(2) |
| HD | | 0.160465 | [325] | 7-15 R(0) |
| HD | | 0.160569 | [289] | 7-15 P(1) |
| HD | | 0.160647 | [325,289] | 5-13 P(3) |
| HD | | 0.160648 | [325,289] | 7-15 P(2) |
| HD | | 0.160674 | [325] | 6-14 R(0) |

| | WAVELENGTH IN VACUUM [MICROMETER] | | TRANSITION |
|---|---|---|---|
| HD | 0.160692 | [325,289] | 7-15 P(3) |
| HD | 0.160747 | [289] | 7-15 P(4) |
| HD | 0.160794 | [325,289] | 6-14 P(1) |
| HD | 0.160827 | [289] | 5-13 P(4) |
| HD | 0.160893 | [325,289] | 6-14 P(2) |
| HD | 0.161005 | [325,289] | 6-14 P(3) |
| HD | 0.161131 | [289] | 6-14 P(4) |

| | WAVELENGTH IN AIR [MICROMETER] | | WAVELENGTH IN AIR [MICROMETER] |
|---|---|---|---|
| HD | 0.917201 | [086] | |

NUMBER OF LINES IN HYDROGEN DEUTERIUM 45

ACTIVE MEDIUM : SODIUM, MOLECULAR
SYMBOL : NA2

OPERATING CONDITIONS : EXCITATION BY OPTICAL PUMPING WITH 473 NM OR 659 NM LASER RADIATION, OR WITH ARGON LASER AT 454 NM, 488 NM [452]. GENERATION OF NA2-VAPOR BY HEATING METALLIC SODIUM WITHIN A STAINLESS STEEL HEAT PIPE TO 500-600 C. HE BUFFER GAS PRESSURE 30 TORR. SUPERFLUORESCENCE.

| | WAVELENGTH IN AIR [MICROMETER] | | | WAVELENGTH IN AIR [MICROMETER] | |
|---|---|---|---|---|---|
| NA2 | 0.5250 | [452] | NA2 | 0.5453 | [452] |
| NA2 | 0.526333 | [434] | NA2 | 0.5459 | [452] |
| NA2 | 0.5279 | [452] | NA2 | 0.5472 | [452] |
| NA2 | 0.529952 | [434] | NA2 | 0.5474 | [452] |
| NA2 | 0.529816 | [434] | NA2 | 0.5485 | [452] |
| NA2 | 0.5338 | [452] | NA2 | 0.5490 | [452] |
| NA2 | 0.5326 | [452] | NA2 | 0.5491 | [452] |
| NA2 | 0.534283 | [434] | NA2 | 0.549158 | [434] |
| NA2 | 0.5345 | [452] | NA2 | 0.5504 | [452] |
| NA2 | 0.534930 | [434] | NA2 | 0.5562 | [452] |
| NA2 | 0.536902 | [434] | NA2 | 0.5568 | [452] |
| NA2 | 0.5375 | [452] | NA2 | 0.5596 | [452] |
| NA2 | 0.5376 | [452] | | | |
| NA2 | 0.537814 | [434] | NA2 | 0.784930 | [434] |
| NA2 | 0.5381 | [452] | NA2 | 0.786590 | [434] |
| NA2 | 0.538497 | [434] | NA2 | 0.789740 | [434] |
| NA2 | 0.538635 | [434] | NA2 | 0.789790 | [434] |
| NA2 | 0.5394 | [452] | NA2 | 0.791783 | [434] |
| NA2 | 0.540244 | [434] | NA2 | 0.792947 | [434] |
| NA2 | 0.541311 | [434] | NA2 | 0.793697 | [434] |
| NA2 | 0.5417 | [452] | NA2 | 0.797474 | [434] |
| NA2 | 0.5421 | [452] | NA2 | 0.797657 | [434] |
| NA2 | 0.544150 | [434] | NA2 | 0.799091 | [434] |
| NA2 | 0.544694 | [434] | NA2 | 0.799660 | [434] |

| | WAVELENGTH IN AIR [MICROMETER] | | | WAVELENGTH IN AIR [MICROMETER] | |
|---|---|---|---|---|---|
| NA2 | 0.80011 | [435] | NA2 | 0.805366 | [434] |
| NA2 | 0.8002 | [434] | NA2 | 0.805611 | [434] |
| NA2 | 0.800840 | [434] | NA2 | 0.8066 | [435] |
| NA2 | 0.80154 | [435] | NA2 | 0.806943 | [434] |
| NA2 | 0.803650 | [434] | NA2 | 0.80715 | [435] |
| NA2 | 0.803931 | [434] | NA2 | 0.80659 | [435] |
| NA2 | 0.804447 | [434] | | | |

NUMBER OF LINES IN SODIUM, MOLECULAR 60

ACTIVE MEDIUM : NITROGEN, MOLECULAR
SYMBOL : N2

OPERATING CONDITIONS : PULSED DISCHARGE WITH HIGH CURRENT DENSITY IN 1-4 TORR N2. WITH MANY LINES, COOLING TO 77 K IS NECESSARY. A TRANSVERSE ELECTRODE CONFIGURATION IS ADVANTAGEOUS [288]. SIMPLE-DESIGN LASER [511]. ALSO EXCITATION BY 300-400 KEV ELECTRON BEAM. TYPICAL PRESSURE 20 TORR [249]. ELECTRON BEAM EXCITATION IN HIGH-PRESSURE AR+N2 MIXTURES [379].
ON THE DOMINATING UV LINES AT 337 NM, PULSE POWERS UP TO 2 MW CAN BE GENERATED [400]. THE MEAN POWER AT 1200 HZ REP. FREQUENCY HAS BEEN 1.5 WATTS [304].
THE N2 ION TRANSITION IS GENERATED BY CHARGE TRANSFER PUMPING FROM ELECTRON BEAM EXCITED HE2+ [378].

| | | WAVELENGTH IN AIR [MICROMETER] | | TRANSITION | |
|---|---|---|---|---|---|
| N2 | | 0.315756 | [248] | ?1-0 | |
| N2 | | 0.315778 | [248] | ?1-0 | |
| N2 | | 0.315798 | [248] | ?1-0 | |
| N2 | | 0.315803 | [248] | ?1-0 | |
| N2 | | 0.315816 | [248] | ?1-0 | |
| N2 | | 0.315827 | [248] | ?1-0 | |
| N2 | | 0.315832 | [248] | ?1-0 | |
| N2 | | 0.315844 | [248] | ?1-0 | |
| N2 | | 0.315853 | [248] | ?1-0 | |
| N2 | > | 0.315861 | [248] | ?1-0 | |
| N2 | | 0.315870 | [248] | ?1-0 | |
| N2 | | 0.315874 | [248] | ?1-0 | |
| N2 | > | 0.315883 | [248] | ?1-0 | |
| N2 | | 0.315891 | [248] | ?1-0 | |
| N2 | | 0.315900 | [248] | ?1-0 | |
| N2 | | 0.315911 | [248] | ?1-0 | |
| N2 | > | 0.315919 | [248] | ?1-0 | |
| N2 | | 0.3364909 | [247] | ?0-0 | R1-7 |
| N2 | | 0.3365425 | [247] | ?0-0 | --- |
| N2 | | 0.3365478 | [247] | 0-0 | R3-6 |
| N2 | | 0.3366913 | [247,254] | 0-0 | R2-4 |
| N2 | | 0.3369541 | [246] | 0-0 | Q3?-2 |
| N2 | | 0.3369552 | [246] | 0-0 | P1?-2 |
| N2 | | 0.3369769 | [247] | ?0-0 | --- |

| | | WAVELENGTH IN AIR [MICROMETER] | | TRANSITION |
|---|---|---|---|---|
| N2 | | 0.3369823 | [246] | 0-0 P1?-3 |
| N2 | | 0.3369835 | [246] | 0-0 P1-3 |
| N2 | | 0.3369907 | [247] | 0-0 P1-17 |
| N2 | | 0.3370027 | [246] | 0-0 P2-2 |
| N2 | | 0.3370075 | [246] | 0-0 P1?-4 |
| N2 | | 0.3370081 | [246] | 0-0 P1-4 |
| N2 | | 0.3370137 | [247] | 0-0 P1-16 |
| N2 | | 0.3370174 | [247] | ?0-0 P2-16 |
| N2 | | 0.3370288 | [246] | 0-0 P1?-5 |
| N2 | | 0.3370295 | [246,231] | 0-0 P1-5 |
| N2 | | 0.3370312 | [246] | 0-0 P2?-3 |
| N2 | | 0.3370381 | [247] | 0-0 P1-15 |
| N2 | > | 0.3370438 | [247,254] | 0-0 P2-15 |
| N2 | | 0.3370466 | [246] | 0-0 P1?-6 |
| N2 | | 0.3370474 | [246] | 0-0 P1-6 |
| N2 | | 0.3370555 | [246] | 0-0 P1-14 |
| N2 | | 0.3370562 | [246] | 0-0 P2-4 |
| N2 | | 0.3370608 | [246] | 0-0 P1?-7 |
| N2 | | 0.3370619 | [246,231] | 0-0 P1-7 |
| N2 | | 0.3370665 | [247] | 0-0 P2-14 |
| N2 | | 0.3370677 | [246] | 0-0 P1-13 |
| N2 | | 0.3370714 | [246,231] | 0-0 P1?-8 |
| N2 | | 0.3370726 | [246] | 0-0 P1-8 |
| N2 | > | 0.3370749 | [246,231] | 0-0 P1?-12 |
| N2 | > | 0.3370758 | [246] | 0-0 P2?-5 |
| N2 | | 0.3370782 | [246] | 0-0 P1?-9 |
| N2 | > | 0.3370797 | [246,231] | 0-0 P1-9 |
| N2 | | 0.3370812 | [246,231] | 0-0 P1?-10 |
| N2 | | 0.3370816 | [246,231] | 0-0 P1-11 |
| N2 | | 0.3370826 | [246] | 0-0 P1-10 |
| N2 | > | 0.3370919 | [246,231] | 0-0 P2-6 |
| N2 | > | 0.3370986 | [246,231] | ?0-0 P3?-4 |
| N2 | > | 0.3371037 | [246,231] | 0-0 P2?-7 |
| N2 | | 0.3371075 | [246] | 0-0 P2?-11 |
| N2 | | 0.3371082 | [246] | 0-0 P2-11 |
| N2 | | 0.3371113 | [246,231] | 0-0 P2-8 |
| N2 | | 0.3371121 | [246] | 0-0 P3?-12 |
| N2 | > | 0.3371135 | [246,231] | 0-0 P2-10 |
| N2 | > | 0.3371143 | [246,231] | 0-0 P2?-9 |
| N2 | > | 0.3371172 | [246,231] | 0-0 P3-5 |
| N2 | | 0.3371266 | [246,231] | 0-0 P3-11 |
| N2 | > | 0.3371307 | [246,231] | 0-0 P3?-6 |
| N2 | > | 0.3371366 | [246,231] | 0-0 P3?-10 |
| N2 | > | 0.3371392 | [246,231] | 0-0 P3-7 |
| N2 | > | 0.3371421 | [246,231] | 0-0 P3-9 |
| N2 | > | 0.3371429 | [246,231] | 0-0 P3?-8 |
| N2 | | 0.3386428 | [247] | ?0-0 P1-21 |
| N2 | | 0.3575460 | [247] | ?0-1 --- |
| N2 | | 0.3575798 | [247] | 0-1 P1-5 |
| N2 | | 0.3575980 | [246] | ?0-1 P1-6 |
| N2 | | 0.3576112 | [247] | 0-1 P1-7 |
| N2 | | 0.3576194 | [246] | 0-1 P1-11 |
| N2 | | 0.3576250 | [246] | 0-1 P1-9 |
| N2 | | 0.3576320 | [246] | ?0-1 P2-5 |
| N2 | | 0.3576571 | [246] | ?0-1 P2-7 |
| N2 | | 0.3576613 | [246] | ?0-1 P3-11 |
| N2 | | 0.3576778 | [246] | ?0-1 P3-5 |

| | | WAVELENGTH IN AIR [MICROMETER] | | TRANSITION |
|---|---|---|---|---|
| N2 | > | 0.3576899 | [246] | ?0-1 P3-9 |
| N2 | > | 0.3576955 | [246] | ?0-1 P3-7 |
| N2 | | 0.3804 | [401] | ?0-2 |
| N2 | | 0.4058 | [401] | ?0-3 |
| N2+ | | 0.4278 | [365,378,472] | |
| N2 | | 0.7482187 | [246] | 4-2 QR23-1 |
| N2 | | 0.748274 | [231] | 4-2 Q1-11 |
| N2 | | 0.7485941 | [246] | ?4-2 --- |
| N2 | | 0.7486135 | [246] | 4-2 PQ23-3 |
| N2 | | 0.7486413 | [246] | 4-2 P2-4 |
| N2 | | 0.7486253 | [246] | 4-2 PQ23-4 |
| N2 | | 0.7487409 | [246,231] | ?4-2 Q1-9 |
| N2 | | 0.7488046 | [246] | ?4-2 --- |
| N2 | > | 0.7488246 | [246] | ?4-2 P2-6 |
| N2 | | 0.7489107 | [246] | ?4-2 PQ23-6 |
| N2 | | 0.7489626 | [246] | ?4-2 Q1-8 |
| N2 | | 0.7489809 | [231,246] | ?4-2 P2-13 |
| N2 | | 0.7490096 | [246] | 4-2 P2-12 |
| N2 | | 0.7490317 | [246] | 4-2 PQ23-8 |
| N2 | | 0.7491510 | [246] | ?4-2 --- |
| N2 | > | 0.7491705 | [231,246] | 4-2 Q1-7 |
| N2 | | 0.7492379 | [246] | 4-2 OP23-4 |
| N2 | | 0.7493082 | [246] | 4-2 QR12-6 |
| N2 | | 0.7493716 | [246] | 4-2 Q1-6 |
| N2 | | 0.7493910 | [246] | ?4-2 --- |
| N2 | | 0.7495086 | [246] | 4-2 OP23-5 |
| N2 | | 0.7495465 | [246] | ?4-2 --- |
| N2 | | 0.7495660 | [231,246] | 4-2 Q1-5 |
| N2 | | 0.7496024 | [246] | ?4-2 --- |
| N2 | | 0.7497256 | [246] | ?4-2 --- |
| N2 | > | 0.7497524 | [246] | 4-2 Q1-4 |
| N2 | | 0.7497728 | [246] | ?4-2 --- |
| N2 | | 0.7498898 | [246] | ?4-2 --- |
| N2 | | 0.7499013 | [246] | ?4-2 --- |
| N2 | > | 0.7499327 | [246] | 4-2 Q1-3 |
| N2 | | 0.7499593 | [246] | ?4-2 --- |
| N2 | | 0.7499825 | [246] | 4-2 P1-13 |
| N2 | | 0.7500071 | [246] | ?4-2 OP23-7 |
| N2 | | 0.7500646 | [246] | ?4-2 --- |
| N2 | | 0.7500734 | [246] | ?4-2 --- |
| N2 | > | 0.7501056 | [246] | 4-2 Q1-2 |
| N2 | | 0.7501295 | [246] | ?4-2 --- |
| N2 | | 0.7501404 | [246] | ?4-2 --- |
| N2 | | 0.7501553 | [231,246] | ?4-2 P1-11 |
| N2 | | 0.7502139 | [246] | 4-2 PQ12-9 |
| N2 | | 0.7502729 | [246] | 4-2 Q1-1 |
| N2 | | 0.7502768 | [246] | 4-2 P1-9 |
| N2 | > | 0.7503035 | [246] | 4-2 PQ12-7 |
| N2 | > | 0.7503371 | [246] | 4-2 PQ12-6 |
| N2 | > | 0.7503418 | [246] | 4-2 PR13-8 |
| N2 | > | 0.7503642 | [246] | 4-2 PQ12-5 |
| N2 | | 0.7503669 | [246] | ?4-2 --- |
| N2 | | 0.7503697 | [246] | ?4-2 PQ12-4 |
| N2 | | 0.7503838 | [246] | 4-2 PR13-7 |
| N2 | | 0.7503960 | [246] | 4-2 PR13-7 |

| | | WAVELENGTH IN AIR [MICROMETER] | | TRANSITION |
|---|---|---|---|---|
| N2 | > | 0.7503994 | [246] | 4-2 PQ12-3 |
| N2 | | 0.7504106 | [246] | 4-2 PQ12-2 |
| N2 | > | 0.7504160 | [246] | 4-2 PQ12-1 |
| N2 | > | 0.7504184 | [246] | 4-2 PR13-6 |
| N2 | | 0.7504274 | [246] | ?4-2 Q1-0 |
| N2 | > | 0.7504598 | [246] | 4-2 P1-3 |
| N2 | | 0.7504768 | [246] | ?4-2 P1-1 |
| N2 | | 0.7505113 | [246] | 4-2 PR13-2 |
| N2 | | 0.7505710 | [246] | ?4-2 --- |
| N2 | | 0.7505903 | [246] | ?4-2 --- |
| N2 | | 0.7506063 | [246] | ?4-2 --- |
| N2 | | 0.7506356 | [246] | 4-2 OP23-10 |
| N2 | | 0.7508145 | [246] | 4-2 OP23-11 |
| N2 | | 0.7509890 | [246] | ?4-2 --- |
| N2 | > | 0.7510133 | [246] | 4-2 OP12-3 |
| N2 | | 0.7510923 | [246] | ?4-2 --- |
| N2 | | 0.7511592 | [246] | 4-2 OP12-4 |
| N2 | | 0.7511799 | [246] | ?4-2 --- |
| N2 | > | 0.7512003 | [246] | 4-2 OP12-5 |
| N2 | | 0.7512569 | [246] | ?4-2 --- |
| N2 | | 0.7513357 | [246] | 4-2 OP12-6 |
| N2 | | 0.7514079 | [246] | ?4-2 --- |
| N2 | | 0.7515446 | [246] | ?4-2 --- |
| N2 | | 0.7515650 | [246] | 4-2 OP12-7 |
| N2 | | 0.7517728 | [246] | 4-2 OQ13-8 |
| N2 | | 0.7518013 | [246] | 4-2 OP12-9 |
| N2 | | 0.7574329 | [250,231] | 3-1 Q2-17 |
| N2 | > | 0.758105 | [231] | 3-1 Q3-13 |
| N2 | | 0.758423 | [231] | 3-1 Q3-11 |
| N2 | | 0.7586439 | [231,246] | 3-1 Q3-9 |
| N2 | | 0.7587693 | [246] | 3-1 Q3-7 |
| N2 | | 0.7589868 | [250] | 3-1 Q2-13 |
| N2 | | 0.7591960 | [250,231] | 3-1 Q2-12 |
| N2 | | 0.7593908 | [250] | 3-1 Q2-11 |
| N2 | | 0.7594941 | [250,231] | ?3-1 Q1-15 |
| N2 | | 0.7597289 | [250] | 3-1 Q2-9 |
| N2 | | 0.759870 | [231] | 3-1 Q1-13 |
| N2 | | 0.7603477 | [246] | ?3-1 --- |
| N2 | | 0.7606374 | [246] | 3-1 PQ23-2 |
| N2 | > | 0.7607626 | [246] | 3-1 PQ23-3 |
| N2 | > | 0.7608801 | [231,246] | ?3-1 PQ23-4 |
| N2 | > | 0.7609853 | [231,246] | 3-1 PQ23-5 |
| N2 | | 0.7613612 | [250] | 3-1 PQ23-7 |
| N2 | | 0.7610759 | [246] | 3-1 PQ23-6 |
| N2 | | 0.7611082 | [246] | 3-1 Q1-8 |
| N2 | > | 0.7611514 | [246] | 3-1 PQ23-7 |
| N2 | | 0.7612105 | [246] | 3-1 PQ23-8 |
| N2 | | 0.7612528 | [246] | 3-1 PQ23-9 |
| N2 | > | 0.7613260 | [231,246] | 3-1 Q1-7 |
| N2 | > | 0.7615347 | [246] | 3-1 Q1-6 |
| N2 | | 0.7616994 | [246] | 3-1 OP23-5 |
| N2 | > | 0.7617357 | [231,246] | 3-1 Q1-5 |
| N2 | | 0.7619288 | [246] | 3-1 Q1-4 |
| N2 | | 0.7620844 | [246] | ?3-1 --- |
| N2 | | 0.7620943 | [246] | ?3-1 --- |
| N2 | > | 0.7621161 | [246] | 3-1 Q1-3 |
| N2 | | 0.7622235 | [246] | 3-1 OP23-7 |

| | | WAVELENGTH IN AIR [MICROMETER] | | TRANSITION |
|---|---|---|---|---|
| N2 | | 0.7622565 | [246] | ?3-1 --- |
| N2 | > | 0.7622959 | [231,246] | ?3-1 Q1-2 |
| N2 | | 0.7623256 | [246] | ?3-1 --- |
| N2 | | 0.7623264 | [250] | 3-1 Q1-3 |
| N2 | | 0.7623311 | [246] | ?3-1 --- |
| N2 | | 0.7623582 | [246] | 3-1 PQ12-10 |
| N2 | | 0.7623686 | [246] | ?3-1 --- |
| N2 | | 0.7623918 | [246] | ?3-1 --- |
| N2 | | 0.7624220 | [231,246] | 3-1 PQ12-9 |
| N2 | > | 0.7624690 | [246] | ?3-1 Q1-1 |
| N2 | > | 0.7624924 | [246] | 3-1 P1-9 |
| N2 | > | 0.7625115 | [231,246] | 3-1 PQ12-7 |
| N2 | > | 0.7625445 | [246] | 3-1 PQ12-6 |
| N2 | | 0.7625709 | [246] | 3-1 PQ12-5 |
| N2 | | 0.7625770 | [246] | ?3-1 --- |
| N2 | | 0.7625812 | [246] | 3-1 P1-7 |
| N2 | | 0.7625906 | [246] | 3-1 PQ12-4 |
| N2 | | 0.7626007 | [246] | 3-1 PR13-7 |
| N2 | | 0.7626044 | [246] | 3-1 PQ12-3 |
| N2 | | 0.7626114 | [246] | 3-1 P1-6 |
| N2 | > | 0.7626180 | [246] | 3-1 PQ12-2 |
| N2 | | 0.7626207 | [246] | 3-1 PQ12-1 |
| N2 | > | 0.7626360 | [246] | 3-1 P1-5 |
| N2 | | 0.7626560 | [246] | 3-1 P1-4 |
| N2 | > | 0.7626700 | [246] | 3-1 P1-3 |
| N2 | > | 0.7626749 | [246] | ?3-1 P1-1 |
| N2 | | 0.7626826 | [246] | ?3-1 OP23-9 |
| N2 | | 0.7627806 | [250] | 3-1 PQ12-5 |
| N2 | | 0.7628854 | [246] | 3-1 OP23-10 |
| N2 | | 0.7629102 | [246] | ?3-1 --- |
| N2 | | 0.7630305 | [246] | ?3-1 --- |
| N2 | | 0.7631880 | [246] | ?3-1 --- |
| N2 | | 0.7632446 | [246] | ?3-1 OP12-3 |
| N2 | | 0.7633348 | [246] | ?3-1 --- |
| N2 | > | 0.7633985 | [246] | ?3-1 OP12-4 |
| N2 | | 0.7634546 | [246] | ?3-1 --- |
| N2 | | 0.7634779 | [246] | ?3-1 --- |
| N2 | > | 0.7635474 | [246] | 3-1 OP12-5 |
| N2 | | 0.7636126 | [246] | ?3-1 --- |
| N2 | | 0.7636904 | [246] | 3-1 OP12-6 |
| N2 | | 0.7637586 | [246] | ?3-1 --- |
| N2 | > | 0.7638274 | [246] | 3-1 OP12-7 |
| N2 | | 0.7639571 | [246] | 3-1 OP12-8 |
| N2 | | 0.7639715 | [246] | ?3-1 --- |
| N2 | | 0.7640383 | [246] | ?3-1 --- |
| N2 | > | 0.7640794 | [246] | 3-1 OP12-9 |
| N2 | | 0.7641929 | [246] | 3-1 OP12-10 |
| N2 | | 0.7642478 | [246] | ?3-1 --- |
| N2 | | 0.7644612 | [246] | ?3-1 --- |
| N2 | > | 0.771206 | [231] | 2-0 Q3-9 |
| N2 | | 0.7724562 | [245] | 2-0 Q1-13 |
| N2 | | 0.7730032 | [245] | 2-0 Q1-11 |
| N2 | | 0.7735040 | [245] | 2-0 Q1-9 |
| N2 | | 0.7739632 | [245] | 2-0 Q1-7 |
| N2 | | 0.7743859 | [231,246] | 2-0 Q1-5 |
| N2 | | 0.775270 | [231] | 2-0 PR13-1 |
| N2 | > | 0.7752354 | [246] | ?2-0 P1-8 |

| | | WAVELENGTH IN AIR [MICROMETER] | | TRANSITION |
|---|---|---|---|---|
| N2 | | 0.7753652 | [246] | ?2-0 P1-2 |
| N2 | | 0.865331 | [231] | 2-1 R3-7 |
| N2 | | 0.865492 | [231] | 2-1 Q3-15 |
| N2 | | 0.866089 | [231] | 2-1 Q3-13 |
| N2 | | 0.866256 | [231] | 2-1 Q2-15 |
| N2 | | 0.866345 | [231] | 2-1 Q3-12 |
| N2 | > | 0.866572 | [231] | 2-1 Q3-11 |
| N2 | | 0.86676 | [231] | 2-1 Q3-10 |
| N2 | | 0.8669223 | [231,246] | 2-1 Q3-9 |
| N2 | | 0.866959 | [231] | 2-1 Q2-13 |
| N2 | | 0.8671332 | [231,246] | 2-1 Q3-7 |
| N2 | | 0.867554 | [231] | 2-1 Q2-11 |
| N2 | | 0.868281 | [231] | 2-1 R1-7 |
| N2 | | 0.8682937 | [250] | 2-1 Q2-9 |
| N2 | | 0.868374 | [231] | 2-1 Q1-13 |
| N2 | | 0.868762 | [231] | 2-1 Q1-12 |
| N2 | > | 0.869136 | [231] | 2-1 Q1-11 |
| N2 | | 0.8692560 | [246] | 2-1 QP23-1 |
| N2 | | 0.869490 | [231] | 2-1 Q1-10 |
| N2 | | 0.8696366 | [246] | 2-1 PQ23-2 |
| N2 | | 0.8697945 | [246] | 2-1 PQ23-3 |
| N2 | | 0.8698263 | [231,246] | 2-1 Q1-9 |
| N2 | | 0.8699397 | [246] | 2-1 PQ23-4 |
| N2 | > | 0.8700670 | [231,246] | 2-1 PQ23-5 |
| N2 | > | 0.8700684 | [246] | 2-1 PQ23-5 |
| N2 | > | 0.8701481 | [231,246] | 2-1 Q1-8 |
| N2 | | 0.8701718 | [246] | 2-1 PQ23-6 |
| N2 | > | 0.8702541 | [231,246] | 2-1 PQ23-7 |
| N2 | | 0.8702681 | [246] | 2-1 OP23-3 |
| N2 | | 0.8703093 | [246] | 2-1 PQ23-8 |
| N2 | | 0.870331 | [231] | 2-1 PQ23-11 |
| N2 | | 0.8703457 | [246] | 2-1 PQ23-9 |
| N2 | > | 0.8704549 | [231,246] | 2-1 Q1-7 |
| N2 | > | 0.8707478 | [231,246] | 2-1 Q1-6 |
| N2 | | 0.8710118 | [246] | 2-1 OP23-5 |
| N2 | > | 0.8710273 | [231,246] | 2-1 Q1-5 |
| N2 | > | 0.8712956 | [231,246] | 2-1 Q1-4 |
| N2 | | 0.8713533 | [246] | ?2-1 OP23-6 |
| N2 | | 0.871450 | [231] | 2-1 P1-13 |
| N2 | > | 0.8715519 | [231,246] | 2-1 Q1-3 |
| N2 | | 0.871644 | [231] | 2-1 PQ12-11 |
| N2 | > | 0.8716718 | [246] | 2-1 OP23-7 |
| N2 | | 0.8717377 | [231,246] | 2-1 P1-11 |
| N2 | > | 0.8717970 | [246] | 2-1 Q1-2 |
| N2 | | 0.8718571 | [246] | 2-1 P1-10 |
| N2 | | 0.8718654 | [231,246] | 2-1 PQ12-9 |
| N2 | | 0.8719537 | [246] | 2-1 PQ12-8 |
| N2 | | 0.8719562 | [231,246] | 2-1 P1-9 |
| N2 | | 0.8719562 | [246] | ?2-1 --- |
| N2 | | 0.8719791 | [246] | ?2-1 --- |
| N2 | > | 0.8720251 | [231,246] | 2-1 PQ12-7 |
| N2 | > | 0.8720284 | [246] | 2-1 Q1-1 |
| N2 | | 0.8720308 | [246] | ?2-1 --- |
| N2 | > | 0.8720419 | [246] | 2-1 P1-8 |
| N2 | | 0.8720848 | [246] | 2-1 PQ12-6 |
| N2 | > | 0.8721155 | [246] | 2-1 P1-7 |

| | | WAVELENGTH IN AIR [MICROMETER] | | TRANSITION |
|---|---|---|---|---|
| N2 | | 0.8721327 | [231,246] | 2-1 PQ12-5 |
| N2 | > | 0.8721718 | [246] | ?2-1 P1-6 |
| N2 | | 0.8721971 | [246] | ?2-1 --- |
| N2 | > | 0.8722007 | [231,246] | 2-1 PQ12-3 |
| N2 | > | 0.8722220 | [246] | ?2-1 P1-5 |
| N2 | > | 0.8722341 | [246] | 2-1 PQ12-1 |
| N2 | | 0.8722569 | [246] | 2-1 P1-4 |
| N2 | | 0.8722836 | [246] | 2-1 P1-3 |
| N2 | | 0.8723057 | [246] | 2-1 P1-1 |
| N2 | | 0.8726333 | [246] | 2-1 OP12-1 |
| N2 | | 0.8728430 | [246] | 2-1 OP12-2 |
| N2 | | 0.8730453 | [246] | 2-1 OP12-3 |
| N2 | | 0.8732394 | [246] | 2-1 OP12-4 |
| N2 | > | 0.8734247 | [246] | 2-1 OP12-5 |
| N2 | | 0.8735995 | [246] | 2-1 OP12-6 |
| N2 | > | 0.8737644 | [246] | 2-1 OP12-7 |
| N2 | | 0.8739162 | [246] | 2-1 OP12-8 |
| N2 | | 0.8740559 | [246] | 2-1 OP12-9 |
| N2 | | 0.8742917 | [246] | 2-1 OP12-11 |
| N2 | | 0.884129 | [231] | 1-0 Q3-15 |
| N2 | | 0.8845349 | [246] | 1-0 SR32-1 |
| N2 | | 0.8846598 | [250,231] | ?1-0 RQ21-9 |
| N2 | | 0.884758 | [231] | 1-0 Q3-13 |
| N2 | | 0.884920 | [231] | 1-0 Q2-15 |
| N2 | | 0.885026 | [231] | 1-0 Q3-12 |
| N2 | | 0.885261 | [231] | 1-0 Q3-11 |
| N2 | | 0.885460 | [231] | 1-0 Q3-10 |
| N2 | | 0.8856271 | [231,246] | 1-0 Q3-9 |
| N2 | | 0.885649 | [231] | 1-0 Q2-13 |
| N2 | | 0.8858470 | [231,246] | 1-0 Q3-7 |
| N2 | | 0.8861195 | [250,231] | ?1-0 R1-9 |
| N2 | | 0.886153 | [231] | 1-0 RQ21-5 |
| N2 | | 0.886256 | [231] | 1-0 Q1-15 |
| N2 | | 0.886278 | [231] | 1-0 Q2-11 |
| N2 | | 0.886697 | [231] | 1-0 Q1-14 |
| N2 | | 0.886799 | [231] | 1-0 Q2-9 |
| N2 | > | 0.887121 | [231] | 1-0 Q1-13 |
| N2 | | 0.887531 | [231] | 1-0 Q1-12 |
| N2 | > | 0.887918 | [231] | 1-0 Q1-11 |
| N2 | | 0.8880521 | [246] | 1-0 QR23-1 |
| N2 | | 0.888288 | [231] | 1-0 Q1-10 |
| N2 | | 0.8884527 | [246] | 1-0 PQ23-2 |
| N2 | | 0.8886204 | [246] | 1-0 PQ23-3 |
| N2 | | 0.8886378 | [231,246] | 1-0 Q1-9 |
| N2 | | 0.8887756 | [246] | 1-0 PQ23-4 |
| N2 | | 0.8889111 | [231,246] | 1-0 PQ23-5 |
| N2 | | 0.8889738 | [231,246] | 1-0 Q1-8 |
| N2 | | 0.8890243 | [246] | 1-0 PQ23-6 |
| N2 | | 0.8891133 | [231,246] | 1-0 PQ23-7 |
| N2 | | 0.8891769 | [246] | 1-0 PQ23-8 |
| N2 | | 0.8892149 | [231,246] | ?1-0 PQ23-9 |
| N2 | | 0.8892940 | [231,246] | 1-0 Q1-7 |
| N2 | | 0.8896001 | [231,246] | 1-0 Q1-6 |
| N2 | | 0.8898930 | [231,246] | 1-0 Q1-5 |
| N2 | | 0.8899078 | [246] | 1-0 OP23-5 |
| N2 | | 0.8901733 | [231,246] | 1-0 Q1-4 |
| N2 | | 0.8902420 | [250] | 1-0 P1-15 |

| | WAVELENGTH IN AIR [MICROMETER] | | TRANSITION |
|---|---|---|---|
| N2 | 0.8902711 | [246] | 1-0 OP23-6 |
| N2 | 0.890372 | [231] | 1-0 P1-13 |
| N2 | 0.8904419 | [231,246] | 1-0 Q1-3 |
| N2 | 0.890566 | [231] | 1-0 PQ12-11 |
| N2 | 0.8906097 | [246] | 1-0 OP23-7 |
| N2 | 0.8906649 | [231,246] | 1-0 P1-11 |
| N2 | 0.8906994 | [246] | 1-0 Q1-2 |
| N2 | 0.8907920 | [231,246] | 1-0 PQ12-9 |
| N2 | 0.8908808 | [246] | 1-0 PQ12-8 |
| N2 | 0.8908878 | [231,246] | 1-0 P1-9 |
| N2 | 0.8909451 | [246] | 1-0 Q1-1 |
| N2 | 0.8909527 | [231,246] | 1-0 PQ12-7 |
| N2 | 0.8909750 | [246] | 1-0 P1-8 |
| N2 | 0.8910132 | [246] | 1-0 PQ12-6 |
| N2 | 0.8910480 | [231,246] | 1-0 P1-7 |
| N2 | 0.8910612 | [246] | 1-0 PQ12-5 |
| N2 | 0.8911001 | [246] | 1-0 PQ12-4 |
| N2 | 0.8911063 | [246] | 1-0 P1-6 |
| N2 | 0.8911280 | [231,246] | 1-0 PQ12-3 |
| N2 | 0.8911502 | [246] | 1-0 PQ12-2 |
| N2 | 0.8911538 | [246] | 1-0 P1-5 |
| N2 | 0.8911608 | [246] | 1-0 PQ12-1 |
| N2 | 0.8911898 | [246] | 1-0 P1-4 |
| N2 | 0.8912139 | [246] | ?1-0 OP23-9 |
| N2 | 0.8918033 | [246] | 1-0 PP12-2 |
| N2 | 0.8920184 | [246] | 1-0 OP12-3 |
| N2 | 0.8922249 | [246] | 1-0 OP12-4 |
| N2 | 0.8924223 | [246] | 1-0 OP12-5 |
| N2 | 0.8926099 | [246] | 1-0 OP12-6 |
| N2 | 0.8927865 | [246] | 1-0 OP12-7 |
| N2 | 0.8929509 | [246] | 1-0 OP12-8 |
| N2 | 0.8931019 | [246] | 1-0 OP12-9 |
| N2 | 0.8933580 | [246] | 1-0 OP12-11 |
| N2 | 0.965389 | [246] | 3-3 P2-11 |
| N2 | 0.965846 | [246] | 3-3 Q1-7 |
| N2 | 0.966599 | [246] | 3-3 Q1-5 |
| N2 | 0.967270 | [246] | 3-3 Q1-3 |
| N2 | 0.967758 | [246] | 3-3 PQ12-7 |
| N2 | 0.967943 | [246] | 3-3 PQ12-5 |
| N2 | 0.968061 | [246] | 3-3 PQ12-3 |
| N2 | 0.969552 | [246] | 3-3 OP12-5 |
| N2 | 0.969879 | [246] | 3-3 OP12-7 |

| | WAVELENGTH IN VACUUM [MICROMETER] | | TRANSITION |
|---|---|---|---|
| N2 | 1.043874 | [231] | 0-0 Q1-15 |
| N2 | 1.044548 | [231] | 0-0 Q1-14 |
| N2 | 1.045189 | [231] | 0-0 Q1-13 |
| N2 | 1.045806 | [231] | 0-0 Q1-12 |
| N2 | 1.046404 | [231] | 0-0 Q1-11 |
| N2 | 1.046956 | [231] | 0-0 Q1-10 |
| N2 | 1.047482 | [231] | 0-0 Q1-9 |
| N2 | 1.047979 | [231] | 0-0 Q1-8 |
| N2 | > 1.048249 | [231] | 0-0 PQ23-7 |
| N2 | 1.048461 | [231] | 0-0 Q1-7 |

| | | WAVELENGTH IN VACUUM [MICROMETER] | | TRANSITION |
|---|---|---|---|---|
| N2 | | 1.048922 | [231] | 0-0 Q1-6 |
| N2 | > | 1.049348 | [231] | 0-0 Q1-5 |
| N2 | | 1.049766 | [231] | 0-0 Q1-4 |
| N2 | | 1.050161 | [231] | 0-0 Q1-3 |
| N2 | > | 1.050519 | [231] | 0-0 PQ12-9 |
| N2 | | 1.050800 | [231] | 0-0 PQ12-7 |
| N2 | | 1.051005 | [231] | 0-0 PQ12-5 |
| N2 | | 1.051118 | [231] | 0-0 PQ12-3 |
| N2 | | 1.052548 | [231] | 0-0 OQ12-3 |
| N2 | | 1.052911 | [231] | 0-0 OP12-5 |
| N2 | | 1.053382 | [231] | 0-0 OP12-7 |
| N2 | | 1.053760 | [231] | 0-0 OP12-9 |
| N2 | | 1.230598 | [231] | 0-1 Q1-11 |
| N2 | > | 1.231430 | [231] | 0-1 Q1-10 |
| N2 | | 1.232219 | [231] | 0-1 Q1-9 |
| N2 | | 1.232962 | [231] | 0-1 Q1-8 |
| N2 | | 1.233671 | [231] | 0-1 Q1-7 |
| N2 | | 1.234332 | [231] | 0-1 Q1-6 |
| N2 | | 1.234969 | [231] | 0-1 Q1-5 |
| N2 | | 3.29462 | [086] | 2-1 Q(14) |
| N2 | | 3.30166 | [086] | 2-1 Q(12) |
| N2 | | 3.30755 | [086] | 2-1 Q(10) |
| N2 | | 3.31239 | [086] | 2-1 Q(08) |
| N2 | | 3.31616 | [086] | 2-1 Q(06) |
| N2 | | 3.31892 | [086] | 2-1 Q(04) |
| N2 | | 3.45212 | [086] | 1-0 Q(12) |
| N2 | | 3.45852 | [086] | 1-0 Q(10) |
| N2 | | 3.46377 | [086] | 1-0 Q(08) |
| N2 | | 3.46804 | [086] | 1-0 Q(06) |
| N2 | | 3.47127 | [086] | 1-0 Q(04) |
| N2 | | 8.18384 | [086] | 0-0 Q(08) |
| N2 | | 8.21106 | [086] | 0-0 Q(06) |

NUMBER OF LINES IN NITROGEN, MOLECULAR 447

ACTIVE MEDIUM : BISMUTH, MOLECULAR
SYMBOL : BI2

OPERATING CONDITIONS : OPTICAL PUMPING WITH AR-LASER AT 514.5 NM AND DYE-LASER [565].

| | WAVELENGTH IN AIR [MICROMETER] | | | WAVELENGTH IN AIR [MICROMETER] | |
|---|---|---|---|---|---|
| BI2 | 0.5929 | [551] | BI2 | 0.6414 | [551] |
| BI2 | 0.6160 | [551] | BI2 | 0.6422 | [551] |
| BI2 | 0.6339 | [551] | BI2 | 0.6603 | [551] |
| BI2 | 0.6576 | [551] | BI2 | 0.6809 | [551] |
| BI2 | 0.6582 | [551] | BI2 | 0.7006 | [551] |
| BI2 | 0.6650 | [551] | BI2 | 0.7013 | [551] |
| | | | | | |
| BI2 | 0.6239 | [551] | BI2 | 0.7292 | [551] |
| BI2 | 0.6300 | [551] | BI2 | 0.7301 | [551] |

| WAVELENGTH IN AIR [MICROMETER] | | | WAVELENGTH IN AIR [MICROMETER] | | |
|---|---|---|---|---|---|
| BI2 | 0.7366 | [551] | BI2 | 0.7408 | [551] |
| BI2 | 0.7376 | [551] | BI2 | 0.7471 | [551] |
| | | | BI2 | 0.7482 | [551] |
| BI2 | 0.7364 | [551] | | | |
| BI2 | 0.7439 | [551] | BI2 | 0.7468 | [551] |
| | | | BI2 | 0.7475 | [551] |
| BI2 | 0.7335 | [551] | BI2 | 0.7543 | [551] |
| BI2 | 0.7398 | [551] | BI2 | 0.7551 | [551] |

NUMBER OF LINES IN BISMUTH, MOLECULAR 29

ACTIVE MEDIUM : SULPHUR, MOLECULAR
SYMBOL : S2

OPERATING CONDITIONS : PUMPING BY UV-PHOTODISSOCATION OR OPTICAL PUMPING WITH SHG-DYE LASER.

| WAVELENGTH IN AIR [MICROMETER] | | | WAVELENGTH IN AIR [MICROMETER] | | |
|---|---|---|---|---|---|
| S2 | 0.365-0.57 | [519] | S2 | 1.0941 | [520] |
| S2 | 1.086 | [520] | S2 | 1.0946 | [520] |
| S2 | 1.0915 | [520] | S2 | 1.099 | [520] |
| S2 | 1.0917 | [520] | S2 | 1.100 | [520] |
| S2 | 1.0920 | [520] | S2 | 1.1567 | [520] |
| S2 | 1.0923 | [520] | | | |

NUMBER OF LINES IN SULPHUR, MOLECULAR 11

ACTIVE MEDIUM : TELLURIUM, MOLECULAR
SYMBOL : TE2

OPERATING CONDITIONS : OPTICAL PUMPING WITH AR-LASER AT 476.5 NM.

| WAVELENGTH IN AIR [MICROMETER] | | | WAVELENGTH IN AIR [MICROMETER] | | |
|---|---|---|---|---|---|
| TE2 | 0.5643 | [551] | TE2 | 0.6004 | [551] |
| TE2 | 0.5647 | [551] | TE2 | 0.6008 | [551] |
| TE2 | 0.5715 | [551] | TE2 | 0.6067 | [551] |
| TE2 | 0.5720 | [551] | TE2 | 0.6168 | [551] |
| TE2 | 0.5767 | [551] | | | |
| TE2 | 0.5784 | [551] | | | |
| TE2 | 0.5790 | [551] | TE2 | 0.5575 | [551] |
| TE2 | 0.5851 | [551] | TE2 | 0.5579 | [551] |
| TE2 | 0.5859 | [551] | TE2 | 0.5650 | [551] |
| TE2 | 0.5927 | [551] | TE2 | 0.5719 | [551] |
| TE2 | 0.5936 | [551] | TE2 | 0.5724 | [551] |

| | WAVELENGTH IN AIR [MICROMETER] | |
|---|---|---|
| TE2 | 0.5794 | [551] |
| TE2 | 0.5798 | [551] |
| TE2 | 0.5870 | [551] |
| TE2 | 0.5874 | [551] |
| | | |
| TE2 | 0.5774 | [551] |
| TE2 | 0.5783 | [551] |
| TE2 | 0.5849 | [551] |
| TE2 | 0.5857 | [551] |
| TE2 | 0.5924 | [551] |
| TE2 | 0.5934 | [551] |
| TE2 | 0.6002 | [551] |
| | | |
| TE2 | 0.5626 | [551] |
| TE2 | 0.5696 | [551] |
| TE2 | 0.5701 | [551] |
| TE2 | 0.5714 | [551] |
| TE2 | 0.5766 | [551] |
| TE2 | 0.5773 | [551] |
| TE2 | 0.5780 | [551] |
| TE2 | 0.5787 | [551] |
| TE2 | 0.5841 | [551] |
| TE2 | 0.5851 | [551] |
| TE2 | 0.6005 | [551] |
| TE2 | 0.6009 | [551] |
| TE2 | 0.6082 | [551] |
| TE2 | 0.6085 | [551] |
| TE2 | 0.6162 | [551] |
| TE2 | 0.6165 | [551] |
| TE2 | 0.6278 | [551] |
| TE2 | 0.6287 | [551] |
| TE2 | 0.6371 | [551] |
| TE2 | 0.6379 | [551] |
| TE2 | 0.6465 | [551] |
| TE2 | 0.6473 | [551] |
| TE2 | 0.6561 | [551] |

| | WAVELENGTH IN AIR [MICROMETER] | |
|---|---|---|
| TE2 | 0.5569 | [551] |
| TE2 | 0.5571 | [551] |
| TE2 | 0.5638 | [551] |
| TE2 | 0.5642 | [551] |
| TE2 | 0.5711 | [551] |
| TE2 | 0.5715 | [551] |
| TE2 | 0.5785 | [551] |
| TE2 | 0.5786 | [551] |
| TE2 | 0.5865 | [551] |
| TE2 | 0.6005 | [551] |
| TE2 | 0.6009 | [551] |
| TE2 | 0.6085 | [551] |
| TE2 | 0.6089 | [551] |
| TE2 | 0.6165 | [551] |
| TE2 | 0.6170 | [551] |
| | | |
| TE2 | 0.5575 | [551] |
| TE2 | 0.5578 | [551] |
| TE2 | 0.5646 | [551] |
| TE2 | 0.5649 | [551] |
| TE2 | 0.5701 | [551] |
| TE2 | 0.5719 | [551] |
| TE2 | 0.5721 | [551] |
| TE2 | 0.5793 | [551] |
| TE2 | 0.5797 | [551] |
| TE2 | 0.5869 | [551] |
| TE2 | 0.6204 | [551] |
| TE2 | 0.6288 | [551] |
| TE2 | 0.6295 | [551] |
| TE2 | 0.6381 | [551] |
| TE2 | 0.6388 | [551] |
| TE2 | 0.6477 | [551] |
| TE2 | 0.6484 | [551] |
| TE2 | 0.6574 | [551] |
| TE2 | 0.6581 | [551] |

NUMBER OF LINES IN TELLURIUM, MOLECULAR 88

ACTIVE MEDIUM : FLUORINE, MOLECULAR
SYMBOL : F2

OPERATING CONDITIONS : EXCITATION BY E-BEAM IN NE+F2 OR HE+F2 MIXTURES.
EXCITATION ALSO BY ELECTRICAL DISCHARGE WITH UV-PREIONISATION [577].

| | WAVELENGTH IN VACUUM [MICROMETER] | | | WAVELENGTH IN VACUUM [MICROMETER] | |
|---|---|---|---|---|---|
| F2 | 0.15671 | [544] | F2 | 0.15759 | [544] |
| F2 | 0.15748 | [475,544] | | | |

NUMBER OF LINES IN FLUORINE, MOLECULAR 3

ACTIVE MEDIUM : BROMINE, MOLECULAR
SYMBOL : BR2

OPERATING CONDITIONS : EXCITATION BY E-BEAM [408] OR ELECTRICAL DISCHARGE [392] IN AR+BR2. OPTICAL PUMPING WITH ND:YAG-SHG LASER [518].

| WAVELENGTH IN AIR [MICROMETER] | | | WAVELENGTH IN AIR [MICROMETER] | | |
|---|---|---|---|---|---|
| BR2 | 0.2915 | [392] | BR2 | 0.63654 | [518] |
| BR2 | 0.55020 | [518] | BR2 | 0.63705 | [518] |
| BR2 | 0.55053 | [518] | BR2 | 0.67408 | [518] |
| BR2 | 0.58048 | [518] | BR2 | 0.67455 | [518] |
| BR2 | 0.58090 | [518] | BR2 | 0.67456 | [518] |
| BR2 | 0.61272 | [518] | BR2 | 0.67506 | [518] |
| BR2 | 0.61316 | [518] | BR2 | 0.74582 | [518] |
| BR2 | 0.61318 | [518] | BR2 | 0.74638 | [518] |
| BR2 | 0.61368 | [518] | BR2 | 0.74641 | [518] |
| BR2 | 0.63612 | [518] | BR2 | 0.74704 | [518] |
| BR2 | 0.63654 | [518] | | | |

NUMBER OF LINES IN BROMINE, MOLECULAR 21

ACTIVE MEDIUM : CHLORINE, MOLECULAR
SYMBOL : CL2

OPERATING CONDITIONS : EXCITATION BY AXIAL ELECTON-BEAM IN A MIXTURE OF 2.0 TORR CL2 AND 9300 TORR HE.

| WAVELENGTH IN AIR [MICROMETER] | | | WAVELENGTH IN AIR [MICROMETER] | | |
|---|---|---|---|---|---|
| CL2 | 0.2580 | [595] | | | |

NUMBER OF LINES IN CHLORINE, MOLECULAR 1

ACTIVE MEDIUM : IODINE, MOLECULAR
SYMBOL : I2

OPERATING CONDITIONS : OPTICAL PUMPING OF I2-VAPOR WITH AR-LASER [499,510,612], SHG-ND:YAG-LASER [264] OR DYE-LASER [521]. THE LISTED LASER LINES ARE BUT A SMALL SECTION OF THE TOTAL EMISSION SPECTRUM. WITH PROPER SELECTION OF PUMPING WAVELENGTH MORE THAN A MILLION LASER LINES MAY BE GENERATED. WAVENUMBERS MAY BE CALCULATED FROM THE B-STATE [613] AND X-STATE [612] CONSTANTS.

IN THE 0.5 - 1.4 MICROMETER REGION IN ANY 0.01 CM-1 INTERVAL A I2-LASER LINE CAN BE FOUND PREDICTED TO PLUS/MINUS 0.003 CM-1 TOGETHER WITH THE OPTIMUM PUMP WAVELENGTH [614].

3.6 MW UV-EMISSION IS GENERATED BY E-BEAM EXCITATION OF AN AR+CI3D MIXTURE [390,391,538]. ALSO OPTICAL PUMPING WITH FLASHLAMPS [477].

| | WAVELENGTH IN AIR [MICROMETER] | | | WAVELENGTH IN AIR [MICROMETER] | |
|---|---|---|---|---|---|
| I2 | 0.3420 | [391,538] | I2 | 0.3424 | [391,538] |
| I2 | 0.3423 | [391,538] | I2 | 0.3428 | [391,538] |

| | WAVELENGTH IN AIR [MICROMETER] | | TRANSITION |
|---|---|---|---|
| I2 | 0.5543 | [264] | |
| I2 | 0.5550 | [264] | |
| I2 | 0.5567 | [264] | |
| I2 | 0.5680 | [264] | |
| I2 | 0.5697 | [264,510] | 43, 9 |
| I2 | 0.5745 | [264] | |
| I2 | 0.5764 | [264] | |
| I2 | 0.5815 | [264] | |
| I2 | 0.5830 | [499,510] | 43,11 |
| I2 | 0.5880 | [264] | |
| I2 | 0.5905 | [264] | |
| I2 | 0.5969 | [499,510] | 43,13 |
| I2 | 0.6025 | [264] | |
| I2 | 0.6048 | [264] | |
| I2 | 0.6110 | [499,510] | 43,15 |
| I2 | 0.617482 | [264] | 34,13 R(84) |
| I2 | 0.617676 | [264] | 35,13 R(107) |
| I2 | 0.617868 | [264] | 33,13 R(58) |
| I2 | 0.617947 | [264] | 34,13 P(86),R(89) |
| I2 | 0.618193 | [264] | 33,13 P(60) |
| I2 | 0.618267 | [264] | 35,13 P(109) |
| I2 | 0.618441 | [264] | 34,13 P(91) |
| I2 | 0.618535 | [264] | 33,13 P(65) |
| I2 | 0.6198 | [264] | |
| I2 | 0.6258 | [499,510] | 43,17 |
| I2 | 0.6330 | [264] | |
| I2 | 0.6352 | [264] | |
| I2 | 0.6490 | [264] | |
| I2 | 0.6511 | [264] | |
| I2 | 0.6592 | [264] | |
| I2 | 0.6645 | [264] | |
| I2 | 0.6936 | [264] | |

| | WAVELENGTH IN AIR [MICROMETER] | | TRANSITION |
|---|---|---|---|
| I2 | 0.7114 | [264] | |
| I2 | 0.8144 | [499] | 43,38 |
| I2 | 0.8358 | [499] | 43,40 |
| I2 | 0.8578 | [499,510] | 43,42 |
| I2 | 0.8804 | [499,510] | 43,44 |
| I2 | 0.8813 | [264] | |
| I2 | 0.9037 | [499,510] | 43,46 |
| I2 | 0.9047 | [264] | |
| I2 | 0.9060 | [264] | |
| I2 | 0.9274 | [499,510] | 43,48 |
| I2 | 0.9288 | [264] | |
| I2 | 0.9295 | [264] | |
| I2 | 0.9305 | [264] | |
| I2 | 0.9518 | [499,510] | 43,50 |
| I2 | 0.9545 | [264] | |
| I2 | 0.9555 | [264] | |
| I2 | 0.9766 | [499,510] | 43,52 |
| I2 | 0.9963 | [264] | |
| I2 | 0.9973 | [264] | |
| I2 | 1.0019 | [510] | 43,54 |
| I2 | 1.0053 | [264] | |
| I2 | 1.0225 | [264] | |
| I2 | 1.0245 | [264] | |
| I2 | 1.0255 | [264] | |
| I2 | 1.0274 | [510] | 43,56 |
| I2 | 1.0534 | [264] | |
| I2 | 1.0775 | [264] | |
| I2 | 1.0788 | [264] | |
| I2 | 1.1068 | [264] | |
| I2 | 1.1073 | [264] | |
| I2 | 1.1207 | [521] | 11,44 R(40) |
| I2 | 1.1214 | [521] | 11,44 P(42) |
| I2 | 1.1215 | [521] | 11,44 R(58) |
| I2 | 1.1224 | [521] | 11,44 P(60) |
| I2 | 1.1255 | [264] | |
| I2 | 1.1327 | [521] | 12,45 R(127) |
| I2 | 1.1336 | [521] | 13,46 R(84) |
| I2 | 1.1347 | [521] | 12,45 P(129) |
| I2 | 1.1349 | [521] | 13,46 P(86) |
| I2 | 1.1350 | [264] | |
| I2 | 1.1454 | [521] | 12,46 R(61) |
| I2 | 1.1464 | [521] | 12,46 P(63) |
| I2 | 1.217 | [499] | 43,71 |
| I2 | 1.274 | [499] | 43,76 |
| I2 | 1.2870 | [264] | |
| I2 | 1.2925 | [264] | |
| I2 | 1.294 | [499] | 43,78 |
| I2 | 1.304 | [499] | 43,79 |
| I2 | 1.3153 | [264] | |
| I2 | 1.3192 | [264] | |
| I2 | 1.320 | [499] | 43,81 |
| I2 | 1.3282 | [264] | |
| I2 | 1.3291 | [264] | |
| I2 | 1.3310 | [264] | |
| I2 | 1.3324 | [264] | |
| I2 | 1.3333 | [264] | |
| I2 | 1.3349 | [264] | |

| | WAVELENGTH IN AIR [MICROMETER] | | TRANSITION |
|---|---|---|---|
| I2 | 1.338 | [499] | 43,83 |

| | WAVELENGTH IN VACUUM [MICROMETER] | | TRANSITION |
|---|---|---|---|
| I2 | 1.26392 | [612] | 62,84 P(26) |
| I2 | 1.25663 | [612] | 62,83 R(26) |
| I2 | 1.25697 | [612] | 62,83 P(28) |
| I2 | 1.26359 | [612] | 62,84 R(26) |
| I2 | 1.27918 | [612] | 64,87 P(39) |
| I2 | 1.27980 | [612] | 64,87 P(41) |
| I2 | 1.28212 | [612] | 62,87 R(26) |
| I2 | 1.28245 | [612] | 62,87 P(28) |
| I2 | 1.28468 | [612] | 64,88 P(39) |
| I2 | 1.28513 | [612] | 64,88 P(41) |
| I2 | 1.28757 | [612] | 62,88 R(26) |
| I2 | 1.28789 | [612] | 62,88 P(28) |
| I2 | 1.28966 | [612] | 64,89 R(39) |
| I2 | 1.29007 | [612] | 64,89 P(41) |
| I2 | 1.30545 | [612] | 62,92 R(26) |
| I2 | 1.30572 | [612] | 62,92 P(28) |
| I2 | 1.30656 | [612] | 49,83 R(101) |
| I2 | 1.30784 | [612] | 49,83 P(103) |
| I2 | 1.30900 | [612] | 62,93 R(26) |
| I2 | 1.30926 | [612] | 62,93 P(28) |
| I2 | 1.31131 | [612] | 64,95 R(39) |
| I2 | 1.31166 | [612] | 64,95 P(41) |
| I2 | 1.31216 | [612] | 62,94 R(26) |
| I2 | 1.31241 | [612] | 62,94 R(28) |
| I2 | 1.31374 | [612] | 64,96 R(39) |
| I2 | 1.31407 | [612] | 64,96 P(41) |
| I2 | 1.31498 | [612] | 62,95 R(26) |
| I2 | 1.31523 | [612] | 62,95 P(28) |
| I2 | 1.32952 | [612] | 47,85 R(86) |
| I2 | 1.33065 | [612] | 47,85 P(88) |
| I2 | 1.33546 | [612] | 49,88 P(101) |
| I2 | 1.33567 | [612] | 50,89 R(44) |
| I2 | 1.33573 | [612] | 47,86 R(86) |
| I2 | 1.33619 | [612] | 50,89 R(46) |
| I2 | 1.33661 | [612] | 49,88 P(103) |
| I2 | 1.33681 | [612] | 47,86 P(88) |
| I2 | 1.34142 | [612] | 43,83 P(17) |
| I2 | 1.34145 | [612] | 47,87 R(86) |
| I2 | 1.34173 | [612] | 43,83 R(11) |
| I2 | 1.34173 | [612] | 43,83 R(15) |
| I2 | 1.34192 | [612] | 43,83 P(13) |
| I2 | 1.34193 | [612] | 44,84 R(46) |
| I2 | 1.34248 | [612] | 47,87 P(88) |
| I2 | 1.34256 | [612] | 44,84 P(48) |

NUMBER OF LINES IN IODINE, MOLECULAR 138

ACTIVE MEDIUM : CHLORINE MONOFLUORIDE
SYMBOL : CLF

OPERATING CONDITIONS : EXCITATION BY ELECTRON-BEAM IN A MIXTURE OF NE/F2/CL2.

| | WAVELENGTH IN AIR [MICROMETER] | | WAVELENGTH IN AIR [MICROMETER] |
|---|---|---|---|
| CLF | 0.285 | [596] | |

NUMBER OF LINES IN CHLORINE MONOFLUORIDE 1

ACTIVE MEDIUM : HYDROGEN FLUORIDE
SYMBOL : HF

OPERATING CONDITIONS : EXCITATION BY CHEMICAL REACTIONS, WITH WHICH VIBRATIONALLY EXCITED HF IS PRODUCED. THE REACTION IS CONTROLLED MOSTLY BY AN ELECTRICAL DISCHARGE [175], BUT ALSO BY FLASH PHOTOLYSIS [176], OR SUCCESSIVE ADDITION OF CHEMICAL COMPONENTS [177]. PULSE ENERGIES UP TO 2.3 KWS [178] AND MEAN POWERS OF 5 KW HAVE BEEN GENERATED [179]. BY MEANS OF OPTICAL PUMPING WITH SINGLE HF LASER LINES, ROTATIONAL TRANSITIONS IN THE FAR IR CAN BE OBTAINED WITH VERY HIGH GAIN [243], BY OPTICAL PUMPING ROTATIONAL-VIBRATIONAL TRANSITIONS HAVE ALSO BEEN GENERATED [242].
BIBLIOGRAPHY [180,187],

| | WAVELENGTH IN VACUUM [MICROMETER] | | TRANSITION |
|---|---|---|---|
| HF | 2.41 | [242] | 1- 0 R( 4) |
| HF | 2.43 | [242] | 1- 0 R( 3) |
| HF | 2.45 | [242] | 1- 0 R( 2) |
| HF | 2.48 | [242] | 1- 0 R( 1) |
| HF | 2.551 | [310] | 1- 0 P( 1) |
| HF | 2.579 | [310] | 1- 0 P( 2) |
| HF | 2.6084 | [176,242] | 1- 0 P( 3) |
| HF | 2.6396 | [176,242] | 1- 0 P( 4) |
| HF | 2.6726 | [176,242] | 1- 0 P( 5) |
| HF | 2.7075238 | [176,181,483] | 1- 0 P( 6) |
| HF | 2.7440 | [176,181] | 1- 0 P( 7) |
| HF | 2.7826 | [176,181] | 1- 0 P( 8) |
| HF | 2.8231 | [181] | 1- 0 P( 9) |
| HF | 2.8657 | [181] | 1- 0 P(10) |
| HF | 2.9103 | [181] | 1- 0 P(11) |
| HF | 2.9573 | [181] | 1- 0 P(12) |
| HF | 3.0064 | [181] | 1- 0 P(13) |
| HF | 3.0582 | [181] | 1- 0 P(14) |
| HF | 3.1125 | [181] | 1- 0 P(15) |
| HF | 3.1695 | [420,421,423] | 1- 0 P(16) |
| HF | 3.2292 | [420,421,423] | 1- 0 P(17) |
| HF | 3.2919 | [420,423] | 1- 0 P(18) |
| HF | 2.6668 | [176] | 2- 1 P( 1) |

| | WAVELENGTH IN VACUUM [MICROMETER] | | TRANSITION |
|---|---|---|---|
| HF | 2.6963 | [176,181] | 2-1 P(2) |
| HF | 2.7275 | [176,181] | 2-1 P(3) |
| HF | 2.7604 | [176,181] | 2-1 P(4) |
| HF | 2.7952 | [176,181] | 2-1 P(5) |
| HF | 2.8319 | [176,181] | 2-1 P(6) |
| HF | 2.8705 | [176,181] | 2-1 P(7) |
| HF | 2.9111 | [181] | 2-1 P(8) |
| HF | 2.9539 | [181] | 2-1 P(9) |
| HF | 2.9989 | [181] | 2-1 P(10) |
| HF | 3.0461 | [181] | 2-1 P(11) |
| HF | 3.0958 | [181] | 2-1 P(12) |
| HF | 3.1480 | [181] | 2-1 P(13) |
| HF | 3.2029 | [181] | 2-1 P(14) |
| HF | 3.2603 | [181] | 2-1 P(15) |
| HF | 3.2206 | [420,421,423] | 2-1 P(16) |
| | | | |
| HF | 2.7902 | [176] | 3-2 P(1) |
| HF | 2.8213 | [176,181] | 3-2 P(2) |
| HF | 2.8542 | [176,181] | 3-2 P(3) |
| HF | 2.8890 | [176,181] | 3-2 P(4) |
| HF | 2.9257 | [176,181] | 3-2 P(5) |
| HF | 2.9644 | [176,181] | 3-2 P(6) |
| HF | 3.0052 | [176,181] | 3-2 P(7) |
| HF | 3.0482 | [176,181] | 3-2 P(8) |
| HF | 3.0935 | [420,423] | 3-2 P(9) |
| HF | 3.1411 | [420,423] | 3-2 P(10) |
| HF | 3.1912 | [420,423] | 3-2 P(11) |
| HF | 3.2438 | [183] | 3-2 P(12) |
| HF | 3.2991 | [420,423] | 3-2 P(13) |
| | | | |
| HF | 2.9221 | [176] | 4-3 P(1) |
| HF | 2.9549 | [176] | 4-3 P(2) |
| HF | 2.9896 | [176,319] | 4-3 P(3) |
| HF | 3.026 | [319] | 4-3 P(4) |
| HF | 3.065 | [319] | 4-3 P(5) |
| HF | 3.1454 | [183] | 4-3 P(8) |
| HF | 3.1492 | [183] | 4-3 P(7) |
| | | | |
| HF | 3.0982 | [183] | 5-4 P(2) |
| HF | 3.1350 | [422,423] | 5-4 P(3) |
| HF | 3.1640 | [422,423] | 5-4 P(4) |
| HF | 3.2151 | [422,423] | 5-4 P(5) |
| HF | 3.256 | [319] | 5-4 P(6) |
| HF | 3.3044 | [422,423] | 5-4 P(7) |
| | | | |
| HF | 3.333 | [319] | 6-5 P(4) |
| HF | 3.377 | [319] | 6-5 P(5) |
| | | | |
| HF | 10.1978 | [205] | 0 (28-27) |
| HF | 10.4578 | [205] | 0 (27-26) |
| HF | 10.7439 | [205] | 0 (26-25) |
| HF | 11.0573 | [205] | 0 (25-24) |
| HF | 11.4033 | [205] | 0 (24-23) |
| HF | 11.7854 | [205] | 0 (23-22) |
| HF | 12.2082 | [205] | 0 (22-21) |
| HF | 12.6781 | [205] | 0 (21-20) |
| HF | 13.2009 | [205] | 0 (20-19) |

| | WAVELENGTH IN VACUUM [MICROMETER] | | TRANSITION | |
|---|---|---|---|---|
| HF | 13.7841 | [205] | 0 | (19-18) |
| HF | 14.4406 | [205] | 0 | (18-17) |
| HF | 15.1744 | [108] | 0 | (17-16) |
| HF | 16.0215 | [205] | 0 | (16-15) |
| HF | 18.09 | [482] | 0 | (14-13) |
| HF | 19.35 | [482] | 0 | (13-12) |
| HF | 20.835 | [482] | 0 | (12-11) |
| HF | 12.2619 | [205] | 1 | (23-22) |
| HF | 12.7006 | [205] | 1 | (22-21) |
| HF | 13.1877 | [205] | 1 | (21-20) |
| HF | 13.7277 | [205] | ?1 | (20-19) |
| HF | 15.0163 | [205] | 1 | (18-17) |
| HF | 18.6010 | [205] | 1 | (14-13) |
| HF | 20.1337 | [205] | 1 | (13-12) |
| HF | 21.6986 | [205] | 1 | (12-11) |
| HF | 36.5 | [243] | 1 | ( 7- 6) |
| HF | 42.4 | [243] | 1 | ( 6- 5) |
| HF | 50.6 | [243] | 1 | ( 5- 4) |
| HF | 63.4 | [243] | 1 | ( 4- 3) |
| HF | 84.4 | [243] | 1 | ( 3- 2) |
| HF | 126.5 | [243] | 1 | ( 2- 1) |
| HF | 10.5819 | [205] | 2 | (30-29) |
| HF | 10.8117 | [205] | 2 | (29-28) |
| HF | 13.2211 | [205] | 2 | (22-21) |
| HF | 14.2861 | [205] | 2 | (20-19) |
| HF | 20.9393 | [205] | 2 | (13-12) |
| HF | 19.55 | [482] | 2 | (14-13) |
| HF | 11.5408 | [205] | ?3 | (28-27) |
| HF | 19.1129 | [205] | 3 | (15-14) |
| HF | 20.3513 | [205] | 3 | (14-13) |
| HF | 21.7865 | [205] | 3 | (13-12) |
| HF | 19.915 | [482] | 4 | (15-14) |

NUMBER OF LINES IN HYDROGEN FLUORIDE 107

ACTIVE MEDIUM : DEUTERIUM FLUORIDE
SYMBOL : DF

OPERATING CONDITIONS : EXCITATION BY CHEMICAL REACTIONS, WITH WHICH VIBRATIONALLY EXCITED DF IS PRODUCED. THE REACTION IS INITIATED MOSTLY BY AN ELECTRICAL DISCHARGE [175], BUT ALSO BY FLASH PHOTOLYSIS [182], OR PURELY CHEMICAL MEANS [177].
WITH A DISPERSIVE RESONATOR OVERTONE EMISSION [182] (DELTA V = 2) IS POSSIBLE.
BIBLIOGRAPHY AT [180,187].

| | WAVELENGTH IN VACUUM [MICROMETER] | | TRANSITION |
|---|---|---|---|
| DF | 1.836 | [182] | 3- 1 P( 4) |
| DF | 1.844 | [182] | 3- 1 P( 5) |
| DF | 1.854 | [182] | 3- 1 P( 6) |

| | WAVELENGTH IN VACUUM [MICROMETER] | | TRANSITION |
|---|---|---|---|
| DF | 3.493 | [309] | 1- 0 P( 2) |
| DF | 3.521 | [309] | 1- 0 P( 3) |
| DF | 3.550 | [309] | 1- 0 P( 4) |
| DF | 3.581 | [309] | 1- 0 P( 5) |
| DF | 3.612 | [309] | 1- 0 P( 6) |
| DF | 3.645 | [309] | 1- 0 P( 7) |
| DF | 3.679 | [309] | 1- 0 P( 8) |
| DF | 3.716 | [376,175] | 1- 0 P( 9) |
| DF | 3.752 | [376] | 1- 0 P(10) |
| DF | 3.7901 | [175] | 1- 0 P(11) |
| DF | 3.8298 | [181,175] | 1- 0 P(12) |
| DF | 3.8707 | [175] | 1- 0 P(13) |
| DF | 3.9133 | [175] | 1- 0 P(14) |
| DF | 3.9572 | [181,175] | 1- 0 P(15) |
| DF | 4.0032 | [181,175] | 1- 0 P(16) |
| DF | 4.0502 | [175] | 1- 0 P(17) |
| | | | |
| DF | 3.6363 | [181,175] | 2- 1 P( 3) |
| DF | 3.6665 | [181,175] | 2- 1 P( 4) |
| DF | 3.6983 | [181,175] | 2- 1 P( 5) |
| DF | 3.7310 | [181,175] | 2- 1 P( 6) |
| DF | 3.7651 | [181,175] | 2- 1 P( 7) |
| DF | 3.8007 | [181,175] | 2- 1 P( 8) |
| DF | 3.8375 | [181,175] | 2- 1 P( 9) |
| DF | 3.8757 | [181,175] | 2- 1 P(10) |
| DF | 3.9155 | [181,175] | 2- 1 P(11) |
| DF | 3.9565 | [181,175] | 2- 1 P(12) |
| DF | 3.9995 | [181,175] | 2- 1 P(13) |
| DF | 4.0435 | [175] | 2- 1 P(14) |
| DF | 4.0893 | [175] | 2- 1 P(15) |
| DF | 4.1369 | [181,175] | 2- 1 P(16) |
| DF | 4.1862 | [181,175] | 2- 1 P(17) |
| | | | |
| DF | 3.7563 | [181,175] | 3- 2 P( 3) |
| DF | 3.7878 | [181,175] | 3- 2 P( 4) |
| DF | 3.8206 | [181,175] | 3- 2 P( 5) |
| DF | 3.8547 | [181,175] | 3- 2 P( 6) |
| DF | 3.8903 | [181,175] | 3- 2 P( 7) |
| DF | 3.9272 | [181,175] | 3- 2 P( 8) |
| DF | 3.9654 | [181,175] | 3- 2 P( 9) |
| DF | 4.0054 | [181,175] | 3- 2 P(10) |
| DF | 4.0464 | [181,175] | 3- 2 P(11) |
| DF | 4.0895 | [181,175] | 3- 2 P(12) |
| DF | 4.1337 | [175] | 3- 2 P(13) |
| DF | 4.1798 | [181,175] | 3- 2 P(14) |
| | | | |
| DF | 3.8503 | [175] | 4- 3 P( 2) |
| DF | 3.8817 | [175] | 4- 3 P( 3) |
| DF | 3.9145 | [175] | 4- 3 P( 4) |
| DF | 3.9487 | [181] | 4- 3 P( 5) |
| DF | 3.9843 | [181] | 4- 3 P( 6) |
| DF | 4.0212 | [181] | 4- 3 P( 7) |
| DF | 4.0595 | [175] | 4- 3 P( 8 |

NUMBER OF LINES IN DEUTERIUM FLUORIDE 53

ACTIVE MEDIUM : HYDROGEN CHLORIDE
SYMBOL : HCL

OPERATING CONDITIONS : EXCITATION BY CHEMICAL REACTIONS IN H2+CL2O OR CL2+HI. THE REACTIONS ARE INITIATED BY DISSOCIATION OF THE COMPONENTS IN A PULSED DISCHARGE [206,276,281], BY FLASH PHOTOLYSIS [277], OR CONTINUOUS RF-DISCHARGE [278,279].ISOTOPIC PUMPING [567].
THE ROTATIONAL LINES ARE GENERATED BY PULSED DISCHARGE IN A MIXTURE OF CL2+CH3CL, CL2+CH3BR OR CL2+H2+CCLF3 [280].
IN A TRANSVERSE DISCHARGE CONFIGURATION, PULSED POWERS OF 2 KW HAVE BEEN OBTAINED [278].

| | WAVELENGTH IN VACUUM [MICROMETER] | | TRANSITION |
|---|---|---|---|
| HCL | 3.5728 | [175] | 1-0 P( 4) CL[35] |
| HCL | 3.6026 | [175] | 1-0 P( 5) CL[35] |
| HCL | 3.6337 | [175] | 1-0 P( 6) CL[35] |
| HCL | 3.6660 | [175] | 1-0 P( 7) CL[35] |
| HCL | 3.6996 | [206,175] | 1-0 P( 8) CL[35] |
| HCL | 3.7341 | [206] | 1-0 P( 9) CL[35] |
| HCL | 3.7707 | [206] | 1-0 P(10) CL[35] |
| HCL | 3.8081 | [206] | 1-0 P(11) CL[35] |
| HCL | 3.6362 | [175] | 1-0 P( 6) CL[37] |
| HCL | 3.6685 | [175] | 1-0 P( 7) CL[37] |
| HCL | 3.7021 | [175] | 1-0 P( 8) CL[37] |
| HCL | 3.7370 | [175] | 1-0 P( 9) CL[37] |
| HCL | 3.7071 | [280,281] | 2-1 P( 4) CL[35] |
| HCL | 3.7383 | [280,281] | 2-1 P( 5) CL[35] |
| HCL | 3.7710 | [280,281] | 2-1 P( 6) CL[35] |
| HCL | 3.8050 | [280,281] | 2-1 P( 7) CL[35] |
| HCL | 3.8401 | [280,281] | 2-1 P( 8) CL[35] |
| HCL | 3.8768 | [280,281] | 2-1 P( 9) CL[35] |
| HCL | 3.9149 | [280,281] | 2-1 P(10) CL[35] |
| HCL | 3.7098 | [175] | 2-1 P( 4) CL[37] |
| HCL | 3.7408 | [280,175] | 2-1 P( 5) CL[37] |
| HCL | 3.7735 | [280,175] | 2-1 P( 6) CL[37] |
| HCL | 3.8074 | [280,175] | 2-1 P( 7) CL[37] |
| HCL | 3.8425 | [280,175] | 2-1 P( 8) CL[37] |
| HCL | 3.8509 | [281,175] | 3-2 P( 4) CL[35] |
| HCL | 3.8840 | [280,281,175] | 3-2 P( 5) CL[35] |
| HCL | 3.9181 | [280,281,175] | 3-2 P( 6) CL[35] |
| HCL | 3.9536 | [280,281,175] | 3-2 P( 7) CL[35] |
| HCL | 3.9909 | [280,281,175] | 3-2 P( 8) CL[35] |
| HCL | 4.0295 | [280,281,175] | 3-2 P( 9) CL[35] |
| HCL | 3.8536 | [175] | 3-2 P( 4) CL[37] |
| HCL | 3.8864 | [175] | 3-2 P( 5) CL[37] |
| HCL | 3.9205 | [280,175] | 3-2 P( 6) CL[37] |
| HCL | 3.9560 | [280] | 3-2 P( 7) CL[37] |
| HCL | 4.0054 | [175] | 4-3 P( 4) CL[35] |
| HCL | 4.0399 | [175] | 4-3 P( 5) CL[35] |
| HCL | 4.0759 | [175] | 4-3 P( 6) CL[35] |

| WAVELENGTH IN VACUUM [MICROMETER] | | | TRANSITION | | |
|---|---|---|---|---|---|
| HCL | 4.1135 | [175] | 4-3 | P( 7) | CL[35] |
| HCL | 13.8720 | [280] | 0 | R(40) | CL[35] |
| HCL | 14.0994 | [280] | 0 | R(39) | CL[35] |
| HCL | 14.3434 | [280] | 0 | R(38) | CL[35] |
| HCL | 16.2125 | [280] | 0 | R(32) | CL[35] |
| HCL | 16.6085 | [280] | 0 | R(31) | CL[35] |
| HCL | 17.0340 | [280,282] | 0 | R(30) | CL[35] |
| HCL | 17.4923 | [280,282] | 0 | R(29) | CL[35] |
| HCL | 17.9874 | [280,282] | 0 | R(28) | CL[35] |
| HCL | 18.522 | [280] | 0 | R(27) | CL[35] |
| HCL | 20.4106 | [280] | 0 | R(24) | CL[35] |
| HCL | 21.1556 | [280] | 0 | R(23) | CL[35] |
| HCL | 21.9706 | [280] | 0 | R(22) | CL[35] |
| HCL | 22.8637 | [280] | 0 | R(21) | CL[35] |
| HCL | 23.8485 | [280] | 0 | R(20) | CL[35] |
| HCL | 24.9367 | [280] | 0 | R(19) | CL[35] |
| HCL | 26.1462 | [280] | 0 | R(18) | CL[35] |
| HCL | 27.508 | [282] | 0 | R(17) | CL[35] |
| HCL | 16.664 | [282] | 0 | R(31) | CL[37] |
| HCL | 17.997 | [282] | 0 | R(28) | CL[37] |
| HCL | 19.122 | [282] | 0 | R(26) | CL[37] |
| HCL | 17.125 | [282] | 1 | R(31) | CL[35] |
| HCL | 17.575 | [282] | 1 | R(30) | CL[35] |
| HCL | 18.035 | [282] | 1 | R(29) | CL[35] |
| HCL | 18.555 | [282] | 1 | R(28) | CL[35] |
| HCL | 19.7002 | [280] | 1 | R(26) | CL[35] |
| HCL | 20.3455 | [280] | 1 | R(25) | CL[35] |
| HCL | 21.0470 | [280] | 1 | R(24) | CL[35] |
| HCL | 21.8127 | [280] | 1 | R(23) | CL[35] |
| HCL | 22.6514 | [280] | 1 | R(22) | CL[35] |
| HCL | 23.5705 | [280] | 1 | R(21) | CL[35] |
| HCL | 24.5833 | [280] | 1 | R(20) | CL[35] |
| HCL | 25.7040 | [280] | 1 | R(19) | CL[35] |
| HCL | 16.765 | [282] | 1 | R(32) | CL[37] |
| HCL | 18.593 | [282] | 1 | R(28) | CL[37] |
| HCL | 19.145 | [282] | 1 | R(27) | CL[37] |
| HCL | 24.6177 | [280] | 1 | R(20) | CL[37] |
| HCL | 19.183 | [282] | 2 | R(28) | CL[35] |
| HCL | 20.9991 | [280] | 2 | R(26) | CL[35] |
| HCL | 24.3178 | [280] | 2 | R(21) | CL[35] |
| HCL | ? 19.783 | [282] | 3 | R(28) | CL[35] |
| HCL | ? 19.821 | [282] | 3 | | |

NUMBER OF LINES IN HYDROGEN CHLORIDE 79

ACTIVE MEDIUM : DEUTERIUM CHLORIDE
SYMBOL : DCL

OPERATING CONDITIONS : EXCITATION BY A CHEMICAL REACTION IN A MIXTURE OF D2+CL2. THE REACTION IS STARTED BY DISSOCIATION OF THE COMPONENTS IN A PULSED DISCHARGE. TYPICAL PRSSURE 1.5 TORR H2 AND 2.4 TORR CL2.

| | WAVELENGTH IN VACUUM [MICROMETER] | | TRANSITION |
|---|---|---|---|
| DCL | 5.0445 | [280] | 2-1 P( 5) CL[35] |
| DCL | 5.0743 | [280] | 2-1 P( 6) CL[35] |
| DCL | 5.1049 | [280,287] | 2-1 P( 7) CL[35] |
| DCL | 5.1363 | [280,287] | 2-1 P( 8) CL[35] |
| DCL | 5.1688 | [280,287] | 2-1 P( 9) CL[35] |
| DCL | 5.0514 | [280] | 2-1 P( 5) CL[37] |
| DCL | 5.0811 | [280] | 2-1 P( 6) CL[37] |
| DCL | 5.1118 | [280] | 2-1 P( 7) CL[37] |
| DCL | 5.1431 | [280] | 2-1 P( 8) CL[37] |
| DCL | 5.1511 | [280,287] | 3-2 P( 4) CL[35] |
| DCL | 5.1811 | [280,287] | 3-2 P( 5) CL[35] |
| DCL | 5.2118 | [280] | 3-2 P( 6) CL[35] |
| DCL | 5.2435 | [280] | 3-2 P( 7) CL[35] |
| DCL | 5.2760 | [280] | 3-2 P( 8) CL[35] |
| DCL | 5.3097 | [280] | 3-2 P( 9) CL[35] |
| DCL | 5.3443 | [280] | 3-2 P(10) CL[35] |
| DCL | 5.3799 | [280] | 3-2 P(11) CL[35] |
| DCL | 5.1879 | [280] | 3-2 P( 5) CL[37] |
| DCL | 5.2186 | [280] | 3-2 P( 6) CL[37] |
| DCL | 5.2503 | [280] | 3-2 P( 7) CL[37] |
| DCL | 5.2829 | [280] | 3-2 P( 8) CL[37] |
| DCL | 5.3244 | [280] | 4-3 P( 5) CL[35] |
| DCL | 5.3562 | [280] | 4-3 P( 6) CL[35] |
| DCL | 5.3889 | [280] | 4-3 P( 7) CL[35] |
| DCL | 5.4577 | [280] | 4-3 P( 9) CL[35] |
| DCL | 5.4935 | [280] | 4-3 P(10) CL[35] |
| DCL | 5.5304 | [280] | 4-3 P(11) CL[35] |
| DCL | 5.3629 | [280] | 4-3 P( 6) CL[37] |
| DCL | 5.3956 | [280] | 4-3 P( 7) CL[37] |
| DCL | 5.4295 | [280] | 4-3 P( 8) CL[37] |
| DCL | 5.5084 | [280] | 5-4 P( 6) CL[35] |
| DCL | 5.5423 | [280] | 5-4 P( 7) CL[35] |
| DCL | 5.5776 | [280] | 5-4 P( 8) CL[35] |
| DCL | 5.6137 | [280] | 5-4 P( 9) CL[35] |

NUMBER OF LINES IN DEUTERIUM CHLORIDE 34

ACTIVE MEDIUM : HYDROGEN BROMIDE
SYMBOL : HBR

OPERATING CONDITIONS : EXCITATION BY CHEMICAL REACTION WITHIN A MIXTURE OF H2+BR2. INITIATION OF THE REACTION BY DISSOCIATION OF THE COMPONENTS IN A PULSED DISCHARGE. TYPICAL PRESSURE 1.5 TORR H2 AND 2.4 TORR BR2. WITH A TRANSVERSE DISCHARGE CONFIGURATION, PULSE POWERS OF APPROXIMATELY 1 KW HAVE BEEN OBTAINED [276].

| | WAVELENGTH IN VACUUM [MICROMETER] | | TRANSITION |
|---|---|---|---|
| HBR | 4.0170 | [280] | 1-0 P(4) BR[79] |
| HBR | 4.0470 | [280] | 1-0 P(5) BR[79] |
| HBR | 4.0783 | [280] | 1-0 P(6) BR[79] |
| HBR | 4.1107 | [280] | 1-0 P(7) BR[79] |
| HBR | 4.1442 | [280] | 1-0 P(8) BR[79] |
| HBR | 4.0176 | [280] | 1-0 P(4) BR[81] |
| HBR | 4.0475 | [280] | 1-0 P(5) BR[81] |
| HBR | 4.0788 | [280] | 1-0 P(6) BR[81] |
| HBR | 4.1112 | [280] | 1-0 P(7) BR[81] |
| HBR | 4.1448 | [280] | 1-0 P(8) BR[81] |
| HBR | 4.1796 | [280] | 1-0 P(9) BR[81] |
| HBR | 4.1653 | [280] | 2-1 P(4) BR[79] |
| HBR | 4.1970 | [280] | 2-1 P(5) BR[79] |
| HBR | 4.2295 | [280] | 2-1 P(6) BR[79] |
| HBR | 4.2633 | [280] | 2-1 P(7) BR[79] |
| HBR | 4.2988 | [280] | 2-1 P(8) BR[79] |
| HBR | 4.3354 | [280] | 2-1 P(9) BR[79] |
| HBR | 4.1658 | [280] | 2-1 P(4) BR[81] |
| HBR | 4.1975 | [280] | 2-1 P(5) BR[81] |
| HBR | 4.2639 | [280] | 2-1 P(7) BR[81] |
| HBR | 4.2994 | [280] | 2-1 P(8) BR[81] |
| HBR | 4.3359 | [280] | 2-1 P(9) BR[81] |
| HBR | 4.3250 | [280] | 3-2 P(4) BR[79] |
| HBR | 4.3579 | [280] | 3-2 P(5) BR[79] |
| HBR | 4.3925 | [280] | 3-2 P(6) BR[79] |
| HBR | 4.4281 | [280] | 3-2 P(7) BR[79] |
| HBR | 4.4652 | [280] | 3-2 P(8) BR[79] |
| HBR | 4.5041 | [280] | 3-2 P(9) BR[79] |
| HBR | 4.3255 | [280] | 3-2 P(4) BR[81] |
| HBR | 4.3585 | [280] | 3-2 P(5) BR[81] |
| HBR | 4.3931 | [280] | 3-2 P(6) BR[81] |
| HBR | 4.4307 | [280] | 3-2 P(7) BR[81] |
| HBR | 4.4658 | [280] | 3-2 P(8) BR[81] |
| HBR | 4.5047 | [280] | 3-2 P(9) BR[81] |
| HBR | 4.5330 | [280] | 4-3 P(5) BR[79] |
| HBR | 4.5691 | [280] | 4-3 P(6) BR[79] |
| HBR | 4.6070 | [280] | 4-3 P(7) BR[79] |
| HBR | 4.6463 | [280] | 4-3 P(8) BR[79] |
| HBR | 4.5335 | [280] | 4-3 P(5) BR[81] |

| | WAVELENGTH IN VACUUM [MICROMETER] | | TRANSITION |
|---|---|---|---|
| HBR | 4.5696 | [280] | 4-3 P(6) BR[81] |
| HBR | 4.6076 | [280] | 4-3 P(7) BR[81] |
| HBR | 4.6467 | [280] | 4-3 P(8) BR[81] |
| HBR | 19.399 | [282] | 0 R(33) BR[81] |
| HBR | 20.360 | [282] | 0 R(31) |
| HBR | 20.896 | [282] | 0 R(30) BR[79] |
| HBR | 20.949 | [282] | 0 R(30) BR[81] |
| HBR | 21.501 | [282] | 0 R(29) |
| HBR | 22.136 | [282] | 0 R(28) BR[81] |
| HBR | 30.948 | [282] | 0 R(19) |
| HBR | 32.469 | [282] | 0 R(18) |
| HBR | 19.988 | [282] | 1 R(33) |
| HBR | 21.546 | [282] | 1 R(30) BR[79] |
| HBR | 30.445 | [282] | 1 R(20) |
| HBR | 31.849 | [282] | 1 R(19) |
| HBR | 33.409 | [282] | 1 R(18) |
| HBR | 22.226 | [282] | 2 R(30) BR[79] |
| HBR | 22.855 | [282] | 2 R(29) |
| HBR | 31.366 | [282] | 2 R(20) |
| HBR | 32.799 | [282] | 2 R(19) |
| HBR | 40.526 | [282] | 2 R(15) |
| HBR | 23.436 | [282] | 3 R(22) BR[81] |
| HBR | 29.786 | [282] | 3 R(22) BR[81] |

NUMBER OF LINES IN HYDROGEN BROMIDE 62

ACTIVE MEDIUM : DEUTERIUM BROMIDE
SYMBOL : DBR

OPERATING CONDITIONS : EXCITATION BY CHEMICAL REACTION WITHIN A MIXTURE OF D2+BR2.
INITIATION OF THE REACTION BY DISSOCIATION OF THE COMPONENTS IN A PULSED DISCHARGE. TYPICAL PRESSURE 0.6 TORR D2 AND 0.3 TORR BR2.

| | WAVELENGTH IN VACUUM [MICROMETER] | | TRANSITION |
|---|---|---|---|
| DBR | 5.8049 | [280] | 2-1 P( 8) |
| DBR | 5.8620 | [280] | 3-2 P( 5) BR[79] |
| DBR | 5.8928 | [280] | 3-2 P( 6) BR[79] |
| DBR | 5.9246 | [280] | 3-2 P( 7) BR[79] |
| DBR | 5.9573 | [280] | 3-2 P( 8) BR[79] |
| DBR | 5.8626 | [280] | 3-2 P( 5) BR[81] |
| DBR | 5.8944 | [280] | 3-2 P( 6) BR[81] |
| DBR | 5.9261 | [280] | 3-2 P( 7) BR[81] |
| DBR | 5.9590 | [280] | 3-2 P( 8) BR[81] |

| | WAVELENGTH IN VACUUM [MICROMETER] | | TRANSITION |
|---|---|---|---|
| DBR | 6.0209 | [280] | 4-3 P( 5) BR[79] |
| DBR | 6.0529 | [280] | 4-3 P( 6) BR[79] |
| DBR | 6.0858 | [280] | 4-3 P( 7) BR[79] |
| DBR | 6.1200 | [280] | 4-3 P( 8) BR[79] |
| DBR | 6.1546 | [280] | 4-3 P( 9) BR[79] |
| DBR | 6.1903 | [280] | 4-3 P(10) BR[79] |
| DBR | 6.2272 | [280] | 4-3 P(11) BR[79] |
| | | | |
| DBR | 6.0225 | [280] | 4-3 P( 5) BR[81] |
| DBR | 6.0544 | [280] | 4-3 P( 6) BR[81] |
| DBR | 6.0873 | [280] | 4-3 P( 7) BR[81] |
| DBR | 6.1216 | [280] | 4-3 P( 8) BR[81] |
| DBR | 6.1562 | [280] | 4-3 P( 9) BR[81] |
| DBR | 6.1918 | [280] | 4-3 P(10) BR[81] |
| DBR | 6.2289 | [280] | 4-3 P(11) BR[81] |
| | | | |
| DBR | 6.2566 | [280] | 5-4 P( 7) BR[79] |
| DBR | 6.2916 | [280] | 5-4 P( 8) BR[79] |
| DBR | 6.3279 | [280] | 5-4 P( 9) BR[79] |
| | | | |
| DBR | 6.2237 | [280] | 5-4 P( 6) BR[81] |
| DBR | 6.2581 | [280] | 5-4 P( 7) BR[81] |
| DBR | 6.2932 | [280] | 5-4 P( 8) BR[81] |
| DBR | 6.3294 | [280] | 5-4 P( 9) BR[81] |

NUMBER OF LINES IN DEUTERIUM BROMIDE 30

ACTIVE MEDIUM : NITROGEN MONOXIDE
SYMBOL : NO

OPERATING CONDITIONS : PULSED DISCHARGE IN A FLOWING MIXTURE OF 3.5 TORR NOCL AND 5.8 TORR HE [233] OR EXCITATION BY PHOTODISSOCIATION OF NOCL [265,266], OR NO AND NO2 [524].

| | | WAVELENGTH IN VACUUM [MICROMETER] | | | WAVELENGTH IN VACUUM [MICROMETER] | |
|---|---|---|---|---|---|---|
| NO | | 1.2237 | [524] | NO | 5.9632 | [233,265] |
| NO | | 1.1069 | [524] | NO | 5.9673 | [233] |
| NO | ? | 2.6072 | [524] | NO | 5.9756 | [233,265] |
| NO | ? | 2.6380 | [524] | NO | 5.9799 | [233] |
| | | | | NO | 5.9862 | [233] |
| NO | | 5.8462 | [233] | NO | 5.9931 | [233] |
| NO | | 5.8549 | [233] | NO | 6.0010 | [233,265] |
| NO | | 5.8584 | [233] | NO | 6.0054 | [233] |
| NO | | 5.8706 | [233] | NO | 6.0192 | [233] |
| NO | | 5.8789 | [233] | NO | 6.0267 | [233,265] |
| NO | | 5.9036 | [233] | NO | 6.0324 | [233,265] |
| NO | | 5.9083 | [233] | NO | 6.0366 | [265] |
| NO | | 5.9423 | [233] | NO | 6.0402 | [233,265] |
| NO | | 5.9546 | [233] | NO | 6.0419 | [233] |
| NO | | 5.9550 | [265] | NO | 6.0543 | [233] |

| | WAVELENGTH IN VACUUM [MICROMETER] | | | WAVELENGTH IN VACUUM [MICROMETER] | |
|---|---|---|---|---|---|
| NO | 6.0628 | [233,265] | NO | 6.2328 | [233] |
| NO | 6.0673 | [233] | NO | 6.2381 | [233] |
| NO | 6.0801 | [233] | NO | 6.2511 | [233] |
| NO | 6.0864 | [233,265] | NO | 6.2602 | [233] |
| NO | 6.0934 | [233,265] | NO | 6.2645 | [233] |
| NO | 6.1015 | [233] | NO | 6.2778 | [233] |
| NO | 6.1204 | [233] | NO | 6.2865 | [233] |
| NO | 6.1417 | [233,265] | NO | 6.2913 | [233] |
| NO | 6.1538 | [233] | NO | 6.2998 | [233] |
| NO | 6.1546 | [265] | NO | 6.3051 | [233] |
| NO | 6.1576 | [233] | NO | 6.3136 | [233] |
| NO | 6.1663 | [233] | NO | 6.3191 | [233] |
| NO | 6.1792 | [233] | NO | 6.3274 | [233] |
| NO | 6.1838 | [233] | NO | 6.3336 | [233] |
| NO | 6.1921 | [233] | NO | 6.3764 | [233] |
| NO | 6.1972 | [233] | NO | 6.3894 | [233] |
| NO | 6.1973 | [265] | NO | 6.3960 | [233] |
| NO | 6.2055 | [233] | NO | 6.4031 | [233] |
| NO | 6.2110 | [233] | NO | 6.4262 | [233] |
| NO | 6.2191 | [233] | NO | 6.4321 | [233] |
| NO | 6.2249 | [233] | | | |

NUMBER OF LINES IN NITROGEN MONOXIDE 70

ACTIVE MEDIUM : CYANOGEN
SYMBOL : CN

OPERATING CONDITIONS : EXCITATION OF MOLECULAR ELECTRONIC AND VIBRATIONAL LASER EMISSION BY PHOTODISSOCIATION AND PREDISSOCIATION OF CN-BEARING COMPOUNDS. ALSO DISCHARGE PUMPING IN HCN VAPOR [444].

| | WAVELENGTH IN AIR [MICROMETER] | | TRANSITION |
|---|---|---|---|
| CN | 1.09966 | [197] | Q1(02) |
| CN | 1.09963 | [197] | Q1(03) |
| CN | 1.09965 | [197] | Q1(04) |
| CN | 1.09974 | [197] | Q1(01) |
| CN | 1.09974 | [197] | Q1(05) |
| CN | 1.09987 | [197] | Q1(06) |
| CN | 1.10007 | [197] | Q1(07) |
| CN | 1.10031 | [197] | Q1(08) |
| CN | 1.10061 | [197] | Q1(09) |
| CN | 1.10096 | [197] | Q1(10) |
| CN | 1.10136 | [197] | Q1(11) |
| CN | 1.10082 | [197] | Q1(12) |
| CN | 1.10232 | [197] | Q1(13) |
| CN | 1.10288 | [197] | Q1(14) |
| CN | 1.10348 | [197] | Q1(15) |
| CN | 1.10414 | [197] | Q1(16) |
| CN | 1.10485 | [197] | Q1(17) |
| CN | 1.10726 | [197] | Q1(20) |

| | WAVELENGTH IN AIR [MICROMETER] | | TRANSITION |
|---|---|---|---|
| CN | 1.10445 | [197] | P1(09) |
| CN | 1.10521 | [197] | P1(10) |
| CN | 1.10603 | [197] | P1(11) |
| CN | 1.10689 | [197] | P1(12) |
| CN | 1.10782 | [197] | P1(13) |
| CN | 1.10879 | [197] | P1(14) |
| CN | 1.10981 | [197] | P1(15) |
| CN | 1.11090 | [197] | P1(16) |
| CN | 1.11200 | [197] | P1(17) |
| CN | 1.11321 | [197] | P1(18) |
| CN | 1.41830 | [197] | Q1(05) |
| CN | 1.41849 | [197] | Q1(06) |
| CN | 1.41876 | [197] | Q1(07) |
| CN | 1.41911 | [197] | Q1(08) |
| CN | 1.41954 | [197] | Q1(09) |
| CN | 1.42005 | [197] | Q1(10) |
| CN | 1.42065 | [197] | Q1(11) |
| CN | 1.42132 | [197] | Q1(12) |
| CN | 1.42207 | [197] | Q1(13) |
| CN | 1.42289 | [197] | Q1(14) |
| CN | 1.42380 | [197] | Q1(15) |
| CN | 1.42478 | [197] | Q1(16) |
| CN | 1.42583 | [197] | Q1(17) |
| CN | 1.42696 | [197] | Q1(18) |
| CN | 1.42808 | [197] | Q1(19) |
| CN | 1.42945 | [197] | Q1(20) |
| CN | 1.42081 | [197] | Q1(21) |

NUMBER OF LINES IN CYANOGEN 45

ACTIVE MEDIUM : HYDROXYL
SYMBOL : OH

OPERATING CONDITIONS : FLASH PHOTOLYSIS IN A MIXTURE OF O3+H2 [259]. PULSED DISCHARGE IN A MIXTURE OF O3+H2+HE [260]. THE ROTATIONAL TRANSITIONS IN A FLOWING MIXTURE OF SF6+H2+O2 WITH A PULSED DISCHARGE [306].

| | WAVELENGTH IN VACUUM [MICROMETER] | | | WAVELENGTH IN VACUUM [MICROMETER] | |
|---|---|---|---|---|---|
| OH | 2.93432 | [259] | OH | 3.15697 | [259] |
| OH | 2.96999 | [259] | OH | 3.23615 | [259] |
| OH | 3.07877 | [259] | OH | 3.27653 | [259] |
| OH | 3.11677 | [259] | | | |

| | WAVELENGTH IN AIR [MICROMETER] | | TRANSITION |
|---|---|---|---|
| OH | 12.273 | [306] | 0 R 1(24) |
| OH | 12.279 | [306] | 0 R 2(24) |
| OH | 12.660 | [306] | 0 R 1(23) |
| OH | 12.663 | [306] | 0 R 2(23) |
| OH | 13.073 | [306] | 0 R'2(22) |
| OH | 13.079 | [306] | 0 R 1(22) |
| OH | 13.088 | [306] | 0 R 2(22) |
| OH | 13.525 | [306] | 0 R'1(21) |
| OH | 13.538 | [306] | 0 R'2(21) |
| OH | 13.547 | [306] | 0 R 1(21) |
| OH | 13.557 | [306] | 0 R 2(21) |
| OH | 14.043 | [306] | 0 R'1(20) |
| OH | 14.059 | [306] | 0 R'2(20) |
| OH | 14.067 | [306,455] | 0 R 1(20) |
| OH | 14.081 | [306] | 0 R 2(20) |
| OH | 14.620 | [306] | 0 R'1(19) |
| OH | 14.640 | [306] | 0 R'2(19) |
| OH | 14.646 | [306,455] | 0 R 1(19) |
| OH | 14.662 | [306] | 0 R 2(19) |
| OH | 15.289 | [306] | 0 R'2(18) |
| OH | 15.294 | [306] | 0 R 1(18) |
| OH | 15.313 | [306] | 0 R 2(18) |
| OH | 18.788 | [306] | 0 R'1(14) |
| OH | 18.828 | [306] | 0 R 1(14) |
| OH | 18.849 | [306] | 0 R'2(14) |
| OH | 18.878 | [306] | 0 R 2(14) |
| OH | 20.05 | [455] | 0 R 1(13) |
| OH | 21.48 | [455] | 0 R 1(12) |
| OH | 21.57 | [455] | 0 R 2[12) |
| OH | 23.14 | [455] | 0 R 1(11) |
| OH | 23.26 | [455] | 0 R 2[11) |
| OH | 25.11 | [455] | 0 R 1(10) |
| OH | 25.28 | [455] | 0 R 2[10) |
| OH | 27.47 | [455] | 0 R 1(09) |
| OH | 27.71 | [455] | 0 R 2[09) |
| | | | |
| OH | 13.632 | [306] | 1 R 1(22) |
| OH | 13.642 | [306] | 1 R 2(22) |
| OH | 14.118 | [306] | 1 R 1(21) |
| OH | 14.129 | [306] | 1 R 2(21) |
| OH | 14.655 | [306] | 1 R 1(20) |
| OH | 14.669 | [306] | 1 R 2(20) |
| OH | 15.256 | [306] | 1 R 1(19) |
| OH | 15.274 | [306] | 1 R 2(19) |
| OH | 18.455 | [306] | 1 R'1(15) |
| OH | 18.492 | [306] | 1 R 1(15) |
| OH | 18.502 | [306] | 1 R'2(15) |
| OH | 18.532 | [306] | 1 R 2(15) |
| OH | 19.555 | [306] | 1 R'1(14) |
| OH | 19.594 | [306,455] | 1 R 1(14) |
| OH | 19.619 | [306] | 1 R'2(14) |
| OH | 19.650 | [306,455] | 1 R 2(14) |
| OH | 20.87 | [455] | 1 R 1(13) |
| OH | 20.93 | [455] | 1 R 2(13) |
| OH | 22.33 | [455] | 1 R 1(12) |
| OH | 22.45 | [455] | 1 R 2(12) |
| OH | 24.07 | [455] | 1 R 1(11) |

| | WAVELENGTH IN AIR [MICROMETER] | | TRANSITION |
|---|---|---|---|
| OH | 24.18 | [455] | 1 R 2(11) |
| OH | 26.12 | [455] | 1 R 1(10) |
| OH | 26.30 | [455] | 1 R 2(10) |
| OH | 19.273 | [306] | 2 R 1(15) |
| OH | 19.321 | [306] | 2 R 2(15) |

NUMBER OF LINES IN HYDROXYL 68

ACTIVE MEDIUM : HYDROXYL (DEUTERIUM)
SYMBOL : OD

OPERATING CONDITIONS : PULSED DISCHARGE WITHIN A FLOWING MIXTURE OF SF6+D2+O2 [306].

| | WAVELENGTH IN AIR [MICROMETER] | | TRANSITION |
|---|---|---|---|
| OD | 18.121 | [306] | 0 R 1(30) |
| OD | 18.138 | [306] | 0 R 2(30) |
| OD | 18.590 | [306] | 0 R'1(29) |
| OD | 18.603 | [306] | 0 R 1(29) |
| OD | 18.624 | [306] | 0 R 2(29) |
| OD | 19.102 | [306] | 0 R'1(28) |
| OD | 19.121 | [306] | 0 R 1(28) |
| OD | 19.141 | [306] | 0 R'2(28) |
| OD | 19.161 | [306] | 0 R 2(28) |
| OD | 19.662 | [306] | 0 R'1(27) |
| OD | 19.681 | [306] | 0 R 1(27) |
| OD | 19.696 | [306] | 0 R'2(27) |
| OD | 19.704 | [306] | 0 R 2(27) |
| OD | 20.271 | [306] | 0 R'1(26) |
| OD | 20.286 | [306] | 0 R 1(26) |
| OD | 20.296 | [306] | 0 R'2(26) |
| OD | 20.313 | [306] | 0 R 2(26) |

NUMBER OF LINES IN HYDROXYL (DEUTERIUM) 17

ACTIVE MEDIUM : CARBON MONOXIDE
SYMBOL : CO

OPERATING CONDITIONS : STIMULATION OF THE ELECTRONIC TRANSITIONS IN A PULSED DISCHARGE OF HIGH CURRENT DENSITY. TYPICAL PRESSURE FOR THE VAC-UV LINES 60 TORR. IN THE COMPILATION, ONLY THE BAND EDGES ARE LISTED. TYPICAL PRESSURE FOR THE VISIBLE LINES 0.7-2 TORR. STIMULATION OF THE ROT.-VIB. TRANSITIONS WITH PULSED OR CONTINUOUS DISCHARGE IN A MIXTURE OF CO AND N2, SOMETIMES WITH HE, XE OR HG. MANY LINES CAN BE GENERATED ONLY WITH COOLING TO 77K. OPERATION WITH FAST GAS FLOW [443], WITH GASDYNAMIC EXPANSION [198], WITH TRANSVERSE ELECTRODE CONFIGURATION [199]. ALSO EXCITATION BY CHEMICAL REACTIONS (E.G. CS2+3O2+O2) [542,194,195,196], IN THIS CASE, EMISSION WITH DELTA V = 2 (OVERTONE EMISSION) IS OBSERVED [237,399], RECENTLY ALSO WITH ELECTRICAL EXCITATION AND SUPERSONIC EXPANSION [509]. MAXIMUM GENERATED POWER (AT 5.4 MICROMETER): 940 WATT [443]. LINES MARKED WITH @ HAVE NOT YET BEEN OBSERVED, SOME OF THESE PRESUMABLY BECAUSE OF WATER VAPOR INTERFERENCE. ADDITIONAL DATA ON CO-ISOTOPE LASER FREQUENCIES AT [336,608,609].

| | | WAVELENGTH IN VACUUM [MICROMETER] | | | | WAVELENGTH IN VACUUM [MICROMETER] | |
|---|---|---|---|---|---|---|---|
| CO | | 0.181085 | [238] | CO | | 0.195006 | [238] |
| CO | | 0.187831 | [238] | CO | | 0.197013 | [238] |
| CO | | 0.189784 | [238] | | | | |

| | | WAVELENGTH IN AIR [MICROMETER] | | | | WAVELENGTH IN AIR [MICROMETER] | |
|---|---|---|---|---|---|---|---|
| CO | | 0.450248 | [214] | CO | | 0.519154 | [214] |
| CO | | 0.450382 | [214] | CO | | 0.519281 | [214] |
| CO | | 0.450501 | [214] | CO | | 0.519363 | [214] |
| CO | | 0.450602 | [214] | CO | | 0.519387 | [214] |
| CO | | 0.450690 | [214] | CO | | 0.519472 | [214] |
| CO | | 0.450762 | [214] | CO | | 0.519488 | [214] |
| CO | | 0.450821 | [214] | CO | | 0.519595 | [214] |
| CO | | 0.451076 | [214] | CO | | 0.519680 | [214] |
| CO | | 0.481947 | [214] | CO | | 0.519743 | [214] |
| CO | | 0.482155 | [214] | CO | | 0.519786 | [214] |
| CO | | 0.482348 | [214] | CO | | 0.519807 | [214] |
| CO | | 0.482524 | [214] | CO | | 0.519807 | [214] |
| CO | | 0.482685 | [214] | CO | | 0.558411 | [214] |
| CO | | 0.482829 | [214] | CO | | 0.558748 | [214] |
| CO | | 0.482956 | [214] | CO | | 0.559058 | [215,214] |
| CO | ? | 0.483067 | [214] | CO | > | 0.559343 | [215,214] |
| CO | | 0.483162 | [214] | CO | > | 0.559602 | [215,214] |
| CO | | 0.483467 | [214] | CO | > | 0.558362 | [215,214] |
| CO | | 0.483503 | [214] | CO | > | 0.560043 | [215,214] |
| CO | | 0.483523 | [214] | CO | > | 0.560224 | [215,214] |
| CO | | 0.517959 | [214] | CO | | 0.560380 | [215,214] |
| CO | | 0.518211 | [214] | CO | | 0.560509 | [214] |
| CO | | 0.518442 | [214] | CO | | 0.560563 | [214] |
| CO | | 0.518653 | [214] | CO | | 0.560613 | [214] |
| CO | | 0.518843 | [214] | CO | ? | 0.560693 | [214] |
| CO | | 0.519001 | [214] | CO | | 0.560700 | [214] |

| | | WAVELENGTH IN AIR [MICROMETER] | |
|---|---|---|---|
| CO | | 0.560811 | [214] |
| CO | | 0.560898 | [214] |
| CO | | 0.560958 | [214] |
| CO | | 0.560993 | [214] |
| CO | | 0.604369 | [214] |
| CO | | 0.604816 | [214] |
| CO | | 0.605232 | [214] |
| CO | | 0.605619 | [214] |
| CO | > | 0.606296 | [215,214] |
| CO | > | 0.606588 | [215,214] |
| CO | > | 0.606848 | [215,214] |
| CO | > | 0.607075 | [215,214] |
| CO | | 0.607270 | [215,214] |
| CO | | 0.607436 | [215,214] |
| CO | | 0.607566 | [214] |
| CO | | 0.607584 | [214] |
| CO | | 0.607663 | [214] |
| CO | | 0.607731 | [214] |
| CO | | 0.607850 | [214] |
| CO | ? | 0.607934 | [214] |
| CO | | 0.607987 | [214] |
| CO | ? | 0.608007 | [214] |
| CO | | 0.658103 | [214] |
| CO | | 0.658623 | [214] |
| CO | | 0.659105 | [214] |
| CO | | 0.659547 | [215,214] |
| CO | > | 0.659948 | [215,214] |
| CO | > | 0.660306 | [215,214] |
| CO | > | 0.660630 | [215,214] |
| CO | > | 0.660910 | [215,214] |
| CO | | 0.660971 | [214] |
| CO | | 0.661151 | [215,214] |
| CO | | 0.661243 | [214] |
| CO | | 0.661353 | [215,214] |
| CO | | 0.661512 | [214] |
| CO | | 0.661634 | [214] |
| CO | | 0.661919 | [214] |
| CO | | 0.662003 | [214] |
| CO | | 0.662033 | [214] |

| | WAVELENGTH IN VACUUM [MICROMETER] | | TRANSITION |
|---|---|---|---|
| CO | 2.3474 | [237] | 2- 0 |
| CO | 2.3769 | [237] | 3- 1 |
| CO | 2.4380 | [237] | 4- 2 |
| CO | 2.4344 | [237] | 5- 3 |
| CO | 2.4696 | [237] | 6- 4 |
| CO | 2.5019 | [237] | 7- 5 |
| CO | 2.5350 | [237] | 8- 6 |
| CO | 2.5689 | [237] | 9- 7 |
| CO | 2.6036 | [237] | 10- 8 |
| CO | 2.6392 | [237] | 11- 9 |
| CO | 2.6756 | [237] | 12-10 |
| CO | 2.6886 | [509] | 12-10 P(05) |
| CO | 2.6914 | [509] | 12-10 P(06) |
| CO | 2.7129 | [237] | 13-11 |
| CO | 2.7262 | [509] | 13-11 P(05) |
| CO | 2.7290 | [509] | 13-11 P(06) |
| CO | 2.7319 | [509] | 13-11 P(07) |
| CO | 2.7511 | [237] | 14-12 |
| CO | 2.7647 | [509] | 14-12 P(05) |
| CO | 2.7676 | [509] | 14-12 P(06) |
| CO | 2.7705 | [509] | 14-12 P(07) |
| CO | 2.7903 | [237] | 15-13 |
| CO | 2.8042 | [509] | 15-13 P(05) |
| CO | 2.8071 | [509] | 15-13 P(06) |
| CO | 2.8101 | [509] | 15-13 P(07) |
| CO | 2.8306 | [237] | 16-14 |
| CO | 2.8446 | [509] | 16-14 P(05) |
| CO | 2.8476 | [509] | 16-14 P(06) |
| CO | 2.8507 | [509] | 16-14 P(07) |
| CO | 2.8892 | [509] | 17-15 P(05) |
| CO | 2.8923 | [509] | 17-15 P(06) |
| CO | 2.9288 | [509] | 18-16 P(05) |

| | | WAVELENGTH IN VACUUM [MICROMETER] | | TRANSITION |
|---|---|---|---|---|
| CO | | 2.9319 | [509] | 18-16 P(06) |
| CO | | 2.9351 | [509] | 18-16 P(07) |
| CO | | 2.9725 | [509] | 19-17 P(05) |
| CO | | 2.9757 | [509] | 19-17 P(06) |
| CO | | 2.9789 | [509] | 19-17 P(07) |
| CO | | 3.0174 | [509] | 20-18 P(05) |
| CO | | 3.0206 | [509] | 20-18 P(06) |
| CO | | 3.0668 | [509] | 21-19 P(06) |
| CO | @ | 4.735872 | [174] | 1- 0 P( 8) |
| CO | | 4.745130 | [174,212,315] | 1- 0 P( 9) |
| CO | | 4.754501 | [174,212,315] | 1- 0 P(10) |
| CO | | 4.763984 | [174,212,315] | 1- 0 P(11) |
| CO | | 4.773582 | [174,212] | 1- 0 P(12) |
| CO | | 4.783295 | [174,212] | 1- 0 P(13) |
| CO | | 4.793123 | [174,212] | 1- 0 P(14) |
| CO | | 4.803067 | [174,212] | 1- 0 P(15) |
| CO | | 4.813129 | [174,212] | 1- 0 P(16) |
| CO | | 4.823310 | [174,212] | 1- 0 P(17) |
| CO | | 4.833609 | [174,212] | 1- 0 P(18) |
| CO | | 4.844029 | [174,212] | 1- 0 P(19) |
| CO | | 4.854569 | [174,212] | 1- 0 P(20) |
| CO | | 4.865231 | [174,212] | 1- 0 P(21) |
| CO | @ | 4.876016 | [174] | 1- 0 P(22) |
| CO | @ | 4.767821 | [174] | 2- 1 P( 5) |
| CO | | 4.776892 | [174,212] | 2- 1 P( 6) |
| CO | | 4.786076 | [174,210,212] | 2- 1 P( 7) |
| CO | | 4.795373 | [174,210,212] | 2- 1 P( 8) |
| CO | | 4.804785 | [174,210,212] | 2- 1 P( 9) |
| CO | | 4.814312 | [174,210,212] | 2- 1 P(10) |
| CO | | 4.823954 | [174,210,212] | 2- 1 P(11) |
| CO | | 4.833714 | [174,210,212] | 2- 1 P(12) |
| CO | | 4.843591 | [174,210,212] | 2- 1 P(13) |
| CO | | 4.853586 | [174,210,212] | 2- 1 P(14) |
| CO | | 4.863700 | [174,210,212] | 2- 1 P(15) |
| CO | | 4.873935 | [174,212] | 2- 1 P(16) |
| CO | | 4.884291 | [174,212] | 2- 1 P(17) |
| CO | | 4.894769 | [174,212] | 2- 1 P(18) |
| CO | | 4.905369 | [174,212] | 2- 1 P(19) |
| CO | | 4.916094 | [174,212] | 2- 1 P(20) |
| CO | @ | 4.926943 | [174] | 2- 1 P(21) |
| CO | @ | 4.846781 | [174] | 3- 2 P( 7) |
| CO | | 4.856233 | [174,212] | 3- 2 P( 8) |
| CO | | 4.865803 | [174,212] | 3- 2 P( 9) |
| CO | | 4.875490 | [174,210,212] | 3- 2 P(10) |
| CO | | 4.885296 | [174,210,212] | 3- 2 P(11) |
| CO | | 4.895221 | [174,210,212] | 3- 2 P(12) |
| CO | | 4.905267 | [174,210,212] | 3- 2 P(13) |
| CO | | 4.915434 | [174,210,212] | 3- 2 P(14) |
| CO | | 4.925723 | [174,212] | 3- 2 P(15) |
| CO | | 4.936136 | [174,212] | 3- 2 P(16) |
| CO | | 4.946672 | [174,212] | 3- 2 P(17) |
| CO | | 4.957333 | [174,212] | 3- 2 P(18) |
| CO | | 4.968120 | [174,212] | 3- 2 P(19) |
| CO | | 4.979035 | [174,212] | 3- 2 P(20) |

| | | WAVELENGTH IN VACUUM [MICROMETER] | | TRANSITION |
|---|---|---|---|---|
| CO | | 4.990016 | [174,212] | 3- 2 P(21) |
| CO | | 5.001277 | [174,212] | 3- 2 P(22) |
| CO | | 5.012578 | [174,212] | 3- 2 P(23) |
| CO | | 5.023976 | [174,212] | 3- 2 P(24) |
| CO | | 5.035544 | [174,212] | 3- 2 P(25) |
| CO | | 5.047242 | [174,212] | 3- 2 P(26) |
| CO | | 5.059073 | [174,212] | 3- 2 P(27) |
| CO | | 5.071040 | [174,212] | 3- 2 P(28) |
| CO | | 5.083144 | [174,212] | 3- 2 P(29) |
| CO | | 5.095386 | [174,212] | 3- 2 P(30) |
| CO | | 5.107766 | [174,212] | 3- 2 P(31) |
| CO | | 5.120267 | [174,212] | 3- 2 P(32) |
| CO | | 5.132949 | [174,212] | 3- 2 P(33) |
| CO | @ | 5.145754 | [174] | 3- 2 P(34) |
| CO | @ | 4.880759 | [174] | 4- 3 P( 4) |
| CO | | 4.890016 | [174,509] | 4- 3 P( 5) |
| CO | | 4.899391 | [174,509] | 4- 3 P( 6) |
| CO | | 4.908883 | [174,509] | 4- 3 P( 7) |
| CO | | 4.918494 | [174,212] | 4- 3 P( 8) |
| CO | | 4.928238 | [174,212] | 4- 3 P( 9) |
| CO | | 4.938078 | [174,210,212] | 4- 3 P(10) |
| CO | | 4.948052 | [174,210,212] | 4- 3 P(11) |
| CO | | 4.958148 | [174,212] | 4- 3 P(12) |
| CO | | 4.968369 | [337,210,212] | 4- 3 P(13) |
| CO | | 4.978711 | [337,210,212] | 4- 3 P(14) |
| CO | | 4.989181 | [337,210,212] | 4- 3 P(15) |
| CO | | 4.999775 | [174,212] | 4- 3 P(16) |
| CO | | 5.010497 | [174,212] | 4- 3 P(17) |
| CO | | 5.021347 | [174,212] | 4- 3 P(18) |
| CO | | 5.032321 | [174,212] | 4- 3 P(19) |
| CO | | 5.043435 | [174,212] | 4- 3 P(20) |
| CO | | 5.054676 | [174,212] | 4- 3 P(21) |
| CO | | 5.066048 | [174,212] | 4- 3 P(22) |
| CO | @ | 5.077554 | [174] | 4- 3 P(23) |
| CO | | 4.943828 | [174,509] | 5- 4 P( 4) |
| CO | | 4.953240 | [174,509] | 5- 4 P( 5) |
| CO | | 4.962772 | [174,509] | 5- 4 P( 6) |
| CO | | 4.972425 | [174] | 5- 4 P( 7) |
| CO | | 4.982220 | [174,212] | 5- 4 P( 8) |
| CO | | 4.992099 | [174,212] | 5- 4 P( 9) |
| CO | | 5.002121 | [337,212] | 5- 4 P(10) |
| CO | | 5.012268 | [337,212] | 5- 4 P(11) |
| CO | | 5.022539 | [337,212] | 5- 4 P(12) |
| CO | | 5.032938 | [337,212] | 5- 4 P(13) |
| CO | | 5.043462 | [337,212] | 5- 4 P(14) |
| CO | | 5.054117 | [337,212] | 5- 4 P(15) |
| CO | | 5.064899 | [337,212] | 5- 4 P(16) |
| CO | | 5.075812 | [337,212] | 5- 4 P(17) |
| CO | | 5.086856 | [337,212] | 5- 4 P(18) |
| CO | | 5.098033 | [174,219,212] | 5- 4 P(19) |
| CO | | 5.109343 | [174,219,212] | 5- 4 P(20) |
| CO | | 5.120787 | [174,219,212] | 5- 4 P(21) |
| CO | | 5.131252 | [174,219,212] | 5- 4 P(22) |
| CO | | 5.144084 | [174,219,212] | 5- 4 P(23) |
| CO | | 5.155938 | [174,219,212] | 5- 4 P(24) |

| | | WAVELENGTH IN VACUUM [MICROMETER] | | TRANSITION |
|---|---|---|---|---|
| CO | | 5.167931 | [174,219,212] | 5- 4 P(25) |
| CO | | 5.180064 | [174,219,212] | 5- 4 P(26) |
| CO | | 5.192338 | [174,219,212] | 5- 4 P(27) |
| CO | | 5.204755 | [174,212] | 5- 4 P(28) |
| CO | | 5.217312 | [174,212] | 5- 4 P(29) |
| CO | | 5.230020 | [174,212] | 5- 4 P(30) |
| CO | | 5.242872 | [174,212] | 5- 4 P(31) |
| CO | | 5.255870 | [174,212] | 5- 4 P(32) |
| CO | | 5.269018 | [174,212] | 5- 4 P(33) |
| CO | @ | 5.282316 | [174] | 5- 4 P(34) |
| | | | | |
| CO | @ | 5.008369 | [174] | 6- 5 P( 4) |
| CO | | 5.017940 | [174,509] | 6- 5 P( 5) |
| CO | | 5.027635 | [174,509] | 6- 5 P( 6) |
| CO | | 5.037454 | [174,212,216] | 6- 5 P( 7) |
| CO | | 5.047397 | [174,216,212] | 6- 5 P( 8) |
| CO | | 5.057467 | [174,216,212] | 6- 5 P( 9) |
| CO | | 5.067663 | [337,216,212] | 6- 5 P(10) |
| CO | > | 5.077988 | [337,216,212] | 6- 5 P(11) |
| CO | | 5.088440 | [337,216,212] | 6- 5 P(12) |
| CO | | 5.099023 | [337,216,212] | 6- 5 P(13) |
| CO | | 5.109734 | [337,216,212] | 6- 5 P(14) |
| CO | | 5.120577 | [337,218,212] | 6- 5 P(15) |
| CO | | 5.131555 | [337,218,219] | 6- 5 P(16) |
| CO | | 5.142663 | [337,218,219] | 6- 5 P(17) |
| CO | | 5.153909 | [174,219,212] | 6- 5 P(18) |
| CO | | 5.165289 | [174,219,212] | 6- 5 P(19) |
| CO | | 5.176606 | [174,219,212] | 6- 5 P(20) |
| CO | | 5.188460 | [174,219,212] | 6- 5 P(21) |
| CO | | 5.200254 | [174,219,212] | 6- 5 P(22) |
| CO | | 5.21187 | [174,219,212] | 6- 5 P(23) |
| CO | | 5.224262 | [174,219,212] | 6- 5 P(24) |
| CO | | 5.236479 | [174,219,212] | 6- 5 P(25) |
| CO | | 5.248840 | [174,219,212] | 6- 5 P(26) |
| CO | | 5.261343 | [174,219,212] | 6- 5 P(27) |
| CO | | 5.273997 | [174,219,212] | 6- 5 P(28) |
| CO | | 5.286796 | [174,212] | 6- 5 P(29) |
| CO | | 5.299744 | [174,212] | 6- 5 P(30) |
| CO | | 5.312842 | [174,212] | 6- 5 P(31) |
| CO | | 5.326091 | [174,212] | 6- 5 P(32) |
| CO | | 5.339493 | [174,212] | 6- 5 P(33) |
| CO | @ | 5.353049 | [174] | 6- 5 P(34) |
| | | | | |
| CO | @ | 5.074432 | [174] | 7- 6 P( 4) |
| CO | | 5.084166 | [174,509] | 7- 6 P( 5) |
| CO | | 5.094028 | [174,509] | 7- 6 P( 6) |
| CO | | 5.104017 | [174,216,212] | 7- 6 P( 7) |
| CO | | 5.114134 | [174,216,212] | 7- 6 P( 8) |
| CO | | 5.124408 | [337,216,212] | 7- 6 P( 9) |
| CO | | 5.134757 | [337,216,212] | 7- 6 P(10) |
| CO | | 5.145264 | [337,216,218] | 7- 6 P(11) |
| CO | | 5.155902 | [337,216,218] | 7- 6 P(12) |
| CO | | 5.166672 | [337,216,218] | 7- 6 P(13) |
| CO | | 5.177575 | [337,216,218] | 7- 6 P(14) |
| CO | | 5.188617 | [337,216,218] | 7- 6 P(15) |
| CO | | 5.199792 | [337,219,212] | 7- 6 P(16) |
| CO | | 5.211102 | [337,219,212] | 7- 6 P(17) |

| | | WAVELENGTH IN VACUUM [MICROMETER] | | TRANSITION |
|---|---|---|---|---|
| CO | | 5.222555 | [174,212,219] | 7- 6 P(18) |
| CO | | 5.234145 | [174,212,219] | 7- 6 P(19) |
| CO | | 5.245874 | [174,212,219] | 7- 6 P(20) |
| CO | | 5.257745 | [174,219,212] | 7- 6 P(21) |
| CO | | 5.269759 | [174,219,212] | 7- 6 P(22) |
| CO | | 5.281916 | [174,219,212] | 7- 6 P(23) |
| CO | | 5.294218 | [174,219,212] | 7- 6 P(24) |
| CO | | 5.306666 | [174,219,212] | 7- 6 P(25) |
| CO | | 5.319261 | [174,219,212] | 7- 6 P(26) |
| CO | | 5.332005 | [174,219,212] | 7- 6 P(27) |
| CO | | 5.344899 | [174,219,212] | 7- 6 P(28) |
| CO | | 5.357945 | [174,212] | 7- 6 P(29) |
| CO | | 5.371143 | [174,212] | 7- 6 P(30) |
| CO | | 5.384494 | [174,212] | 7- 6 P(31) |
| CO | | 5.398001 | [174,212] | 7- 6 P(32) |
| CO | @ | 5.411665 | [174] | 7- 6 P(33) |
| | | | | |
| CO | @ | 5.142062 | [174] | 8- 7 P( 4) |
| CO | | 5.151996 | [174,509] | 8- 7 P( 5) |
| CO | | 5.162000 | [174,509] | 8- 7 P( 6) |
| CO | | 5.172164 | [174,212,216] | 8- 7 P( 7) |
| CO | | 5.182459 | [174,216,212] | 8- 7 P( 8) |
| CO | | 5.192888 | [337,216,212] | 8- 7 P( 9) |
| CO | | 5.203447 | [337,216,217] | 8- 7 P(10) |
| CO | | 5.214142 | [337,216,212] | 8- 7 P(11) |
| CO | | 5.224972 | [337,216,217] | 8- 7 P(12) |
| CO | | 5.235937 | [337,216,212] | 8- 7 P(13) |
| CO | | 5.247038 | [337,216,218] | 8- 7 P(14) |
| CO | | 5.258279 | [337,218,219] | 8- 7 P(15) |
| CO | | 5.269659 | [337,218,219] | 8- 7 P(16) |
| CO | | 5.281183 | [174,218,219] | 8- 7 P(17) |
| CO | | 5.292846 | [174,218,219] | 8- 7 P(18) |
| CO | | 5.304651 | [174,218,219] | 8- 7 P(19) |
| CO | | 5.316600 | [174,218,219] | 8- 7 P(20) |
| CO | | 5.328694 | [174,218,219] | 8- 7 P(21) |
| CO | | 5.340935 | [174,218,219] | 8- 7 P(22) |
| CO | | 5.353322 | [174,218,219] | 8- 7 P(23) |
| CO | | 5.365859 | [174,218,219] | 8- 7 P(24) |
| CO | | 5.378545 | [174,218,219] | 8- 7 P(25) |
| CO | | 5.391382 | [174,218,219] | 8- 7 P(26) |
| CO | | 5.404372 | [174,218,219] | 8- 7 P(27) |
| CO | | 5.417516 | [174,218,219] | 8- 7 P(28) |
| CO | | 5.430815 | [174,212] | 8- 7 P(29) |
| CO | | 5.444270 | [174,212] | 8- 7 P(30) |
| CO | | 5.457884 | [174,212] | 8- 7 P(31) |
| CO | | 5.471632 | [174,212] | 8- 7 P(32) |
| CO | | 5.485591 | [174,212] | 8- 7 P(33) |
| CO | @ | 5.499688 | [174] | 8- 7 P(34) |
| | | | | |
| CO | | 5.108980 | [174,212] | 9- 8 R( 6) |
| CO | | 5.100430 | [174,212] | 9- 8 R( 7) |
| CO | | 5.092002 | [174,212] | 9- 8 R( 8) |
| CO | @ | 5.201370 | [174] | 9- 8 P( 3) |
| CO | | 5.211315 | [174] | 9- 8 P( 4) |
| CO | | 5.221392 | [174,509] | 9- 8 P( 5) |
| CO | | 5.231603 | [174,212,509] | 9- 8 P( 6) |
| CO | | 5.241946 | [174,216,212] | 9- 8 P( 7) |

| | | WAVELENGTH IN VACUUM [MICROMETER] | | TRANSITION |
|---|---|---|---|---|
| CO | | 5.252424 | [337,216,212] | 9- 8 P( 8) |
| CO | | 5.263039 | [337,216,217] | 9- 8 P( 9) |
| CO | | 5.273789 | [337,216,217] | 9- 8 P(10) |
| CO | | 5.284678 | [337,216,217] | 9- 8 P(11) |
| CO | | 5.295704 | [337,216,217] | 9- 8 P(12) |
| CO | | 5.306870 | [337,216,217] | 9- 8 P(13) |
| CO | | 5.318176 | [337,216,217] | 9- 8 P(14) |
| CO | | 5.329624 | [337,218,219] | 9- 8 P(15) |
| CO | | 5.341216 | [337,218,219] | 9- 8 P(16) |
| CO | | 5.352955 | [174,218,219] | 9- 8 P(17) |
| CO | | 5.364837 | [174,218,219] | 9- 8 P(18) |
| CO | | 5.376865 | [174,218,219] | 9- 8 P(19) |
| CO | | 5.389039 | [174,218,219] | 9- 8 P(20) |
| CO | | 5.401364 | [174,218,219] | 9- 8 P(21) |
| CO | | 5.413837 | [174,218,219] | 9- 8 P(22) |
| CO | | 5.426463 | [174,218,219] | 9- 8 P(23) |
| CO | | 5.439219 | [174,218,219] | 9- 8 P(24) |
| CO | | 5.452172 | [174,218,219] | 9- 8 P(25) |
| CO | | 5.465259 | [174,218,219] | 9- 8 P(26) |
| CO | | 5.478502 | [174,218,219] | 9- 8 P(27) |
| CO | | 5.491904 | [174,218,219] | 9- 8 P(28) |
| CO | | 5.505464 | [174,212] | 9- 8 P(29) |
| CO | | 5.519186 | [174,212] | 9- 8 P(30) |
| CO | | 5.533070 | [174,212] | 9- 8 P(31) |
| CO | | 5.547118 | [174,212] | 9- 8 P(32) |
| CO | | 5.561330 | [174,212] | 9- 8 P(33) |
| CO | @ | 5.575710 | [174] | 9- 8 P(34) |
| | | | | |
| CO | | 5.096828 | [174,212] | 10- 9 R(16) |
| CO | | 5.104413 | [174,212] | 10- 9 R(15) |
| CO | | 5.112118 | [174,212] | 10- 9 R(14) |
| CO | | 5.119944 | [174,212] | 10- 9 R(13) |
| CO | | 5.127892 | [174,212] | 10- 9 R(12) |
| CO | | 5.135961 | [174,212] | 10- 9 R(11) |
| CO | @ | 5.272122 | [174] | 10- 9 P( 3) |
| CO | | 5.282243 | [174] | 10- 9 P( 4) |
| CO | | 5.292498 | [174,509] | 10- 9 P( 5) |
| CO | | 5.302890 | [174,212,509] | 10- 9 P( 6) |
| CO | | 5.313418 | [174,212] | 10- 9 P( 7) |
| CO | | 5.324085 | [337,216,212] | 10- 9 P( 8) |
| CO | | 5.334893 | [337,216,212] | 10- 9 P( 9) |
| CO | | 5.345838 | [337,216,217] | 10- 9 P(10) |
| CO | | 5.356924 | [337,216,217] | 10- 9 P(11) |
| CO | | 5.368153 | [337,216,217] | 10- 9 P(12) |
| CO | | 5.379526 | [337,216,217] | 10- 9 P(13) |
| CO | | 5.391045 | [337,216,218] | 10- 9 P(14) |
| CO | | 5.402707 | [337,218,219] | 10- 9 P(15) |
| CO | | 5.414516 | [337,218,219] | 10- 9 P(16) |
| CO | | 5.426477 | [174,218,219] | 10- 9 P(17) |
| CO | | 5.438584 | [174,218,219] | 10- 9 P(18) |
| CO | | 5.450840 | [174,218,219] | 10- 9 P(19) |
| CO | | 5.463249 | [174,218,219] | 10- 9 P(20) |
| CO | | 5.475810 | [174,218,219] | 10- 9 P(21) |
| CO | | 5.488524 | [174,218,219] | 10- 9 P(22) |
| CO | | 5.501394 | [174,218,219] | 10- 9 P(23) |
| CO | | 5.514421 | [174,218,219] | 10- 9 P(24) |
| CO | | 5.527606 | [174,218,219] | 10- 9 P(25) |

| | | WAVELENGTH IN VACUUM [MICROMETER] | | TRANSITION |
|---|---|---|---|---|
| CO | | 5.540950 | [174,218,219] | 10- 9 P(26) |
| CO | | 5.554455 | [174,218,219] | 10- 9 P(27) |
| CO | @ | 5.568122 | [174] | 10- 9 P(28) |
| CO | | 5.122753 | [174,212] | 11-10 R(22) |
| CO | | 5.129726 | [174,212] | 11-10 R(21) |
| CO | | 5.136819 | [174,212] | 11-10 R(20) |
| CO | | 5.144032 | [174,212] | 11-10 R(19) |
| CO | | 5.151366 | [174,212] | 11-10 R(18) |
| CO | | 5.158822 | [174,212] | 11-10 R(17) |
| CO | | 5.166400 | [174,212] | 11-10 R(16) |
| CO | | 5.174100 | [174,212] | 11-10 R(15) |
| CO | | 5.181924 | [174,212] | 11-10 R(14) |
| CO | | 5.189871 | [174,212] | 11-10 R(13) |
| CO | @ | 5.344601 | [174] | 11-10 P( 3) |
| CO | | 5.354901 | [174] | 11-10 P( 4) |
| CO | | 5.365340 | [174,509] | 11-10 P( 5) |
| CO | | 5.375920 | [174] | 11-10 P( 6) |
| CO | | 5.386640 | [174,212,509] | 11-10 P( 7) |
| CO | | 5.397499 | [337,212] | 11-10 P( 8) |
| CO | | 5.408504 | [337,212,217] | 11-10 P( 9) |
| CO | | 5.419652 | [337,212,217] | 11-10 P(10) |
| CO | | 5.430943 | [337,212,217] | 11-10 P(11) |
| CO | | 5.442381 | [337,212,217] | 11-10 P(12) |
| CO | | 5.453966 | [337,218,219] | 11-10 P(13) |
| CO | | 5.465699 | [337,218,219] | 11-10 P(14) |
| CO | | 5.477582 | [337,218,219] | 11-10 P(15) |
| CO | | 5.489622 | [174,218,219] | 11-10 P(16) |
| CO | | 5.501808 | [174,218,219] | 11-10 P(17) |
| CO | | 5.514147 | [174,218,219] | 11-10 P(18) |
| CO | | 5.526640 | [174,218,219] | 11-10 P(19) |
| CO | | 5.539288 | [174,218,219] | 11-10 P(20) |
| CO | | 5.552093 | [174,218,219] | 11-10 P(21) |
| CO | | 5.565057 | [174,218,219] | 11-10 P(22) |
| CO | | 5.578179 | [174,218,219] | 11-10 P(23) |
| CO | | 5.591463 | [174,218,219] | 11-10 P(24) |
| CO | | 5.604909 | [174,218,219] | 11-10 P(25) |
| CO | | 5.618519 | [174,212] | 11-10 P(26) |
| CO | | 5.632293 | [174,212] | 11-10 P(27) |
| CO | @ | 5.646235 | [174] | 11-10 P(28) |
| CO | | 5.179608 | [174,212] | 12-11 R(24) |
| CO | | 5.186476 | [174,212] | 12-11 R(23) |
| CO | | 5.193387 | [174,212] | 12-11 R(22) |
| CO | | 5.200459 | [174,212] | 12-11 R(21) |
| CO | | 5.207655 | [174,212] | 12-11 R(20) |
| CO | | 5.214974 | [174,212] | 12-11 R(19) |
| CO | | 5.222417 | [174,212] | 12-11 R(18) |
| CO | | 5.229984 | [174,212] | 12-11 R(17) |
| CO | | 5.237677 | [174,212] | 12-11 R(16) |
| CO | | 5.245496 | [174,212] | 12-11 R(15) |
| CO | | 5.429350 | [174] | 12-11 P( 4) |
| CO | | 5.439978 | [174] | 12-11 P( 5) |
| CO | | 5.450750 | [174] | 12-11 P( 6) |
| CO | | 5.461666 | [174] | 12-11 P( 7) |
| CO | | 5.472735 | [337,212] | 12-11 P( 8) |
| CO | | 5.483932 | [337,212,217] | 12-11 P( 9) |

| | | WAVELENGTH IN VACUUM [MICROMETER] | | TRANSITION |
|---|---|---|---|---|
| CO | | 5.495289 | [337,212,217] | 12-11 P(10) |
| CO | | 5.506793 | [337,212,217] | 12-11 P(11) |
| CO | | 5.518447 | [337,212,217] | 12-11 P(12) |
| CO | | 5.530252 | [337,212,217] | 12-11 P(13) |
| CO | | 5.542208 | [337,212,217] | 12-11 P(14) |
| CO | | 5.554315 | [337,212] | 12-11 P(15) |
| CO | | 5.566589 | [174,212] | 12-11 P(16) |
| CO | | 5.579010 | [174,218,219] | 12-11 P(17) |
| CO | | 5.591588 | [174,218,219] | 12-11 P(18) |
| CO | | 5.604325 | [174,218,219] | 12-11 P(19) |
| CO | | 5.617221 | [174,219,212] | 12-11 P(20) |
| CO | | 5.630278 | [174,219,212] | 12-11 P(21) |
| CO | | 5.643498 | [174,219,212] | 12-11 P(22) |
| CO | | 5.656881 | [174,219,212] | 12-11 P(23) |
| CO | | 5.670430 | [174,219,212] | 12-11 P(24) |
| CO | | 5.684145 | [174,219,212] | 12-11 P(25) |
| CO | | 5.698029 | [174,212] | 12-11 P(26) |
| CO | @ | 5.712083 | [174] | 12-11 P(27) |
| | | | | |
| CO | | 5.343257 | [174,212] | 13-12 R(12) |
| CO | | 5.351719 | [174,212] | 13-12 R(11) |
| CO | | 5.360314 | [174,212] | 13-12 R(10) |
| CO | | 5.369042 | [174,212] | 13-12 R( 9) |
| CO | | 5.377903 | [174,212] | 13-12 R( 8) |
| CO | | 5.386899 | [174,212] | 13-12 R( 7) |
| CO | | 5.396031 | [174,212] | 13-12 R( 6) |
| CO | | 5.405298 | [174,212] | 13-12 R( 5) |
| CO | | 5.414703 | [174,212] | 13-12 R( 4) |
| CO | | 5.424245 | [174,212] | 13-12 R( 3) |
| CO | | 5.433926 | [174,212] | 13-12 R( 2) |
| CO | | 5.516474 | [174] | 13-12 P( 5) |
| CO | | 5.527444 | [174] | 13-12 P( 6) |
| CO | @ | 5.505651 | [174] | 13-12 P( 4) |
| CO | | 5.538556 | [337] | 13-12 P( 7) |
| CO | | 5.549827 | [337] | 13-12 P( 8) |
| CO | | 5.561252 | [337,212] | 13-12 P( 9) |
| CO | | 5.572816 | [337,212] | 13-12 P(10) |
| CO | | 5.584536 | [337,212] | 13-12 P(11) |
| CO | | 5.596412 | [337,212,217] | 13-12 P(12) |
| CO | | 5.608443 | [337,212,217] | 13-12 P(13) |
| CO | | 5.620630 | [337,212] | 13-12 P(14) |
| CO | | 5.632975 | [337,212,219] | 13-12 P(15) |
| CO | | 5.645485 | [174,219,212] | 13-12 P(16) |
| CO | | 5.658148 | [174,219,212] | 13-12 P(17) |
| CO | | 5.670974 | [174,219,212] | 13-12 P(18) |
| CO | | 5.683961 | [174,219,212] | 13-12 P(19) |
| CO | | 5.697113 | [174,219,212] | 13-12 P(20) |
| CO | | 5.710431 | [174,219,212] | 13-12 P(21) |
| CO | | 5.723916 | [174,219,212] | 13-12 P(22) |
| CO | | 5.737568 | [174,219,212] | 13-12 P(23) |
| CO | | 5.751390 | [174,219,212] | 13-12 P(24) |
| CO | | 5.765384 | [174,212] | 13-12 P(25) |
| CO | | 5.779551 | [174,212] | 13-12 P(26) |
| CO | | 5.793891 | [174,212] | 13-12 P(27) |
| CO | | 5.808408 | [174,212] | 13-12 P(28) |
| CO | @ | 5.823103 | [174] | 13-12 P(29) |

| | | WAVELENGTH IN VACUUM [MICROMETER] | | TRANSITION |
|---|---|---|---|---|
| CO | | 5.354648 | [174,212] | 14-13 R(20) |
| CO | | 5.362186 | [174,212] | 14-13 R(19) |
| CO | | 5.369854 | [174,212] | 14-13 R(18) |
| CO | | 5.377654 | [174,212] | 14-13 R(17) |
| CO | | 5.385585 | [174,212] | 14-13 R(16) |
| CO | | 5.393649 | [174,212] | 14-13 R(15) |
| CO | | 5.401846 | [174,212] | 14-13 R(14) |
| CO | | 5.410177 | [174,212] | 14-13 R(13) |
| CO | | 5.418642 | [174,212] | 14-13 R(12) |
| CO | | 5.427242 | [174,212] | 14-13 R(11) |
| CO | | 5.435978 | [174,212] | 14-13 R(10) |
| CO | | 5.444850 | [174,212] | 14-13 R( 9) |
| CO | @ | 5.583870 | [174] | 14-13 P( 4) |
| CO | | 5.594893 | [174] | 14-13 P( 5) |
| CO | | 5.606068 | [174] | 14-13 P( 6) |
| CO | | 5.617391 | [337] | 14-13 P( 7) |
| CO | | 5.628872 | [337] | 14-13 P( 8) |
| CO | @ | 5.640505 | [337] | 14-13 P( 9) |
| CO | | 5.652298 | [337,212,217] | 14-13 P(10) |
| CO | | 5.664243 | [337,212,217] | 14-13 P(11) |
| CO | | 5.676348 | [337,212,217] | 14-13 P(12) |
| CO | | 5.688612 | [337,212] | 14-13 P(13) |
| CO | | 5.701037 | [337,212] | 14-13 P(14) |
| CO | | 5.713623 | [337,212,219] | 14-13 P(15) |
| CO | | 5.726379 | [174,219,212] | 14-13 P(16) |
| CO | | 5.739293 | [174,219,212] | 14-13 P(17) |
| CO | | 5.752373 | [174,219,212] | 14-13 P(18) |
| CO | | 5.765620 | [174,219] | 14-13 P(19) |
| CO | | 5.779036 | [174,219] | 14-13 P(20) |
| CO | | 5.792622 | [174,219] | 14-13 P(21) |
| CO | | 5.806380 | [174,219] | 14-13 P(22) |
| CO | | 5.820310 | [174,219] | 14-13 P(23) |
| CO | | 5.834415 | [174,219] | 14-13 P(24) |
| CO | | 5.848696 | [174,219] | 14-13 P(25) |
| CO | @ | 5.863154 | [174] | 14-13 P(26) |
| CO | | 5.374479 | [174,212] | 15-14 R(28) |
| CO | | 5.381078 | [174,212] | 15-14 R(27) |
| CO | | 5.387807 | [174,212] | 15-14 R(26) |
| CO | | 5.394666 | [174,212] | 15-14 R(25) |
| CO | | 5.401656 | [174,212] | 15-14 R(24) |
| CO | | 5.408780 | [174,212] | 15-14 R(23) |
| CO | | 5.416030 | [174,212] | 15-14 R(22) |
| CO | | 5.423415 | [174,212] | 15-14 R(21) |
| CO | | 5.430932 | [174,212] | 15-14 R(20) |
| CO | | 5.438584 | [174,212] | 15-14 R(19) |
| CO | | 5.446369 | [174,212] | 15-14 R(18) |
| CO | | 5.454289 | [174,212] | 15-14 R(17) |
| CO | | 5.462344 | [174,212] | 15-14 R(16) |
| CO | | 5.470535 | [174,212] | 15-14 R(15) |
| CO | | 5.686687 | [337] | 15-14 P( 6) |
| CO | | 5.698230 | [337] | 15-14 P( 7) |
| CO | @ | 5.709930 | [337] | 15-14 P( 8) |
| CO | | 5.721786 | [337] | 15-14 P( 9) |
| CO | | 5.733804 | [337] | 15-14 P(10) |
| CO | | 5.745984 | [337] | 15-14 P(11) |
| CO | | 5.758322 | [337,212] | 15-14 P(12) |

| | | WAVELENGTH IN VACUUM [MICROMETER] | | TRANSITION |
|---|---|---|---|---|
| CO | | 5.770827 | [337,212] | 15-14 P(13) |
| CO | | 5.783496 | [337,212,219] | 15-14 P(14) |
| CO | | 5.796340 | [174,219,212] | 15-14 P(15) |
| CO | | 5.809343 | [174,219,212] | 15-14 P(16) |
| CO | | 5.822516 | [174,219] | 15-14 P(17) |
| CO | | 5.835859 | [174,219] | 15-14 P(18) |
| CO | @ | 5.583870 | [174] | 14-13 P( 4) |
| CO | | 5.594893 | [174] | 14-13 P( 5) |
| CO | | 5.606068 | [174] | 14-13 P( 6) |
| CO | | 5.849374 | [174,219] | 15-14 P(19) |
| CO | | 5.863063 | [174,219] | 15-14 P(20) |
| CO | | 5.876926 | [174,219] | 15-14 P(21) |
| CO | | 5.890965 | [174,219] | 15-14 P(22) |
| CO | | 5.905182 | [174,219] | 15-14 P(23) |
| CO | | 5.919579 | [174,219] | 15-14 P(24) |
| CO | @ | 5.934157 | [174] | 15-14 P(25) |
| CO | @ | 5.948917 | [174] | 15-14 P(26) |
| | | | | |
| CO | @ | 5.746341 | [174] | 16-15 P( 4) |
| CO | | 5.757944 | [337] | 16-15 P( 5) |
| CO | @ | 5.769382 | [337] | 16-15 P( 6) |
| CO | | 5.781150 | [337,174] | 16-15 P( 7) |
| CO | | 5.793071 | [337] | 16-15 P( 8) |
| CO | | 5.817413 | [337] | 16-15 P(10) |
| CO | | 5.829830 | [337] | 16-15 P(11) |
| CO | | 5.842415 | [337] | 16-15 P(12) |
| CO | | 5.855168 | [337] | 16-15 P(13) |
| CO | @ | 5.868096 | [174] | 16-15 P(14) |
| CO | @ | 5.881189 | [174] | 16-15 P(15) |
| CO | | 5.894455 | [174,219] | 16-15 P(16) |
| CO | | 5.907895 | [174,219] | 16-15 P(17) |
| CO | | 5.921510 | [174,219] | 16-15 P(18) |
| CO | | 5.935301 | [174,219] | 16-15 P(19) |
| CO | | 5.949271 | [174,219] | 16-15 P(20) |
| CO | | 5.963420 | [174,219] | 16-15 P(21) |
| CO | | 5.977751 | [174,219] | 16-15 P(22) |
| CO | | 5.992264 | [174,219] | 16-15 P(23) |
| CO | @ | 6.006963 | [174] | 16-15 P(24) |
| | | | | |
| CO | @ | 5.842407 | [174] | 17-16 P( 5) |
| CO | | 5.854229 | [337] | 17-16 P( 6) |
| CO | | 5.866221 | [337] | 17-16 P( 7) |
| CO | | 5.878384 | [337,174] | 17-16 P( 8) |
| CO | | 5.890704 | [337] | 17-16 P( 9) |
| CO | | 5.903202 | [337] | 17-16 P(10) |
| CO | | 5.915865 | [337] | 17-16 P(11) |
| CO | | 5.928701 | [337] | 17-16 P(12) |
| CO | @ | 5.941711 | [337] | 17-16 P(13) |
| CO | | 5.954841 | [337] | 17-16 P(14) |
| CO | @ | 5.968257 | [174] | 17-16 P(15) |
| CO | | 5.981799 | [174,219] | 17-16 P(16) |
| CO | | 5.995510 | [174,219] | 17-16 P(17) |
| CO | | 6.009406 | [174,219] | 17-16 P(18) |
| CO | | 6.023483 | [174,219] | 17-16 P(19) |
| CO | | 6.037744 | [174,219] | 17-16 P(20) |
| CO | @ | 6.052189 | [174] | 17-16 P(21) |

| | | WAVELENGTH IN VACUUM [MICROMETER] | | TRANSITION |
|---|---|---|---|---|
| CO | @ | 5.941314 | [174] | 18-17 P( 6) |
| CO | | 5.953562 | [337] | 18-17 P( 7) |
| CO | | 5.965936 | [337] | 18-17 P( 8) |
| CO | | 5.978508 | [337] | 18-17 P( 9) |
| CO | | 5.991256 | [337,174] | 18-17 P(10) |
| CO | | 6.004170 | [337] | 18-17 P(11) |
| CO | | 6.017267 | [337] | 18-17 P(12) |
| CO | | 6.030541 | [337] | 18-17 P(13) |
| CO | @ | 6.043400 | [174] | 18-17 P(14) |
| CO | | 6.05760 | [174,219] | 18-17 P(15) |
| CO | | 6.07142 | [174,219] | 18-17 P(16) |
| CO | | 6.08542 | [174,219] | 18-17 P(17) |
| CO | | 6.09960 | [174,219] | 18-17 P(18) |
| CO | | 6.11397 | [174,219] | 18-17 P(19) |
| CO | | 6.12852 | [174,219] | 18-17 P(20) |
| CO | @ | 6.14327 | [174] | 18-17 P(21) |
| | | | | |
| CO | @ | 6.030715 | [174] | 19-18 P( 6) |
| CO | | 6.043187 | [174] | 19-18 P( 7) |
| CO | | 6.055827 | [337] | 19-18 P( 8) |
| CO | | 6.068653 | [337] | 19-18 P( 9) |
| CO | | 6.081656 | [337] | 19-18 P(10) |
| CO | | 6.094841 | [337] | 19-18 P(11) |
| CO | | 6.108206 | [337] | 19-18 P(12) |
| CO | | 6.121753 | [337] | 19-18 P(13) |
| CO | | 6.135488 | [174] | 19-18 P(14) |
| CO | | 6.149406 | [174,219] | 19-18 P(15) |
| CO | | 6.163511 | [174,219] | 19-18 P(16) |
| CO | | 6.177806 | [174,219] | 19-18 P(17) |
| CO | | 6.192291 | [174,219] | 19-18 P(18) |
| CO | | 6.206969 | [174,219] | 19-18 P(19) |
| CO | @ | 6.221841 | [174] | 19-18 P(20) |
| | | | | |
| CO | @ | 6.097625 | [174] | 20-19 P( 4) |
| CO | | 6.109989 | [174] | 20-19 P( 5) |
| CO | | 6.122525 | [337] | 20-19 P( 6) |
| CO | | 6.135248 | [337] | 20-19 P( 7) |
| CO | | 6.148152 | [337] | 20-19 P( 8) |
| CO | | 6.161240 | [337] | 20-19 P( 9) |
| CO | | 6.174509 | [337] | 20-19 P(10) |
| CO | | 6.187966 | [337] | 20-19 P(11) |
| CO | | 6.201608 | [337] | 20-19 P(12) |
| CO | | 6.215438 | [337] | 20-19 P(13) |
| CO | | 6.229461 | [174] | 20-19 P(14) |
| CO | | 6.243673 | [174,219] | 20-19 P(15) |
| CO | | 6.258077 | [174,219] | 20-19 P(16) |
| CO | | 6.272677 | [174,219] | 20-19 P(17) |
| CO | | 6.287472 | [174,219] | 20-19 P(18) |
| CO | @ | 6.302466 | [174] | 20-19 P(19) |
| | | | | |
| CO | @ | 6.191453 | [174] | 21-20 P( 4) |
| CO | | 6.204067 | [174] | 21-20 P( 5) |
| CO | | 6.216856 | [337] | 21-20 P( 6) |
| CO | | 6.229843 | [337] | 21-20 P( 7) |
| CO | | 6.243012 | [337] | 21-20 P( 8) |
| CO | | 6.256366 | [337] | 21-20 P( 9) |
| CO | | 6.269915 | [337] | 21-20 P(10) |

| | | WAVELENGTH IN VACUUM [MICROMETER] | | TRANSITION |
|---|---|---|---|---|
| CO | | 6.283649 | [337] | 21-20 P(11) |
| CO | | 6.297578 | [337] | 21-20 P(12) |
| CO | | 6.311705 | [174] | 21-20 P(13) |
| CO | | 6.326023 | [174,219] | 21-20 P(14) |
| CO | | 6.340538 | [174,219] | 21-20 P(15) |
| CO | | 6.355252 | [174,219] | 21-20 P(16) |
| CO | | 6.370176 | [174,219] | 21-20 P(17) |
| CO | @ | 6.385283 | [174] | 21-20 P(18) |
| CO | | 6.300749 | [337] | 22-21 P( 5) |
| CO | | 6.313813 | [337] | 22-21 P( 6) |
| CO | | 6.327064 | [337] | 22-21 P( 7) |
| CO | | 6.340507 | [337] | 22-21 P( 8) |
| CO | | 6.354144 | [337] | 22-21 P( 9) |
| CO | | 6.367975 | [337] | 22-21 P(10) |
| CO | | 6.382004 | [337] | 22-21 P(11) |
| CO | @ | 6.410658 | [174] | 22-21 P(13) |
| CO | | 6.425284 | [174,219] | 22-21 P(14) |
| CO | | 6.440113 | [174,219] | 22-21 P(15) |
| CO | | 6.455148 | [174,219] | 22-21 P(16) |
| CO | | 6.470388 | [174,219] | 22-21 P(17) |
| CO | @ | 6.485838 | [174] | 22-21 P(18) |
| CO | | 6.400168 | [337] | 23-22 P( 5) |
| CO | | 6.413510 | [337,174] | 23-22 P( 6) |
| CO | | 6.427033 | [337] | 23-22 P( 7) |
| CO | | 6.440760 | [337] | 23-22 P( 8) |
| CO | | 6.454686 | [337] | 23-22 P( 9) |
| CO | | 6.468812 | [337] | 23-22 P(10) |
| CO | @ | 6.497679 | [174] | 23-22 P(12) |
| CO | | 6.512417 | [174,219] | 23-22 P(13) |
| CO | | 6.527363 | [174,219] | 23-22 P(14) |
| CO | | 6.542518 | [174,219] | 23-22 P(15) |
| CO | | 6.557884 | [174,219] | 23-22 P(16) |
| CO | @ | 6.573463 | [174] | 23-22 P(17) |
| CO | | 6.502437 | [337] | 24-23 P( 5) |
| CO | | 6.516055 | [337] | 24-23 P( 6) |
| CO | | 6.529870 | [337] | 24-23 P( 7) |
| CO | | 6.543890 | [337] | 24-23 P( 8) |
| CO | | 6.558117 | [337] | 24-23 P( 9) |
| CO | @ | 6.572552 | [337] | 24-23 P(10) |
| CO | | 6.587194 | [337] | 24-23 P(11) |
| CO | | 6.602046 | [337] | 24-23 P(12) |
| CO | | 6.617109 | [337] | 24-23 P(13) |
| CO | @ | 6.632387 | [174] | 24-23 P(14) |
| CO | | 6.647880 | [174,219] | 24-23 P(15) |
| CO | | 6.663590 | [174,219] | 24-23 P(16) |
| CO | @ | 6.679519 | [174] | 24-23 P(17) |
| CO | | 6.607691 | [337] | 25-24 P( 5) |
| CO | | 6.621596 | [337] | 25-24 P( 6) |
| CO | | 6.635713 | [337] | 25-24 P( 7) |
| CO | | 6.650037 | [337] | 25-24 P( 8) |
| CO | | 6.664574 | [337] | 25-24 P( 9) |
| CO | | 6.679326 | [337] | 25-24 P(10) |
| CO | | 6.694287 | [337] | 25-24 P(11) |

| | | WAVELENGTH IN VACUUM [MICROMETER] | | TRANSITION |
|---|---|---|---|---|
| CO | | 6.709469 | [337] | 25-24 P(12) |
| CO | | 6.702067 | [337] | 26-25 P( 4) |
| CO | @ | 6.716066 | [337] | 26-25 P( 5) |
| CO | | 6.730277 | [337] | 26-25 P( 6) |
| CO | | 6.744698 | [337] | 26-25 P( 7) |
| CO | | 6.759342 | [337] | 26-25 P( 8) |
| CO | | 6.774209 | [337,174] | 26-25 P( 9) |
| CO | | 6.789276 | [337] | 26-25 P(10) |
| CO | | 6.804577 | [337] | 26-25 P(11) |
| CO | | 6.820100 | [337] | 26-25 P(12) |
| CO | | 6.835854 | [337] | 26-25 P(13) |
| CO | | 6.842243 | [337] | 27-26 P( 6) |
| CO | @ | 6.854639 | [337] | 27-26 P( 7) |
| CO | | 6.871970 | [337,174] | 27-26 P( 8) |
| CO | | 6.887152 | [337] | 27-26 P( 9) |
| CO | | 6.902569 | [337] | 27-26 P(10) |
| CO | | 6.918218 | [337] | 27-26 P(11) |
| CO | | 6.934101 | [337] | 27-26 P(12) |
| CO | | 6.942824 | [337] | 28-27 P( 5) |
| CO | @ | 6.957669 | [337] | 28-27 P( 6) |
| CO | | 6.972747 | [337] | 28-27 P( 7) |
| CO | | 6.988057 | [337] | 28-27 P( 8) |
| CO | | 7.003596 | [337] | 28-27 P( 9) |
| CO | | 7.019371 | [337] | 28-27 P(10) |
| CO | | 7.035382 | [337] | 28-27 P(11) |
| CO | | 7.076735 | [337] | 29-28 P( 6) |
| CO | | 7.092158 | [337] | 29-28 P( 7) |
| CO | | 7.107821 | [337] | 29-28 P( 8) |
| CO | | 7.123720 | [337] | 29-28 P( 9) |
| CO | | 7.139864 | [337] | 29-28 P(10) |
| CO | | 7.156250 | [337] | 29-28 P(11) |
| CO | | 7.199637 | [337] | 30-29 P( 6) |
| CO | | 7.215418 | [337] | 30-29 P( 7) |
| CO | | 7.231448 | [337] | 30-29 P( 8) |
| CO | | 7.247721 | [337] | 30-29 P( 9) |
| CO | | 7.264248 | [337] | 30-29 P(10) |
| CO | | 7.310709 | [174] | 31-30 P( 5) |
| CO | | 7.326587 | [337] | 31-30 P( 6) |
| CO | | 7.342742 | [337] | 31-30 P( 7) |
| CO | | 7.359153 | [337] | 31-30 P( 8) |
| CO | | 7.375822 | [337] | 31-30 P( 9) |
| CO | | 7.392748 | [337] | 31-30 P(10) |
| CO | | 7.441563 | [174] | 32-31 P( 5) |
| CO | | 7.457822 | [337] | 32-31 P( 6) |
| CO | | 7.474367 | [337] | 32-31 P( 7) |
| CO | | 7.491175 | [337] | 32-31 P( 8) |
| CO | | 7.508252 | [337] | 32-31 P( 9) |
| CO | | 7.525593 | [337] | 32-31 P(10) |
| CO | | 7.543207 | [337] | 32-31 P(11) |

| | | WAVELENGTH IN VACUUM [MICROMETER] | | TRANSITION |
|---|---|---|---|---|
| CO | | 7.593623 | [174] | 33-32 P( 6) |
| CO | | 7.610553 | [337] | 33-32 P( 7) |
| CO | | 7.627777 | [337] | 33-32 P( 8) |
| CO | | 7.645277 | [337] | 33-32 P( 9) |
| CO | | 7.663057 | [337] | 33-32 P(10) |
| CO | | 7.681105 | [337] | 33-32 P(11) |
| CO | | 7.734191 | [337] | 34-33 P( 6) |
| CO | | 7.751571 | [337] | 34-33 P( 7) |
| CO | | 7.769230 | [337] | 34-33 P( 8) |
| CO | | 7.787173 | [337] | 34-33 P( 9) |
| CO | | 7.805408 | [337] | 34-33 P(10) |
| CO | | 7.823924 | [337] | 34-33 P(11) |
| CO | | 7.879916 | [337] | 35-34 P( 6) |
| CO | | 7.897740 | [337] | 35-34 P( 7) |
| CO | | 7.915858 | [337] | 35-34 P( 8) |
| CO | @ | 7.934266 | [337] | 35-34 P( 9) |
| CO | | 7.952969 | [337] | 35-34 P(10) |
| CO | | 8.031109 | [337] | 36-35 P( 6) |
| CO | | 8.049391 | [337] | 36-35 P( 7) |
| CO | | 8.067991 | [337] | 36-35 P( 8) |
| CO | | 8.086885 | [337] | 36-35 P( 9) |
| CO | | 8.106093 | [337] | 36-35 P(10) |
| CO | | 8.206919 | [337] | 37-36 P( 7) |
| CO | | 8.226011 | [337] | 37-36 P( 8) |
| CO | | 8.245417 | [337] | 37-36 P( 9) |
| CO | | 8.265146 | [337] | 37-36 P(10) |
| CO | | 5.303284 | [380] | 7- 6 P(16) C[13]O[16] |
| CO | | 5.314516 | [380] | 7- 6 P(17) C[13]O[16] |
| CO | | 5.325877 | [380] | 7- 6 P(18) C[13]O[16] |
| CO | | 5.337376 | [380] | 7- 6 P(19) C[13]O[16] |
| CO | | 5.328437 | [380] | 8- 7 P(12) C[13]O[16] |
| CO | | 5.339330 | [380] | 8- 7 P(13) C[13]O[16] |
| CO | @ | 5.350355 | [380] | 8- 7 P(14) C[13]O[16] |
| CO | | 5.367272 | [380] | 8- 7 P(15) C[13]O[16] |
| CO | | 5.372811 | [380] | 8- 7 P(16) C[13]O[16] |
| CO | | 5.384241 | [380] | 8- 7 P(17) C[13]O[16] |
| CO | | 5.395810 | [380] | 8- 7 P(18) C[13]O[16] |
| CO | | 5.407516 | [380] | 8- 7 P(19) C[13]O[16] |
| CO | | 5.377032 | [380] | 9- 8 P(10) C[13]O[16] |
| CO | | 5.387850 | [380] | 9- 8 P(11) C[13]O[16] |
| CO | | 5.398801 | [380] | 9- 8 P(12) C[13]O[16] |
| CO | | 5.409888 | [380] | 9- 8 P(13) C[13]O[16] |
| CO | @ | 5.421112 | [380] | 9- 8 P(14) C[13]O[16] |
| CO | | 5.4322471 | [380] | 9- 8 P(15) C[13]O[16] |
| CO | | 5.4439721 | [380] | 9- 8 P(16) C[13]O[16] |
| CO | | 5.4556001 | [380] | 9- 8 P(17) C[13]O[16] |
| CO | | 5.4673911 | [380] | 9- 8 P(18) C[13]O[16] |
| CO | @ | 5.4793111 | [380] | 9- 8 P(19) C[13]O[16] |
| CO | | 5.4913771 | [380] | 9- 8 P(20) C[13]O[16] |
| CO | | 5.5035831 | [380] | 9- 8 P(21) C[13]O[16] |

| | | WAVELENGTH IN VACUUM [MICROMETER] | | TRANSITION |
|---|---|---|---|---|
| CO | | 5.437800 | [380] | 10- 9 P(09) C[13]O[16] |
| CO | | 5.448674 | [380] | 10- 9 P(10) C[13]O[16] |
| CO | | 5.459684 | [380] | 10- 9 P(11) C[13]O[16] |
| CO | | 5.470831 | [380] | 10- 9 P(12) C[13]O[16] |
| CO | | 5.482117 | [380] | 10- 9 P(13) C[13]O[16] |
| CO | | 5.493543 | [380] | 10- 9 P(14) C[13]O[16] |
| CO | | 5.505110 | [380] | 10- 9 P(15) C[13]O[16] |
| CO | | 5.516821 | [380] | 10- 9 P(16) C[13]O[16] |
| CO | | 5.528674 | [380] | 10- 9 P(17) C[13]O[16] |
| CO | | 5.540673 | [380] | 10- 9 P(18) C[13]O[16] |
| CO | | 5.552816 | [380] | 10- 9 P(19) C[13]O[16] |
| CO | | 5.500027 | [380] | 11-10 P(08) C[13]O[16] |
| CO | | 5.510954 | [380] | 11-10 P(09) C[13]O[16] |
| CO | | 5.522022 | [380] | 11-10 P(10) C[13]O[16] |
| CO | | 5.533232 | [380] | 11-10 P(11) C[13]O[16] |
| CO | | 5.544580 | [380] | 11-10 P(12) C[13]O[16] |
| CO | | 5.556074 | [380] | 11-10 P(13) C[13]O[16] |
| CO | | 5.567715 | [380] | 11-10 P(14) C[13]O[16] |
| CO | | 5.579476 | [380] | 11-10 P(15) C[13]O[16] |
| CO | | 5.591415 | [380] | 11-10 P(16) C[13]O[16] |
| CO | | 5.603490 | [380] | 11-10 P(17) C[13]O[16] |
| CO | @ | 5.615709 | [380] | 11-10 P(18) C[13]O[16] |
| CO | | 5.628080 | [380] | 11-10 P(19) C[13]O[16] |
| CO | @ | 5.640600 | [380] | 11-10 P(20) C[13]O[16] |
| CO | @ | 5.653272 | [380] | 11-10 P(21) C[13]O[16] |
| CO | | 5.666098 | [380] | 11-10 P(22) C[13]O[16] |
| CO | | 5.574745 | [380] | 12-11 P(08) C[13]O[16] |
| CO | | 5.585871 | [380] | 12-11 P(09) C[13]O[16] |
| CO | | 5.597139 | [380] | 12-11 P(10) C[13]O[16] |
| CO | | 5.608550 | [380] | 12-11 P(11) C[13]O[16] |
| CO | | 5.620109 | [380] | 12-11 P(12) C[13]O[16] |
| CO | @ | 5.631813 | [380] | 12-11 P(13) C[13]O[16] |
| CO | | 5.643666 | [380] | 12-11 P(14) C[13]O[16] |
| CO | @ | 5.655664 | [380] | 12-11 P(15) C[13]O[16] |
| CO | | 5.667813 | [380] | 12-11 P(16) C[13]O[16] |
| CO | | 5.680114 | [380] | 12-11 P(17) C[13]O[16] |
| CO | @ | 5.692566 | [380] | 12-11 P(18) C[13]O[16] |
| CO | | 5.705170 | [380] | 12-11 P(19) C[13]O[16] |
| CO | @ | 5.717932 | [380] | 12-11 P(20) C[13]O[16] |
| CO | | 5.730846 | [380] | 12-11 P(21) C[13]O[16] |
| CO | | 5.651276 | [380] | 13-12 P(08) C[13]O[16] |
| CO | | 5.662607 | [380] | 13-12 P(09) C[13]O[16] |
| CO | | 5.674081 | [380] | 13-12 P(10) C[13]O[16] |
| CO | | 5.685704 | [380] | 13-12 P(11) C[13]O[16] |
| CO | | 5.697477 | [380] | 13-12 P(12) C[13]O[16] |
| CO | @ | 5.709398 | [380] | 13-12 P(13) C[13]O[16] |
| CO | | 5.721472 | [380] | 13-12 P(14) C[13]O[16] |
| CO | | 5.733698 | [380] | 13-12 P(15) C[13]O[16] |
| CO | | 5.746076 | [380] | 13-12 P(16) C[13]O[16] |
| CO | | 5.758611 | [380] | 13-12 P(17) C[13]O[16] |
| CO | | 5.771303 | [380] | 13-12 P(18) C[13]O[16] |
| CO | | 5.784148 | [380] | 13-12 P(19) C[13]O[16] |
| CO | | 5.797155 | [380] | 13-12 P(20) C[13]O[16] |

| | | WAVELENGTH IN VACUUM [MICROMETER] | | TRANSITION |
|---|---|---|---|---|
| CO | | 5.810319 | [380] | 13-12 P(21) C[13]O[16] |
| CO | | 5.729687 | [380] | 14-13 P(08) C[13]O[16] |
| CO | | 5.741227 | [380] | 14-13 P(09) C[13]O[16] |
| CO | | 5.752912 | [380] | 14-13 P(10) C[13]O[16] |
| CO | @ | 5.764755 | [380] | 14-13 P(11) C[13]O[16] |
| CO | | 5.776747 | [380] | 14-13 P(12) C[13]O[16] |
| CO | | 5.788893 | [380] | 14-13 P(13) C[13]O[16] |
| CO | | 5.801198 | [380] | 14-13 P(14) C[13]O[16] |
| CO | | 5.813656 | [380] | 14-13 P(15) C[13]O[16] |
| CO | | 5.826267 | [380] | 14-13 P(16) C[13]O[16] |
| CO | | 5.839055 | [380] | 14-13 P(17) C[13]O[16] |
| CO | | 5.851981 | [380] | 14-13 P(18) C[13]O[16] |
| CO | | 5.865079 | [380] | 14-13 P(19) C[13]O[16] |
| CO | | 5.798443 | [380] | 15-14 P(07) C[13]O[16] |
| CO | | 5.810042 | [380] | 15-14 P(08) C[13]O[16] |
| CO | @ | 5.821796 | [380] | 15-14 P(09) C[13]O[16] |
| CO | | 5.833703 | [380] | 15-14 P(10) C[13]O[16] |
| CO | | 5.845769 | [380] | 15-14 P(11) C[13]O[16] |
| CO | | 5.857987 | [380] | 15-14 P(12) C[13]O[16] |
| CO | | 5.870367 | [380] | 15-14 P(13) C[13]O[16] |
| CO | @ | 5.882903 | [380] | 15-14 P(14) C[13]O[16] |
| CO | @ | 5.895604 | [380] | 15-14 P(15) C[13]O[16] |
| CO | | 5.908465 | [380] | 15-14 P(16) C[13]O[16] |
| CO | | 5.921487 | [380] | 15-14 P(17) C[13]O[16] |
| CO | @ | 5.934676 | [380] | 15-14 P(18) C[13]O[16] |
| CO | | 5.948030 | [380] | 15-14 P(19) C[13]O[16] |
| CO | | 5.961582 | [380] | 15-14 P(20) C[13]O[16] |
| CO | | 5.975240 | [380] | 15-14 P(21) C[13]O[16] |
| CO | | 5.892408 | [380] | 16-15 P(08) C[13]O[16] |
| CO | | 5.904387 | [380] | 16-15 P(09) C[13]O[16] |
| CO | @ | 5.916519 | [380] | 16-15 P(10) C[13]O[16] |
| CO | | 5.928814 | [380] | 16-15 P(11) C[13]O[16] |
| CO | @ | 5.941269 | [380] | 16-15 P(12) C[13]O[16] |
| CO | | 5.953887 | [380] | 16-15 P(13) C[13]O[16] |
| CO | | 5.966669 | [380] | 16-15 P(14) C[13]O[16] |
| CO | | 5.979617 | [380] | 16-15 P(15) C[13]O[16] |
| CO | | 5.992728 | [380] | 16-15 P(16) C[13]O[16] |
| CO | | 6.006059 | [380] | 16-15 P(17) C[13]O[16] |
| CO | | 6.019459 | [380] | 16-15 P(18) C[13]O[16] |
| CO | | 6.033077 | [380] | 16-15 P(19) C[13]O[16] |
| CO | | 5.964829 | [380] | 17-16 P(07) C[13]O[16] |
| CO | | 5.976868 | [380] | 17-16 P(08) C[13]O[16] |
| CO | @ | 5.989075 | [380] | 17-16 P(09) C[13]O[16] |
| CO | | 6.001439 | [380] | 17-16 P(10) C[13]O[16] |
| CO | @ | 6.013970 | [380] | 17-16 P(11) C[13]O[16] |
| CO | | 6.026667 | [380] | 17-16 P(12) C[13]O[16] |
| CO | | 6.039534 | [380] | 17-16 P(13) C[13]O[16] |
| CO | @ | 6.052565 | [380] | 17-16 P(14) C[13]O[16] |
| CO | | 6.065767 | [380] | 17-16 P(15) C[13]O[16] |
| CO | | 6.079138 | [380] | 17-16 P(16) C[13]O[16] |
| CO | | 6.092683 | [380] | 17-16 P(17) C[13]O[16] |
| CO | @ | 6.106404 | [380] | 17-16 P(18) C[13]O[16] |
| CO | | 6.120296 | [380] | 17-16 P(19) C[13]O[16] |

| | | WAVELENGTH IN VACUUM [MICROMETER] | | TRANSITION |
|---|---|---|---|---|
| CO | | 6.063494 | [380] | 18-17 P(08) C[13]O[16] |
| CO | @ | 6.075932 | [380] | 18-17 P(09) C[13]O[16] |
| CO | | 6.088540 | [380] | 18-17 P(10) C[13]O[16] |
| CO | | 6.101318 | [380] | 18-17 P(11) C[13]O[16] |
| CO | @ | 6.114263 | [380] | 18-17 P(12) C[13]O[16] |
| CO | | 6.127383 | [380] | 18-17 P(13) C[13]O[16] |
| CO | | 6.140673 | [380] | 18-17 P(14) C[13]O[16] |
| CO | | 6.154138 | [380] | 18-17 P(15) C[13]O[16] |
| CO | | 6.167780 | [380] | 18-17 P(16) C[13]O[16] |
| CO | | 6.181597 | [380] | 18-17 P(17) C[13]O[16] |
| CO | | 6.195595 | [380] | 18-17 P(18) C[13]O[16] |
| CO | | 6.209772 | [380] | 18-17 P(19) C[13]O[16] |
| | | | | |
| CO | | 6.139855 | [380] | 19-18 P(07) C[13]O[16] |
| CO | | 6.152366 | [380] | 19-18 P(08) C[13]O[16] |
| CO | | 6.165049 | [380] | 19-18 P(09) C[13]O[16] |
| CO | | 6.177904 | [380] | 19-18 P(10) C[13]O[16] |
| CO | | 6.190935 | [380] | 19-18 P(11) C[13]O[16] |
| CO | | 6.204140 | [380] | 19-18 P(12) C[13]O[16] |
| CO | | 6.217521 | [380] | 19-18 P(13) C[13]O[16] |
| CO | | 6.231077 | [380] | 19-18 P(14) C[13]O[16] |
| CO | | 6.244817 | [380] | 19-18 P(15) C[13]O[16] |
| CO | | 6.258735 | [380] | 19-18 P(16) C[13]O[16] |
| CO | | 6.272833 | [380] | 19-18 P(17) C[13]O[16] |
| CO | @ | 6.287116 | [380] | 19-18 P(18) C[13]O[16] |
| CO | | 6.301586 | [380] | 19-18 P(19) C[13]O[16] |
| | | | | |
| CO | | 6.230817 | [380] | 20-19 P(07) C[13]O[16] |
| CO | | 6.243577 | [380] | 20-19 P(08) C[13]O[16] |
| CO | | 6.256511 | [380] | 20-19 P(09) C[13]O[16] |
| CO | | 6.269624 | [380] | 20-19 P(10) C[13]O[16] |
| CO | | 6.282915 | [380] | 20-19 P(11) C[13]O[16] |
| CO | | 6.296385 | [380] | 20-19 P(12) C[13]O[16] |
| CO | | 6.310036 | [380] | 20-19 P(13) C[13]O[16] |
| CO | | 6.323871 | [380] | 20-19 P(14) C[13]O[16] |
| CO | | 6.349964 | [380] | 20-19 P(15) C[13]O[16] |
| CO | | 6.352094 | [380] | 20-19 P(16) C[13]O[16] |
| CO | | 6.366487 | [380] | 20-19 P(17) C[13]O[16] |
| CO | | 6.381067 | [380] | 20-19 P(18) C[13]O[16] |
| | | | | |
| CO | | 6.337216 | [380] | 21-20 P(08) C[13]O[16] |
| CO | | 6.350412 | [380] | 21-20 P(09) C[13]O[16] |
| CO | | 6.363789 | [380] | 21-20 P(10) C[13]O[16] |
| CO | | 6.377348 | [380] | 21-20 P(11) C[13]O[16] |
| CO | @ | 6.391091 | [380] | 21-20 P(12) C[13]O[16] |
| CO | | 6.405026 | [380] | 21-20 P(13) C[13]O[16] |
| CO | @ | 6.419144 | [380] | 21-20 P(14) C[13]O[16] |
| CO | @ | 6.433454 | [380] | 21-20 P(15) C[13]O[16] |
| CO | | 6.447956 | [380] | 21-20 P(16) C[13]O[16] |
| CO | | 6.462649 | [380] | 21-20 P(17) C[13]O[16] |
| | | | | |
| CO | | 6.446846 | [380] | 22-21 P(09) C[13]O[16] |
| CO | | 6.460499 | [380] | 22-21 P(10) C[13]O[16] |
| CO | | 6.474336 | [380] | 22-21 P(11) C[13]O[16] |
| CO | | 6.488362 | [380] | 22-21 P(12) C[13]O[16] |
| CO | | 6.502585 | [380] | 22-21 P(13) C[13]O[16] |

| | | WAVELENGTH IN VACUUM [MICROMETER] | | TRANSITION |
|---|---|---|---|---|
| CO | | 6.517002 | [380] | 22-21 P(14) C[13]O[16] |
| CO | | 6.531610 | [380] | 22-21 P(15) C[13]O[16] |
| CO | @ | 6.546418 | [380] | 22-21 P(16) C[13]O[16] |
| CO | | 6.561421 | [380] | 22-21 P(17) C[13]O[16] |
| CO | | 6.559859 | [380] | 23-22 P(10) C[13]O[16] |
| CO | @ | 6.573982 | [380] | 23-22 P(11) C[13]O[16] |
| CO | @ | 6.588305 | [380] | 23-22 P(12) C[13]O[16] |
| CO | | 6.602826 | [380] | 23-22 P(13) C[13]O[16] |
| CO | | 6.617547 | [380] | 23-22 P(14) C[13]O[16] |

NUMBER OF LINES IN CARBON MONOXIDE 905

ACTIVE MEDIUM : OZONE
SYMBOL : O3

OPERATING CONDITIONS : OPTICAL PUMPING WITH SINGLE CO2-LASER LINES.

| | WAVELENGTH IN VACUUM [MICROMETER] | | PUMP TRANSITION |
|---|---|---|---|
| O3 | 121 | [317] | 001-020 P(14) CO2 |
| O3 | 163.61 | [317] | 001-020 P(40) CO2 |
| O3 | 171.5 | [317] | 001-020 P(30) CO2 |

NUMBER OF LINES IN OZONE 3

ACTIVE MEDIUM : CARBON DIOXIDE
SYMBOL : CO2

OPERATING CONDITIONS : CONTINUOUS DISCHARGE IN A MIXTURE OF CO2, N2, AND HE. TYPICAL MIXING RATIO 1:2.5:10. WITH THE CONVENTIONAL CO2 LASER, THE OUTPUT POWER IS LIMITED TO 50-60 WATTS PER METER LENGTH OF THE DISCHARGE TUBE. THE LASING GAS IN THIS CASE IS COOLED BY HEAT CONDUCTION THROUGH WALLS OF THE DISCHARGE TUBE. ONLY A LIMITED AMOUNT OF HEAT CAN BE REMOVED IN THIS WAY WITHOUT THE GAS REACHING AN UNDULY HIGH TEMPERATURE.
SOME MODERN CONSTRUCTION PRINCIPLES OF THE CO2 LASER LEAD TO AN IMPROVED CONTROL OF THE HEAT PRODUCTION:
WITH THE GAS-TRANSPORT LASER (GTL), THE HEAT IS REMOVED FROM THE RESONATOR WITH THE LASER GAS, BY FAST FORCED CONVECTION. THE GAS IS THEN COOLED IN EXTERNAL COOLERS [184].
WITH THE GAS-DYNAMIC LASER (GDL) A HOT-GAS MIXTURE IS COOLED BY ADIABATIC EXPANSION IN A SUPERSONIC NOZZLE [185].
WITH THE CHEMICAL CO2 LASER (TRANSFER CHEMICAL LASER, TCL) THE EXCITATION ENERGY OF VIBRATING DIATOMIC MOLECULES (E.G. HF OR DF) IS TRANSFERRED TO THE CO2 MOLECULE BY COLLISIONS OF THE SECOND KIND. THE PRIMARY EXCITATION IN THIS CASE IS PRODUCED BY CHEMICAL REACTIONS [186,187].
A PULSED EXCITATION IN A CO2-LASER MIXTURE WITH TRANSVERSE DISCHARGE CONFIGURATION AND HIGH PRESSURES (1 ATM. AND MORE) DELIVERS HIGH PEAK POWERS (TEA LASER) [188,189].
MAXIMUM GENERATED POWER AND ENERGY (AT 10.6 MICROMETER):
250 KILOWATT (GDL), 20 KILOWATT (GTL), 1000 JOULE (TEA).
WITH OPTICAL PUMPING (HBR-, HF-LASER [417,425,492]) AND WITH GDL [419,530,610] EMISSION AT 4.3, 14,1 AND 16-17 MICRON IS POSSIBLE.
EXCITATION BY ENERGY TRANSFER FROM BR* [506].
THE COMPILATION OF LASER LINES IN CO2 CONTAINS LINES THAT HAVE BEEN OBTAINED WITH OTHER ISOTOPES THAN C[12]O[16]2.
A BIBLIOGRAPHY UNTIL 1969 AT [190].
A TCL BIBLIOGRAPHY AT [187].
THE LINES MARKED WITH @ HAVE NOT YET BEEN OBSERVED.

| WAVELENGTH IN VACUUM [MICROMETER] | | | TRANSITION |
|---|---|---|---|
| CO2 | 4.3203 | [251] | 102-101 R(17) |
| CO2 | 4.3249 | [251] | 102-101 R(13) |
| CO2 | 4.3276 | [251] | 102-101 R(11) |
| CO2 | 4.3549 | [251] | 102-101 P(07) |
| CO2 | 4.3580 | [251] | 102-101 P(09) |
| CO2 | 4.3612 | [251] | 102-101 P(11) |
| CO2 | 4.3644 | [251] | 102-101 P(13) |
| CO2 | 4.3677 | [251] | 102-101 P(15) |
| CO2 | 4.3711 | [251] | 102-101 P(17) |
| CO2 | 4.3745 | [251] | 102-101 P(19) |
| CO2 | 4.3779 | [251] | 102-101 P(21) |
| CO2 | 4.3814 | [251] | 102-101 P(23) |
| CO2 | 4.3849 | [251] | 102-101 P(25) |
| CO2 | 4.314 | [417] | 101-100 R(08) C[12]O[16,18] |
| CO2 | 4.340 | [417] | 101-100 P(10) C[12]O[16,18] |
| CO2 | 4.354 | [417] | 101-100 P(19) C[12]O[16,18] |
| CO2 | 4.346 | [417] | 101-100 R(08) C[12]O[18]2 |
| CO2 | 4.371 | [417] | 101-100 P(10) C[12]O[18]2 |

| | | WAVELENGTH IN VACUUM [MICROMETER] | | TRANSITION |
|---|---|---|---|---|
| CO2 | | 4.377 | [417] | 101-100 P(14) C[12]O[18]2 |
| CO2 | | 4.382 | [417] | 101-100 P(18) C[12]O[18]2 |
| CO2 | | 4.385 | [417] | 101-100 P(20) C[12]O[18]2 |
| CO2 | | 4.392 | [417] | 101-100 P(24) C[12]O[18]2 |
| CO2 | | 4.398 | [417] | 101-100 P(28) C[12]O[18]2 |
| | | | | |
| CO2 | @ | 9.0702655 | [220] | 001-020 R(70) |
| CO2 | @ | 9.0757663 | [220] | 001-020 R(68) |
| CO2 | @ | 9.0814571 | [220] | 001-020 R(66) |
| CO2 | @ | 9.0873410 | [220] | 001-020 R(64) |
| CO2 | | 9.0934211 | [220,193] | 001-020 R(62) |
| CO2 | | 9.0997003 | [220,193] | 001-020 R(60) |
| CO2 | | 9.1061815 | [220,193] | 001-020 R(58) |
| CO2 | | 9.1128676 | [220,193] | 001-020 R(56) |
| CO2 | | 9.1197615 | [220,193] | 001-020 R(54) |
| CO2 | | 9.1268660 | [220,223] | 001-020 R(52) |
| CO2 | | 9.1341839 | [220,223] | 001-020 R(50) |
| CO2 | | 9.1417179 | [220,223] | 001-020 R(48) |
| CO2 | | 9.1494708 | [220,223] | 001-020 R(46) |
| CO2 | | 9.1574453 | [220,223] | 001-020 R(44) |
| CO2 | | 9.1656440 | [220,223] | 001-020 R(42) |
| CO2 | | 9.1740695 | [220,223] | 001-020 R(40) |
| CO2 | | 9.1827244 | [220,223] | 001-020 R(38) |
| CO2 | | 9.1916114 | [220,223] | 001-020 R(36) |
| CO2 | | 9.2007329 | [220,223] | 001-020 R(34) |
| CO2 | | 9.2100915 | [220,223] | 001-020 R(32) |
| CO2 | | 9.2196895 | [220,223] | 001-020 R(30) |
| CO2 | | 9.2295296 | [220,223] | 001-020 R(28) |
| CO2 | | 9.2396141 | [220,223] | 001-020 R(26) |
| CO2 | | 9.2499453 | [220,223] | 001-020 R(24) |
| CO2 | | 9.2605258 | [220,223] | 001-020 R(22) |
| CO2 | | 9.2713577 | [220,223] | 001-020 R(20) |
| CO2 | | 9.2824434 | [220,223] | 001-020 R(18) |
| CO2 | | 9.2937852 | [220,223] | 001-020 R(16) |
| CO2 | | 9.3053853 | [220,223] | 001-020 R(14) |
| CO2 | | 9.3172460 | [220,223] | 001-020 R(12) |
| CO2 | | 9.3293695 | [220,223] | 001-020 R(10) |
| CO2 | | 9.3417579 | [220,223] | 001-020 R(08) |
| CO2 | | 9.3544134 | [220,223] | 001-020 R(06) |
| CO2 | | 9.3673380 | [220,223] | 001-020 R(04) |
| CO2 | | 9.3805340 | [220,193] | 001-020 R(02) |
| CO2 | | 9.3940033 | [220,193] | 001-020 R(00) |
| CO2 | | 9.4147242 | [220,193] | 001-020 P(02) |
| CO2 | | 9.4288857 | [220,223] | 001-020 P(04) |
| CO2 | | 9.4433275 | [220,223] | 001-020 P(06) |
| CO2 | | 9.4580515 | [220,223] | 001-020 P(08) |
| CO2 | | 9.4730598 | [220,223] | 001-020 P(10) |
| CO2 | | 9.4883540 | [220,223] | 001-020 P(12) |
| CO2 | | 9.5039361 | [220,223] | 001-020 P(14) |
| CO2 | | 9.5198079 | [220,223] | 001-020 P(16) |
| CO2 | | 9.5359711 | [220,223] | 001-020 P(18) |
| CO2 | | 9.5524275 | [220,223] | 001-020 P(20) |
| CO2 | | 9.5691788 | [220,224,223] | 001-020 P(22) |
| CO2 | | 9.5862267 | [220,224,223] | 001-020 P(24) |
| CO2 | | 9.6035727 | [220,224,223] | 001-020 P(26) |
| CO2 | | 9.6212185 | [220,224,223] | 001-020 P(28) |
| CO2 | | 9.6391656 | [220,224,223] | 001-020 P(30) |

| | | WAVELENGTH IN VACUUM [MICROMETER] | | TRANSITION |
|---|---|---|---|---|
| CO2 | | 9.6574156 | [220,224,223] | 001-020 P(32) |
| CO2 | | 9.6750700 | [220,224,223] | 001-020 P(34) |
| CO2 | | 9.6948301 | [220,223] | 001-020 P(36) |
| CO2 | | 9.7139973 | [220,223] | 001-020 P(38) |
| CO2 | | 9.7334730 | [220,223] | 001-020 P(40) |
| CO2 | | 9.7532586 | [220,223] | 001-020 P(42) |
| CO2 | | 9.7733552 | [220,223] | 001-020 P(44) |
| CO2 | | 9.7937640 | [220,223] | 001-020 P(46) |
| CO2 | | 9.8144862 | [220,223] | 001-020 P(48) |
| CO2 | | 9.8355229 | [220,223] | 001-020 P(50) |
| CO2 | | 9.8568751 | [220,223] | 001-020 P(52) |
| CO2 | | 9.8785439 | [220,223] | 001-020 P(54) |
| CO2 | | 9.9005300 | [220,223] | 001-020 P(56) |
| CO2 | | 9.9228344 | [220,223] | 001-020 P(58) |
| CO2 | | 9.9454579 | [220,223] | 001-020 P(60) |
| CO2 | | 9.9684012 | [220,193] | 001-020 P(62) |
| CO2 | | 9.9916650 | [220,193] | 001-020 P(64) |
| CO2 | | 10.0152498 | [220,193] | 001-020 P(66) |
| CO2 | @ | 10.0391561 | [220] | 001-020 P(68) |
| CO2 | @ | 10.0633844 | [220] | 001-020 P(70) |
| CO2 | @ | 10.0879349 | [220] | 001-020 P(72) |
| | | | | |
| CO2 | @ | 9.9985568 | [220] | 001-100 R(70) |
| CO2 | @ | 10.0049238 | [220] | 001-100 R(68) |
| CO2 | @ | 10.0115934 | [220] | 001-100 R(66) |
| CO2 | @ | 10.0185643 | [220] | 001-100 R(64) |
| CO2 | | 10.0258352 | [220,193] | 001-100 R(62) |
| CO2 | | 10.0334048 | [220,193] | 001-100 R(60) |
| CO2 | | 10.0412720 | [220,193] | 001-100 R(58) |
| CO2 | | 10.0494358 | [220,193] | 001-100 R(56) |
| CO2 | | 10.0578953 | [220,223] | 001-100 R(54) |
| CO2 | | 10.0666497 | [220,223] | 001-100 R(52) |
| CO2 | | 10.0756984 | [220,223] | 001-100 R(50) |
| CO2 | | 10.0850408 | [220,223] | 001-100 R(48) |
| CO2 | | 10.0946764 | [220,223] | 001-100 R(46) |
| CO2 | | 10.1046049 | [220,223] | 001-100 R(44) |
| CO2 | | 10.1148262 | [220,223] | 001-100 R(42) |
| CO2 | | 10.1253400 | [220,223] | 001-100 R(40) |
| CO2 | | 10.1361464 | [220,223] | 001-100 R(38) |
| CO2 | | 10.1472454 | [220,223] | 001-100 R(36) |
| CO2 | | 10.1586374 | [220,223] | 001-100 R(34) |
| CO2 | | 10.1703225 | [220,223] | 001-100 R(32) |
| CO2 | | 10.1823014 | [220,223] | 001-100 R(30) |
| CO2 | | 10.1945745 | [220,223] | 001-100 R(28) |
| CO2 | | 10.2071425 | [220,225,223] | 001-100 R(26) |
| CO2 | | 10.2200062 | [220,225,223] | 001-100 R(24) |
| CO2 | | 10.2331666 | [220,225,223] | 001-100 R(22) |
| CO2 | | 10.2466246 | [220,225,223] | 001-100 R(20) |
| CO2 | | 10.2603814 | [220,225,223] | 001-100 R(18) |
| CO2 | | 10.2744384 | [220,225,223] | 001-100 R(16) |
| CO2 | | 10.2887967 | [220,225,223] | 001-100 R(14) |
| CO2 | | 10.3034581 | [220,223] | 001-100 R(12) |
| CO2 | | 10.3184241 | [220,223] | 001-100 R(10) |
| CO2 | | 10.3336965 | [220,223] | 001-100 R(08) |
| CO2 | | 10.3492772 | [220,223] | 001-100 R(06) |
| CO2 | | 10.3651683 | [220,223] | 001-100 R(04) |
| CO2 | | 10.3813718 | [220,193] | 001-100 R(02) |

| | | WAVELENGTH IN VACUUM [MICROMETER] | | TRANSITION |
|---|---|---|---|---|
| CO2 | | 10.3978901 | [220,193] | 001-100 R(00) |
| CO2 | | 10.4232632 | [220,193] | 001-100 P(02) |
| CO2 | | 10.4405795 | [220,223] | 001-100 P(04) |
| CO2 | | 10.4582196 | [220,223] | 001-100 P(06) |
| CO2 | | 10.4761860 | [220,223] | 001-100 P(08) |
| CO2 | | 10.4944835 | [220,223] | 001-100 P(10) |
| CO2 | | 10.5131136 | [220,224,223] | 001-100 P(12) |
| CO2 | > | 10.5320802 | [220,223,226] | 001-100 P(14) |
| CO2 | > | 10.5513868 | [220,226,223] | 001-100 P(16) |
| CO2 | > | 10.5710372 | [220,226,223] | 001-100 P(18) |
| CO2 | > | 10.5910352 | [220,226,223] | 001-100 P(20) |
| CO2 | > | 10.6113848 | [220,226,223] | 001-100 P(22) |
| CO2 | > | 10.6320902 | [220,226,223] | 001-100 P(24) |
| CO2 | > | 10.6531558 | [220,226,223] | 001-100 P(26) |
| CO2 | | 10.6745861 | [220,224,223] | 001-100 P(28) |
| CO2 | | 10.6963859 | [220,224,223] | 001-100 P(30) |
| CO2 | | 10.7185600 | [220,224,223] | 001-100 P(32) |
| CO2 | | 10.7411135 | [220,224,223] | 001-100 P(34) |
| CO2 | | 10.7640517 | [220,224,223] | 001-100 P(36) |
| CO2 | | 10.7873802 | [220,224,223] | 001-100 P(38) |
| CO2 | | 10.8111046 | [220,223] | 001-100 P(40) |
| CO2 | | 10.8352307 | [220,223] | 001-100 P(42) |
| CO2 | | 10.8597648 | [220,223] | 001-100 P(44) |
| CO2 | | 10.8847131 | [220,223] | 001-100 P(46) |
| CO2 | | 10.9100823 | [220,223] | 001-100 P(48) |
| CO2 | | 10.9358790 | [220,223] | 001-100 P(50) |
| CO2 | | 10.9621103 | [220,223] | 001-100 P(52) |
| CO2 | | 10.9887835 | [220,223] | 001-100 P(54) |
| CO2 | | 11.0159060 | [220,223] | 001-100 P(56) |
| CO2 | | 11.0434858 | [220,193] | 001-100 P(58) |
| CO2 | | 11.0715308 | [220,193] | 001-100 P(60) |
| CO2 | | 11.1000493 | [220,193] | 001-100 P(62) |
| CO2 | | 11.1290499 | [220,193] | 001-100 P(64) |
| CO2 | | 11.1585415 | [220,193] | 001-100 P(66) |
| CO2 | | 11.1885334 | [220,193] | 001-100 P(68) |
| CO2 | @ | 11.2190349 | [220] | 001-100 P(70) |
| CO2 | @ | 11.2500559 | [220] | 001-100 P(72) |
| | | | | |
| CO2 | | 9.209171 | [479] | 002-021 R(39) |
| CO2 | | 9.217773 | [479] | 002-021 R(37) |
| CO2 | | 9.226615 | [479] | 002-021 R(35) |
| CO2 | | 9.235699 | [479] | 002-021 R(33) |
| CO2 | | 9.245029 | [479] | 002-021 R(31) |
| CO2 | | 9.254607 | [411,479] | 002-021 R(29) |
| CO2 | | 9.264436 | [411,479] | 002-021 R(27) |
| CO2 | | 9.274517 | [411,479] | 002-021 R(25) |
| CO2 | | 9.284854 | [411,479] | 002-021 R(23) |
| CO2 | | 9.295448 | [411,479] | 002-021 R(21) |
| CO2 | | 9.306302 | [411,479] | 002-021 R(19) |
| CO2 | | 9.316821 | [411,479] | 002-021 R(17) |
| CO2 | | 9.328800 | [411,479] | 002-021 R(15) |
| CO2 | | 9.340448 | [411,479] | 002-021 R(13) |
| CO2 | | 9.352366 | [411,479] | 002-021 R(11) |
| CO2 | | 9.364555 | [411,479] | 002-021 R(09) |
| CO2 | | 9.377018 | [479] | 002-021 R(07) |
| CO2 | | 9.389757 | [479] | 002-021 R(05) |
| CO2 | @ | 9.402774 | [479] | 002-021 R(03) |

| | | WAVELENGTH IN VACUUM [MICROMETER] | | TRANSITION |
|---|---|---|---|---|
| CO2 | @ | 9.450554 | [479] | 002-021 P(03) |
| CO2 | | 9.464848 | [479] | 002-021 P(05) |
| CO2 | | 9.479432 | [479] | 002-021 P(07) |
| CO2 | | 9.494307 | [411,479] | 002-021 P(09) |
| CO2 | | 9.509476 | [411,479] | 002-021 P(11) |
| CO2 | | 9.524939 | [411,479] | 002-021 P(13) |
| CO2 | | 9.540700 | [411,479] | 002-021 P(15) |
| CO2 | > | 9.556760 | [411,479] | 002-021 P(17) |
| CO2 | > | 9.573121 | [411,479] | 002-021 P(19) |
| CO2 | > | 9.589785 | [411,479] | 002-021 P(21) |
| CO2 | > | 9.606753 | [411,479] | 002-021 P(23) |
| CO2 | > | 9.624027 | [411,479] | 002-021 P(25) |
| CO2 | | 9.641609 | [411,479] | 002-021 P(27) |
| CO2 | | 9.655900 | [411,479] | 002-021 P(29) |
| CO2 | | 9.677702 | [411,479] | 002-021 P(31) |
| CO2 | | 9.696217 | [479] | 002-021 P(33) |
| CO2 | | 9.715046 | [479] | 002-021 P(35) |
| CO2 | | 9.734191 | [479] | 002-021 P(37) |
| CO2 | | 9.753653 | [479] | 002-021 P(39) |
| CO2 | @ | 9.773433 | [479] | 002-021 P(41) |
| CO2 | | 9.793533 | [479] | 002-021 P(43) |
| CO2 | | 9.813954 | [479] | 002-021 P(45) |
| CO2 | @ | 10.146624 | [479] | 002-101 R(41) |
| CO2 | | 10.157295 | [411,479] | 002-101 R(39) |
| CO2 | | 10.168257 | [411,479] | 002-101 R(37) |
| CO2 | | 10.179508 | [411,479] | 002-101 R(35) |
| CO2 | | 10.191050 | [411,479] | 002-101 R(33) |
| CO2 | | 10.202883 | [411,479] | 002-101 R(31) |
| CO2 | | 10.215008 | [411,479] | 002-101 R(29) |
| CO2 | | 10.227424 | [411,479] | 002-101 R(27) |
| CO2 | | 10.240133 | [411,479] | 002-101 R(25) |
| CO2 | | 10.253135 | [411,479] | 002-101 R(23) |
| CO2 | | 10.266431 | [411,479] | 002-101 R(21) |
| CO2 | | 10.280023 | [411,479] | 002-101 R(19) |
| CO2 | | 10.293911 | [411,479] | 002-101 R(17) |
| CO2 | > | 10.308097 | [411,479] | 002-101 R(15) |
| CO2 | > | 10.322582 | [411,479] | 002-101 R(13) |
| CO2 | | 10.337367 | [411,479] | 002-101 R(11) |
| CO2 | | 10.352455 | [411,479] | 002-101 R(09) |
| CO2 | | 10.367847 | [411,479] | 002-101 R(07) |
| CO2 | | 10.383545 | [411,479] | 002-101 R(05) |
| CO2 | | 10.399550 | [411,479] | 002-101 R(03) |
| CO2 | @ | 10.415866 | [479] | 002-101 R(01) |
| CO2 | @ | 10.458029 | [411,479] | 002-101 P(03) |
| CO2 | | 10.475449 | [411,479] | 002-101 P(05) |
| CO2 | | 10.493192 | [411,479] | 002-101 P(07) |
| CO2 | | 10.511259 | [411,479] | 002-101 P(09) |
| CO2 | | 10.529654 | [411,479] | 002-101 P(11) |
| CO2 | | 10.548380 | [411,479] | 002-101 P(13) |
| CO2 | | 10.567440 | [411,479] | 002-101 P(15) |
| CO2 | | 10.586838 | [411,479] | 002-101 P(17) |
| CO2 | | 10.606578 | [411,479] | 002-101 P(19) |
| CO2 | > | 10.626664 | [411,479] | 002-101 P(21) |
| CO2 | > | 10.647099 | [411,479] | 002-101 P(23) |
| CO2 | > | 10.667888 | [411,479] | 002-101 P(25) |
| CO2 | > | 10.689036 | [411,479] | 002-101 P(27) |

| | | WAVELENGTH IN VACUUM [MICROMETER] | | TRANSITION |
|---|---|---|---|---|
| CO2 | > | 10.710547 | [411,479] | 002-101 P(29) |
| CO2 | | 10.732425 | [411,479] | 002-101 P(31) |
| CO2 | | 10.754676 | [411,479] | 002-101 P(33) |
| CO2 | | 10.777305 | [411,479] | 002-101 P(35) |
| CO2 | | 10.800317 | [411,479] | 002-101 P(37) |
| CO2 | | 10.823718 | [411,479] | 002-101 P(39) |
| CO2 | | 10.847513 | [411,479] | 002-101 P(41) |
| CO2 | | 10.871709 | [411,479] | 002-101 P(43) |
| CO2 | | 10.896312 | [411,479] | 002-101 P(45) |
| CO2 | @ | 10.921327 | [479] | 002-101 P(47) |
| | | | | |
| CO2 | @ | 10.286987 | [479] | 003-102 R(22) |
| CO2 | | 10.302426 | [479] | 003-102 R(20) |
| CO2 | | 10.316157 | [479] | 003-102 R(18) |
| CO2 | | 10.330184 | [479] | 003-102 R(16) |
| CO2 | | 10.344505 | [479] | 003-102 R(14) |
| CO2 | | 10.359124 | [479] | 003-102 R(12) |
| CO2 | | 10.374040 | [479] | 003-102 R(10) |
| CO2 | | 10.389256 | [479] | 003-102 R(08) |
| CO2 | | 10.404773 | [479] | 003-102 R(06) |
| CO2 | @ | 10.420594 | [479] | 003-102 R(04) |
| CO2 | | 10.53097 | [527] | 003-102 P(08) |
| CO2 | | 10.54916 | [527] | 003-102 P(10) |
| CO2 | | 10.56762 | [527] | 003-102 P(12) |
| CO2 | | 10.58646 | [527] | 003-102 P(14) |
| CO2 | | 10.60562 | [527] | 003-102 P(16) |
| CO2 | @ | 10.665124 | [479] | 003-102 P(22) |
| CO2 | | 10.685646 | [479] | 003-102 P(24) |
| CO2 | | 10.706519 | [479] | 003-102 P(26) |
| CO2 | | 10.727749 | [479] | 003-102 P(28) |
| CO2 | | 10.749339 | [479] | 003-102 P(30) |
| CO2 | | 10.771295 | [479] | 003-102 P(32) |
| CO2 | | 10.793621 | [479] | 003-102 P(34) |
| CO2 | | 10.816324 | [479] | 003-102 P(36) |
| CO2 | | 10.839408 | [479] | 003-102 P(38) |
| CO2 | @ | 10.862879 | [479] | 003-102 P(40) |
| | | | | |
| CO2 | | 10.55376 | [527] | 004-103 P(07) |
| CO2 | | 10.57170 | [527] | 004-103 P(09) |
| CO2 | | 10.5900 | [527] | 004-103 P(11) |
| CO2 | | 10.60858 | [527] | 004-103 P(13) |
| | | | | |
| CO2 | | 10.591025 | [487] | 001-110 R(23) |
| CO2 | | 10.789077 | [487] | 001-110 Q(11) |
| CO2 | | 10.890184 | [487] | 001-110 P(11) |
| CO2 | | 10.900964 | [487] | 001-110 P(12) |
| CO2 | | 10.921469 | [487] | 001-110 P(14) |
| CO2 | | 10.930707 | [487] | 001-110 P(15) |
| CO2 | | 10.942351 | [487] | 001-110 P(16) |
| CO2 | | 10.951486 | [487] | 001-110 P(17) |
| CO2 | | 10.972615 | [487] | 001-110 P(19) |
| CO2 | | 10.985266 | [487] | 001-110 P(20) |
| CO2 | | 11.007301 | [487] | 001-110 P(22) |
| CO2 | | 11.015934 | [487] | 001-110 P(23) |
| CO2 | | 11.029744 | [487] | 001-110 P(24) |
| CO2 | | 11.083630 | [487] | 001-110 P(29) |

| | WAVELENGTH IN VACUUM [MICROMETER] | | TRANSITION |
|---|---|---|---|
| CO2 | 10.50816 | [527,531] | 011-110 R(36) |
| CO2 | 10.51001 | [527,531] | 011-110 R(35) |
| CO2 | 10.52029 | [527,531] | 011-110 R(34) |
| CO2 | 10.52277 | [527,531] | 011-110 R(33) |
| CO2 | 10.53273 | [527,531] | 011-110 R(32) |
| CO2 | 10.54550 | [527,531] | 011-110 R(30) |
| CO2 | 10.54919 | [527,531] | 011-110 R(29) |
| CO2 | 10.55859 | [527,531] | 011-110 R(28) |
| CO2 | 10.56284 | [527,531] | 011-110 R(27) |
| CO2 | 10.57201 | [527,531] | 011-110 R(26) |
| CO2 | 10.57678 | [527,531] | 011-110 R(25) |
| CO2 | 10.58575 | [527,531] | 011-110 R(24) |
| CO2 | 10.59982 | [527,531] | 011-110 R(22) |
| CO2 | 10.60556 | [527,531] | 011-110 R(21) |
| CO2 | 10.61421 | [527,531] | 011-110 R(20) |
| | | | |
| CO2 | 10.92146 | [531] | 011-110 P(14) |
| CO2 | 10.93070 | [531] | 011-110 P(15) |
| CO2 | 10.94235 | [531] | 011-110 P(16) |
| CO2 | 10.95148 | [531] | 011-110 P(17) |
| CO2 | @ 10.96361 | [531] | 011-110 P(18) |
| CO2 | 10.97261 | [531] | 011-110 P(19) |
| CO2 | 10.98526 | [531] | 011-110 P(20) |
| CO2 | 10.99409 | [531] | 011-110 P(21) |
| CO2 | 11.00730 | [531] | 011-110 P(22) |
| CO2 | 11.01593 | [531] | 011-110 P(23) |
| CO2 | 11.02974 | [531] | 011-110 P(24) |
| CO2 | 11.03813 | [531] | 011-110 P(25) |
| CO2 | 11.05258 | [531] | 011-110 P(26) |
| CO2 | 11.06069 | [531] | 011-110 P(27) |
| CO2 | 11.07582 | [531] | 011-110 P(28) |
| CO2 | 11.08363 | [531] | 011-110 P(29) |
| CO2 | 11.09947 | [531] | 011-110 P(30) |
| CO2 | 11.10693 | [531] | 011-110 P(31) |
| CO2 | 11.12354 | [531] | 011-110 P(32) |
| CO2 | 11.13062 | [531] | 011-110 P(33) |
| CO2 | 11.14803 | [531] | 011-110 P(34) |
| CO2 | 11.15468 | [531] | 011-110 P(35) |
| CO2 | 11.17295 | [531] | 011-110 P(36) |
| CO2 | 11.17914 | [531] | 011-110 P(37) |
| CO2 | 11.19830 | [531] | 011-110 P(38) |
| CO2 | 11.20398 | [531] | 011-110 P(39) |
| CO2 | 11.22408 | [531] | 011-110 P(40) |
| | | | |
| CO2 | 10.9735 | [252,223] | 011-030 P(19) |
| CO2 | 10.9950 | [252,223] | 011-030 P(21) |
| CO2 | 11.0165 | [252,223] | 011-030 P(23) |
| CO2 | 11.0300 | [252,223] | 011-030 P(24) |
| CO2 | 11.0385 | [252,223] | 011-030 P(25) |
| CO2 | 11.0535 | [252,223] | 011-030 P(26) |
| CO2 | 11.0610 | [252,223] | 011-030 P(27) |
| CO2 | 11.0760 | [252,223] | 011-030 P(28) |
| CO2 | 11.0850 | [252,223] | 011-030 P(29) |
| CO2 | 11.1000 | [252,223] | 011-030 P(30) |
| CO2 | 11.1070 | [252,223] | 011-030 P(31) |
| CO2 | 11.1235 | [252,223] | 011-030 P(32) |
| CO2 | 11.1315 | [252,223] | 011-030 P(33) |

| | WAVELENGTH IN VACUUM [MICROMETER] | | TRANSITION |
|---|---|---|---|
| CO2 | 11.1465 | [252,223] | 011-030 P(34) |
| CO2 | 11.1555 | [252,223] | 011-030 P(35) |
| CO2 | 11.1736 | [252,223] | 011-030 P(36) |
| CO2 | 11.1790 | [252,223] | 011-030 P(37) |
| CO2 | 11.1960 | [252,223] | 011-030 P(38) |
| CO2 | 11.2035 | [252,223] | 011-030 P(39) |
| CO2 | 11.2235 | [252,223] | 011-030 P(40) |
| CO2 | 11.2295 | [252,223] | 011-030 P(41) |
| CO2 | 11.2495 | [252,223] | 011-030 P(42) |
| CO2 | 11.2545 | [252,223] | 011-030 P(43) |
| CO2 | 11.2770 | [252,223] | 011-030 P(44) |
| CO2 | 11.2805 | [252,223] | 011-030 P(45) |
| | | | |
| CO2 | 10.53907 | [527] | 012-111 R(35) |
| CO2 | 10.54173 | [527] | 012-111 R(34) |
| CO2 | 10.55455 | [527] | 012-111 R(32) |
| CO2 | 10.5639 | [527] | 012-111 R(31) |
| CO2 | 10.58112 | [527] | 012-111 R(28) |
| CO2 | 10.60362 | [527] | 012-111 R(25) |
| CO2 | 10.60885 | [527] | 012-111 R(24) |
| | | | |
| CO2 | 10.51027 | [527] | 101-200 R(12) |
| CO2 | 10.54271 | [527] | 101-200 R(08) |
| | | | |
| CO2 | ? 13.144 | [253] | |
| CO2 | 13.154 | [253] | ?140-050 Q |
| CO2 | ? 13.159 | [253] | |
| CO2 | 13.541 | [253] | ?210-120 Q |
| CO2 | 13.87 | [616] | 100-010 Q |
| CO2 | 14.1 | [418] | 101-011 Q |
| CO2 | 14.16 | [616] | 100-010 P(19) |
| CO2 | 14.19 | [616] | 100-010 P(21) |
| CO2 | 14.21 | [616] | 100-010 P(23) |
| CO2 | 16.586 | [253] | ?140-130 Q |
| CO2 | ? 16.597 | [253] | |
| CO2 | ? 17.023 | [253] | |
| CO2 | 17.029 | [253] | ?031-021 Q |
| CO2 | ? 17.036 | [253] | |
| CO2 | ? 17.048 | [253] | |
| CO2 | ? 17.370 | [253] | |
| CO2 | 17.376 | [253] | ?240-230 Q |
| CO2 | ? 17.390 | [253] | |
| | | | |
| CO2 | @ 8.98770094 | [338] | 001-020 R(50) C[12]O[18]2 |
| CO2 | 8.99496997 | [338] | 001-020 R(48) C[12]O[18]2 |
| CO2 | 9.00240216 | [338] | 001-020 R(46) C[12]O[18]2 |
| CO2 | 9.00999899 | [338] | 001-020 R(44) C[12]O[18]2 |
| CO2 | 9.01776190 | [338] | 001-020 R(42) C[12]O[18]2 |
| CO2 | 9.02569233 | [338] | 001-020 R(40) C[12]O[18]2 |
| CO2 | 9.03379171 | [338] | 001-020 R(38) C[12]O[18]2 |
| CO2 | 9.04206146 | [338] | 001-020 R(36) C[12]O[18]2 |
| CO2 | 9.05050297 | [338] | 001-020 R(34) C[12]O[18]2 |
| CO2 | 9.05911764 | [338] | 001-020 R(32) C[12]O[18]2 |
| CO2 | 9.06790685 | [338] | 001-020 R(30) C[12]O[18]2 |
| CO2 | 9.07687197 | [338] | 001-020 R(28) C[12]O[18]2 |
| CO2 | 9.08601437 | [338] | 001-020 R(26) C[12]O[18]2 |
| CO2 | 9.09533538 | [338] | 001-020 R(24) C[12]O[18]2 |

| | | WAVELENGTH IN VACUUM [MICROMETER] | TRANSITION |
|---|---|---|---|
| CO2 | | 9.10483634 [338] | 001-020 R(22) C[12]O[18]2 |
| CO2 | | 9.11451859 [338] | 001-020 R(20) C[12]O[18]2 |
| CO2 | | 9.12438343 [338] | 001-020 R(18) C[12]O[18]2 |
| CO2 | | 9.13443217 [338] | 001-020 R(16) C[12]O[18]2 |
| CO2 | | 9.14466611 [338] | 001-020 R(14) C[12]O[18]2 |
| CO2 | | 9.15508653 [338] | 001-020 R(12) C[12]O[18]2 |
| CO2 | | 9.16569469 [338] | 001-020 R(10) C[12]O[18]2 |
| CO2 | | 9.17649187 [338] | 001-020 R(08) C[12]O[18]2 |
| CO2 | | 9.18747931 [338] | 001-020 R(06) C[12]O[18]2 |
| CO2 | | 9.19865824 [338] | 001-020 R(04) C[12]O[18]2 |
| CO2 | | 9.21002991 [338] | 001-020 R(02) C[12]O[18]2 |
| CO2 | @ | 9.22159552 [338] | 001-020 R(00) C[12]O[18]2 |
| CO2 | @ | 9.23931022 [338] | 001-020 P(02) C[12]O[18]2 |
| CO2 | | 9.25136594 [338] | 001-020 P(04) C[12]O[18]2 |
| CO2 | | 9.26361978 [338] | 001-020 P(06) C[12]O[18]2 |
| CO2 | | 9.27607290 [338] | 001-020 P(08) C[12]O[18]2 |
| CO2 | | 9.28872646 [338] | 001-020 P(10) C[12]O[18]2 |
| CO2 | | 9.30158160 [338] | 001-020 P(12) C[12]O[18]2 |
| CO2 | | 9.31463943 [338] | 001-020 P(14) C[12]O[18]2 |
| CO2 | | 9.32790109 [338] | 001-020 P(16) C[12]O[18]2 |
| CO2 | | 9.34136768 [338,192] | 001-020 P(18) C[12]O[18]2 |
| CO2 | | 9.35504028 [338,192] | 001-020 P(20) C[12]O[18]2 |
| CO2 | | 9.36891998 [338,192] | 001-020 P(22) C[12]O[18]2 |
| CO2 | | 9.38300763 [338,192] | 001-020 P(24) C[12]O[18]2 |
| CO2 | | 9.39730489 [338,192] | 001-020 P(26) C[12]O[18]2 |
| CO2 | | 9.41181219 [338] | 001-020 P(28) C[12]O[18]2 |
| CO2 | | 9.42653076 [338] | 001-020 P(30) C[12]O[18]2 |
| CO2 | | 9.44146159 [338] | 001-020 P(32) C[12]O[18]2 |
| CO2 | | 9.45660568 [338] | 001-020 P(34) C[12]O[18]2 |
| CO2 | | 9.47196399 [338] | 001-020 P(36) C[12]O[18]2 |
| CO2 | | 9.48753750 [338] | 001-020 P(38) C[12]O[18]2 |
| CO2 | | 9.50332713 [338] | 001-020 P(40) C[12]O[18]2 |
| CO2 | | 9.51933381 [338] | 001-020 P(42) C[12]O[18]2 |
| CO2 | | 9.53555845 [338] | 001-020 P(44) C[12]O[18]2 |
| CO2 | | 9.55200192 [338] | 001-020 P(46) C[12]O[18]2 |
| CO2 | | 9.56866511 [338] | 001-020 P(48) C[12]O[18]2 |
| CO2 | | 9.58554884 [338] | 001-020 P(50) C[12]O[18]2 |
| CO2 | | 9.60265396 [338] | 001-020 P(52) C[12]O[18]2 |
| CO2 | | 9.61998126 [338] | 001-020 P(54) C[12]O[18]2 |
| CO2 | @ | 9.63753153 [338] | 001-020 P(56) C[12]O[18]2 |
| | | | |
| CO2 | | 10.10434660 [338] | 001-100 R(40) C[12]O[18]2 |
| CO2 | | 10.11295490 [338] | 001-100 R(38) C[12]O[18]2 |
| CO2 | | 10.12186285 [338] | 001-100 R(36) C[12]O[18]2 |
| CO2 | | 10.13107088 [338] | 001-100 R(34) C[12]O[18]2 |
| CO2 | | 10.14057947 [338] | 001-100 R(32) C[12]O[18]2 |
| CO2 | | 10.15038920 [338] | 001-100 R(30) C[12]O[18]2 |
| CO2 | | 10.16050076 [338] | 001-100 R(28) C[12]O[18]2 |
| CO2 | | 10.17091494 [338] | 001-100 R(26) C[12]O[18]2 |
| CO2 | | 10.18163264 [338] | 001-100 R(24) C[12]O[18]2 |
| CO2 | | 10.19265484 [338] | 001-100 R(22) C[12]O[18]2 |
| CO2 | | 10.20398263 [338] | 001-100 R(20) C[12]O[18]2 |
| CO2 | | 10.21561722 [338] | 001-100 R(18) C[12]O[18]2 |
| CO2 | | 10.22755988 [338] | 001-100 R(16) C[12]O[18]2 |
| CO2 | | 10.23981204 [338] | 001-100 R(14) C[12]O[18]2 |
| CO2 | | 10.25237519 [338] | 001-100 R(12) C[12]O[18]2 |
| CO2 | | 10.26525095 [338] | 001-100 R(10) C[12]O[18]2 |

| | | WAVELENGTH IN VACUUM [MICROMETER] | TRANSITION |
|---|---|---|---|
| CO2 | | 10.27644106 [338] | 001-100 R(08) C[12]O[18]2 |
| CO2 | | 10.29194733 [338] | 001-100 R(06) C[12]O[18]2 |
| CO2 | @ | 10.30577171 [338] | 001-100 R(04) C[12]O[18]2 |
| CO2 | | 10.38758981 [338] | 001-100 P(06) C[12]O[18]2 |
| CO2 | | 10.40353724 [338] | 001-100 P(08) C[12]O[18]2 |
| CO2 | | 10.41982107 [338] | 001-100 P(10) C[12]O[18]2 |
| CO2 | | 10.43644425 [338] | 001-100 P(12) C[12]O[18]2 |
| CO2 | | 10.45340986 [338] | 001-100 P(14) C[12]O[18]2 |
| CO2 | | 10.47072112 [338] | 001-100 P(16) C[12]O[18]2 |
| CO2 | | 10.48838137 [338] | 001-100 P(18) C[12]O[18]2 |
| CO2 | | 10.50639411 [338] | 001-100 P(20) C[12]O[18]2 |
| CO2 | | 10.52476294 [338] | 001-100 P(22) C[12]O[18]2 |
| CO2 | | 10.54349163 [338] | 001-100 P(24) C[12]O[18]2 |
| CO2 | | 10.56258410 [338] | 001-100 P(26) C[12]O[18]2 |
| CO2 | | 10.58204439 [338] | 001-100 P(28) C[12]O[18]2 |
| CO2 | | 10.60187671 [338] | 001-100 P(30) C[12]O[18]2 |
| CO2 | | 10.62208541 [338] | 001-100 P(32) C[12]O[18]2 |
| CO2 | | 10.64267502 [338] | 001-100 P(34) C[12]O[18]2 |
| CO2 | | 10.66365020 [338] | 001-100 P(36) C[12]O[18]2 |
| CO2 | | 10.68501579 [338] | 001-100 P(38) C[12]O[18]2 |
| CO2 | | 10.70677681 [338] | 001-100 P(40) C[12]O[18]2 |
| CO2 | @ | 10.72893842 [338] | 001-100 P(42) C[12]O[18]2 |
| CO2 | @ | 9.59239947 [338] | 001-020 R(38) C[13]O[16]2 |
| CO2 | | 9.60168999 [338] | 001-020 R(36) C[13]O[16]2 |
| CO2 | | 9.61125719 [338] | 001-020 R(34) C[13]O[16]2 |
| CO2 | | 9.62110404 [338] | 001-020 R(32) C[13]O[16]2 |
| CO2 | | 9.63123346 [338] | 001-020 R(30) C[13]O[16]2 |
| CO2 | | 9.64164836 [338] | 001-020 R(28) C[13]O[16]2 |
| CO2 | | 9.65235161 [338] | 001-020 R(26) C[13]O[16]2 |
| CO2 | | 9.66334607 [338] | 001-020 R(24) C[13]O[16]2 |
| CO2 | | 9.67463455 [338] | 001-020 R(22) C[13]O[16]2 |
| CO2 | | 9.68621987 [338] | 001-020 R(20) C[13]O[16]2 |
| CO2 | | 9.69810480 [338] | 001-020 R(18) C[13]O[16]2 |
| CO2 | | 9.71029210 [338] | 001-020 R(16) C[13]O[16]2 |
| CO2 | | 9.72278448 [338] | 001-020 R(14) C[13]O[16]2 |
| CO2 | | 9.73558466 [338] | 001-020 R(12) C[13]O[16]2 |
| CO2 | | 9.74869533 [338] | 001-020 R(10) C[13]O[16]2 |
| CO2 | | 9.76211913 [338] | 001-020 R(08) C[13]O[16]2 |
| CO2 | | 9.77585872 [338] | 001-020 R(06) C[13]O[16]2 |
| CO2 | | 9.78991669 [338] | 001-020 R(04) C[13]O[16]2 |
| CO2 | @ | 9.80429565 [338] | 001-020 R(02) C[13]O[16]2 |
| CO2 | @ | 9.85718656 [338] | 001-020 P(04) C[13]O[16]2 |
| CO2 | | 9.87304135 [338] | 001-020 P(06) C[13]O[16]2 |
| CO2 | | 9.88923095 [338] | 001-020 P(08) C[13]O[16]2 |
| CO2 | | 9.90575779 [338] | 001-020 P(10) C[13]O[16]2 |
| CO2 | | 9.92262428 [338,191] | 001-020 P(12) C[13]O[16]2 |
| CO2 | | 9.93983281 [338,191] | 001-020 P(14) C[13]O[16]2 |
| CO2 | | 9.95738572 [338,191] | 001-020 P(16) C[13]O[16]2 |
| CO2 | | 9.97528536 [338,191] | 001-020 P(18) C[13]O[16]2 |
| CO2 | | 9.99353402 [338,191] | 001-020 P(20) C[13]O[16]2 |
| CO2 | | 10.01213398 [338,191] | 001-020 P(22) C[13]O[16]2 |
| CO2 | | 10.03108747 [338,191] | 001-020 P(24) C[13]O[16]2 |
| CO2 | | 10.05039672 [338,191] | 001-020 P(26) C[13]O[16]2 |
| CO2 | | 10.07006390 [338,191] | 001-020 P(28) C[13]O[16]2 |
| CO2 | | 10.09009116 [338] | 001-020 P(30) C[13]O[16]2 |
| CO2 | | 10.11048062 [338] | 001-020 P(32) C[13]O[16]2 |

| | | WAVELENGTH IN VACUUM [MICROMETER] | | TRANSITION | | |
|---|---|---|---|---|---|---|
| CO2 | | 10.13123435 | [338] | 001-020 | P(34) | C[13]O[16]2 |
| CO2 | | 10.15235439 | [338] | 001-020 | P(36) | C[13]O[16]2 |
| CO2 | | 10.17384275 | [338] | 001-020 | P(38) | C[13]O[16]2 |
| CO2 | @ | 10.19570139 | [338] | 001-020 | P(40) | C[13]O[16]2 |
| | | | | | | |
| CO2 | | 10.58841900 | [338] | 001-100 | R(46) | C[13]O[16]2 |
| CO2 | | 10.60062644 | [338] | 001-100 | R(44) | C[13]O[16]2 |
| CO2 | | 10.61310290 | [338] | 001-100 | R(42) | C[13]O[16]2 |
| CO2 | | 10.62584867 | [338] | 001-100 | R(40) | C[13]O[16]2 |
| CO2 | | 10.63886409 | [338] | 001-100 | R(38) | C[13]O[16]2 |
| CO2 | | 10.65214962 | [338] | 001-100 | R(36) | C[13]O[16]2 |
| CO2 | | 10.66570577 | [338] | 001-100 | R(34) | C[13]O[16]2 |
| CO2 | | 10.67953319 | [338] | 001-100 | R(32) | C[13]O[16]2 |
| CO2 | | 10.69363257 | [338] | 001-100 | R(30) | C[13]O[16]2 |
| CO2 | | 10.70800473 | [338] | 001-100 | R(28) | C[13]O[16]2 |
| CO2 | | 10.72265056 | [338] | 001-100 | R(26) | C[13]O[16]2 |
| CO2 | | 10.73757106 | [338] | 001-100 | R(24) | C[13]O[16]2 |
| CO2 | | 10.75276730 | [338,191] | 001-100 | R(22) | C[13]O[16]2 |
| CO2 | | 10.76824046 | [338,191] | 001-100 | R(20) | C[13]O[16]2 |
| CO2 | | 10.78399183 | [338,191] | 001-100 | R(18) | C[13]O[16]2 |
| CO2 | | 10.80002276 | [338,191] | 001-100 | R(16) | C[13]O[16]2 |
| CO2 | | 10.81633473 | [338,191] | 001-100 | R(14) | C[13]O[16]2 |
| CO2 | | 10.83292931 | [338] | 001-100 | R(12) | C[13]O[16]2 |
| CO2 | | 10.84980818 | [338] | 001-100 | R(10) | C[13]O[16]2 |
| CO2 | | 10.86697309 | [338] | 001-100 | R(08) | C[13]O[16]2 |
| CO2 | | 10.88442594 | [338] | 001-100 | R(06) | C[13]O[16]2 |
| CO2 | | 10.90216870 | [338] | 001-100 | R(04) | C[13]O[16]2 |
| CO2 | @ | 11.92020346 | [338] | 001-100 | R(02) | C[13]O[16]2 |
| CO2 | @ | 11.93853242 | [338] | 001-100 | R(00) | C[13]O[16]2 |
| CO2 | @ | 11.96658256 | [338] | 001-100 | P(02) | C[13]O[16]2 |
| CO2 | | 11.98565737 | [338] | 001-100 | P(04) | C[13]O[16]2 |
| CO2 | | 11.00503495 | [338] | 001-100 | P(06) | C[13]O[16]2 |
| CO2 | | 11.02471801 | [338] | 001-100 | P(08) | C[13]O[16]2 |
| CO2 | | 11.04470940 | [338] | 001-100 | P(10) | C[13]O[16]2 |
| CO2 | | 11.06501204 | [338,191] | 001-100 | P(12) | C[13]O[16]2 |
| CO2 | | 11.08562903 | [338,191] | 001-100 | P(14) | C[13]O[16]2 |
| CO2 | | 11.10656357 | [338,191] | 001-100 | P(16) | C[13]O[16]2 |
| CO2 | | 11.12781897 | [338,191] | 001-100 | P(18) | C[13]O[16]2 |
| CO2 | | 11.14939871 | [338,191] | 001-100 | P(20) | C[13]O[16]2 |
| CO2 | | 11.17130639 | [338,191] | 001-100 | P(22) | C[13]O[16]2 |
| CO2 | | 11.19354572 | [338,191] | 001-100 | P(24) | C[13]O[16]2 |
| CO2 | | 11.21612059 | [338,191] | 001-100 | P(26) | C[13]O[16]2 |
| CO2 | | 11.23903501 | [338,191] | 001-100 | P(28) | C[13]O[16]2 |
| CO2 | | 11.26229313 | [338] | 001-100 | P(30) | C[13]O[16]2 |
| CO2 | | 11.28589928 | [338] | 001-100 | P(32) | C[13]O[16]2 |
| CO2 | | 11.30985791 | [338] | 001-100 | P(34) | C[13]O[16]2 |
| CO2 | | 11.33417365 | [338] | 001-100 | P(36) | C[13]O[16]2 |
| CO2 | | 11.35885125 | [338] | 001-100 | P(38) | C[13]O[16]2 |
| CO2 | | 11.38389568 | [338] | 001-100 | P(40) | C[13]O[16]2 |
| CO2 | @ | 11.40931204 | [338] | 001-100 | P(42) | C[13]O[16]2 |
| CO2 | @ | 10.49282306 | [338] | 001-100 | R(42) | C[13]O[18]2 |
| CO2 | | 10.50270688 | [338] | 001-100 | R(40) | C[13]O[18]2 |
| CO2 | | 10.51287542 | [338] | 001-100 | R(38) | C[13]O[18]2 |
| CO2 | | 10.52332884 | [338] | 001-100 | R(36) | C[13]O[18]2 |
| CO2 | | 10.53406738 | [338] | 001-100 | R(34) | C[13]O[18]2 |
| CO2 | | 10.54509140 | [338] | 001-100 | R(32) | C[13]O[18]2 |
| CO2 | | 10.55640135 | [338] | 001-100 | R(30) | C[13]O[18]2 |

| | | WAVELENGTH IN VACUUM [MICROMETER] | | TRANSITION |
|---|---|---|---|---|
| CO2 | | 10.56799780 | [338] | 001-100 R(28) C[13]O[18]2 |
| CO2 | | 10.57988138 | [338] | 001-100 R(26) C[13]O[18]2 |
| CO2 | | 10.59205288 | [338] | 001-100 R(24) C[13]O[18]2 |
| CO2 | | 10.60451316 | [338] | 001-100 R(22) C[13]O[18]2 |
| CO2 | | 10.61726318 | [338] | 001-100 R(20) C[13]O[18]2 |
| CO2 | | 10.63030402 | [338] | 001-100 R(18) C[13]O[18]2 |
| CO2 | | 10.64363687 | [338] | 001-100 R(16) C[13]O[18]2 |
| CO2 | | 10.65726303 | [338] | 001-100 R(14) C[13]O[18]2 |
| CO2 | | 10.67118389 | [338] | 001-100 R(12) C[13]O[18]2 |
| CO2 | | 10.68540097 | [338] | 001-100 R(10) C[13]O[18]2 |
| CO2 | | 10.69991589 | [338] | 001-100 R(08) C[13]O[18]2 |
| CO2 | | 10.71473039 | [338] | 001-100 R(06) C[13]O[18]2 |
| CO2 | @ | 11.72984633 | [338] | 001-100 R(04) C[13]O[18]2 |
| CO2 | @ | 11.81846619 | [338] | 001-100 P(06) C[13]O[18]2 |
| CO2 | | 11.83559484 | [338] | 001-100 P(08) C[13]O[18]2 |
| CO2 | | 11.85304268 | [338] | 001-100 P(10) C[13]O[18]2 |
| CO2 | | 11.87081258 | [338] | 001-100 P(12) C[13]O[18]2 |
| CO2 | | 11.88890760 | [338] | 001-100 P(14) C[13]O[18]2 |
| CO2 | | 11.90733089 | [338] | 001-100 P(16) C[13]O[18]2 |
| CO2 | | 11.92608578 | [338] | 001-100 P(18) C[13]O[18]2 |
| CO2 | | 11.94517571 | [338] | 001-100 P(20) C[13]O[18]2 |
| CO2 | | 11.96460427 | [338] | 001-100 P(22) C[13]O[18]2 |
| CO2 | | 11.98437521 | [338] | 001-100 P(24) C[13]O[18]2 |
| CO2 | | 11.00449241 | [338] | 001-100 P(26) C[13]O[18]2 |
| CO2 | | 11.02495992 | [338] | 001-100 P(28) C[13]O[18]2 |
| CO2 | | 11.04578194 | [338] | 001-100 P(30) C[13]O[18]2 |
| CO2 | | 11.06696282 | [338] | 001-100 P(32) C[13]O[18]2 |
| CO2 | | 11.08850708 | [338] | 001-100 P(34) C[13]O[18]2 |
| CO2 | | 11.11041941 | [338] | 001-100 P(36) C[13]O[18]2 |
| CO2 | | 11.13270467 | [338] | 001-100 P(38) C[13]O[18]2 |
| CO2 | | 11.15536788 | [338] | 001-100 P(40) C[13]O[18]2 |
| CO2 | @ | 11.17841426 | [338] | 001-100 P(42) C[13]O[18]2 |
| | | | | |
| CO2 | @ | 9.49185112 | [338] | 001-020 R(46) C[13]O[18]2 |
| CO2 | | 9.49995782 | [338] | 001-020 R(44) C[13]O[18]2 |
| CO2 | | 9.50826101 | [338] | 001-020 R(42) C[13]O[18]2 |
| CO2 | | 9.51676274 | [338] | 001-020 R(40) C[13]O[18]2 |
| CO2 | | 9.52546505 | [338] | 001-020 R(38) C[13]O[18]2 |
| CO2 | | 9.53436995 | [338] | 001-020 R(36) C[13]O[18]2 |
| CO2 | | 9.54347943 | [338] | 001-020 R(34) C[13]O[18]2 |
| CO2 | | 9.55279546 | [338] | 001-020 R(32) C[13]O[18]2 |
| CO2 | | 9.56231997 | [338] | 001-020 R(30) C[13]O[18]2 |
| CO2 | | 9.57205489 | [338] | 001-020 R(28) C[13]O[18]2 |
| CO2 | | 9.58200213 | [338] | 001-020 R(26) C[13]O[18]2 |
| CO2 | | 9.59216355 | [338] | 001-020 R(24) C[13]O[18]2 |
| CO2 | | 9.60254101 | [338] | 001-020 R(22) C[13]O[18]2 |
| CO2 | | 9.61313635 | [338] | 001-020 R(20) C[13]O[18]2 |
| CO2 | | 9.62395138 | [338] | 001-020 R(18) C[13]O[18]2 |
| CO2 | | 9.63498789 | [338] | 001-020 R(16) C[13]O[18]2 |
| CO2 | | 9.64624766 | [338] | 001-020 R(14) C[13]O[18]2 |
| CO2 | | 9.65773242 | [338] | 001-020 R(12) C[13]O[18]2 |
| CO2 | | 9.66944390 | [338] | 001-020 R(10) C[13]O[18]2 |
| CO2 | | 9.68138381 | [338] | 001-020 R(08) C[13]O[18]2 |
| CO2 | | 9.69355383 | [338] | 001-020 R(06) C[13]O[18]2 |
| CO2 | | 9.70595561 | [338] | 001-020 R(04) C[13]O[18]2 |
| CO2 | @ | 9.71859080 | [338] | 001-020 R(02) C[13]O[18]2 |
| CO2 | @ | 9.75121048 | [338] | 001-020 P(02) C[13]O[18]2 |

| | WAVELENGTH IN VACUUM [MICROMETER] | | TRANSITION |
|---|---|---|---|
| CO2 | 9.76467517 | [338] | 001-020 P(04) C[13]O[18]2 |
| CO2 | 9.77638035 | [338] | 001-020 P(06) C[13]O[18]2 |
| CO2 | 9.79232753 | [338] | 001-020 P(08) C[13]O[18]2 |
| CO2 | 9.80651821 | [338] | 001-020 P(10) C[13]O[18]2 |
| CO2 | 9.82095385 | [338] | 001-020 P(12) C[13]O[18]2 |
| CO2 | 9.83563588 | [338] | 001-020 P(14) C[13]O[18]2 |
| CO2 | 9.85056572 | [338] | 001-020 P(16) C[13]O[18]2 |
| CO2 | 9.86574476 | [338] | 001-020 P(18) C[13]O[18]2 |
| CO2 | 9.88117435 | [338] | 001-020 P(20) C[13]O[18]2 |
| CO2 | 9.89685583 | [338] | 001-020 P(22) C[13]O[18]2 |
| CO2 | 9.91279050 | [338] | 001-020 P(24) C[13]O[18]2 |
| CO2 | 9.92897965 | [338] | 001-020 P(26) C[13]O[18]2 |
| CO2 | 9.94542451 | [338] | 001-020 P(28) C[13]O[18]2 |
| CO2 | 9.96212632 | [338] | 001-020 P(30) C[13]O[18]2 |
| CO2 | 9.97908626 | [338] | 001-020 P(32) C[13]O[18]2 |
| CO2 | 9.99630549 | [338] | 001-020 P(34) C[13]O[18]2 |
| CO2 | 10.01378513 | [338] | 001-020 P(36) C[13]O[18]2 |
| CO2 | 10.03152628 | [338] | 001-020 P(38) C[13]O[18]2 |
| CO2 | 10.04952999 | [338] | 001-020 P(40) C[13]O[18]2 |
| CO2 | 10.06779729 | [338] | 001-020 P(42) C[13]O[18]2 |
| CO2 | 10.08632916 | [338] | 001-020 P(44) C[13]O[18]2 |
| CO2 | 10.10512655 | [338] | 001-020 P(46) C[13]O[18]2 |
| CO2 | 10.12419038 | [338] | 001-020 P(48) C[13]O[18]2 |
| CO2 | @ 10.14352152 | [338] | 001-020 P(50) C[13]O[18]2 |
| | | | |
| CO2 | 11.329 | [191] | 001-100 R(22) C[14]O[16]2 |
| CO2 | 11.346 | [191] | 001-100 R(20) C[14]O[16]2 |
| CO2 | 11.364 | [191] | 001-100 R(18) C[14]O[16]2 |
| CO2 | 11.382 | [191] | 001-100 R(16) C[14]O[16]2 |
| CO2 | 11.400 | [191] | 001-100 R(14) C[14]O[16]2 |
| CO2 | @ 11.677 | [191] | 001-100 P(12) C[14]O[16]2 |
| CO2 | 11.700 | [191] | 001-100 P(14) C[14]O[16]2 |
| CO2 | 11.723 | [191] | 001-100 P(16) C[14]O[16]2 |
| CO2 | 11.746 | [191] | 001-100 P(18) C[14]O[16]2 |
| CO2 | 11.770 | [191] | 001-100 P(20) C[14]O[16]2 |
| CO2 | 11.794 | [191] | 001-100 P(22) C[14]O[16]2 |
| CO2 | 11.819 | [191] | 001-100 P(24) C[14]O[16]2 |
| CO2 | 11.843 | [191] | 001-100 P(26) C[14]O[16]2 |
| CO2 | @ 11.868 | [191] | 001-100 P(28) C[14]O[16]2 |
| | | | |
| CO2 | 16.596 | [417] | 100-010 R(06) C[12]O[16,18] |
| CO2 | 16.780 | [417] | 101-011 R(08) C[12]O[16,18] |
| CO2 | 16.927 | [417] | 100-010 P(09) C[12]O[16,18] |
| CO2 | 16.970 | [417] | 100-010 P(11) C[12]O[16,18] |
| | | | |
| CO2 | 17.280 | [417] | 100-010 R(07) C[12]O[18]2 |
| CO2 | 17.463 | [417] | 101-011 R(08) C[12]O[18]2 |
| CO2 | 17.596 | [417] | 100-010 P(07) C[12]O[18]2 |
| CO2 | 17.639 | [417] | 100-010 P(09) C[12]O[18]2 |
| CO2 | 17.684 | [417] | 100-010 P(11) C[12]O[18]2 |
| CO2 | 17.730 | [417] | 100-010 P(13) C[12]O[18]2 |
| CO2 | 17.775 | [417] | 100-010 P(15) C[12]O[18]2 |
| CO2 | 17.821 | [417] | 100-010 P(17) C[12]O[18]2 |
| CO2 | 17.915 | [417] | 100-010 P(21) C[12]O[18]2 |
| CO2 | 17.962 | [417] | 100-010 P(23) C[12]O[18]2 |
| CO2 | 18.010 | [417] | 100-010 P(25) C[12]O[18]2 |
| CO2 | 18.053 | [417] | 100-010 P(27) C[12]O[18]2 |

| WAVELENGTH IN VACUUM [MICROMETER] | TRANSITION |
|---|---|
| NUMBER OF LINES IN CARBON DIOXIDE | 632 |

ACTIVE MEDIUM : DINITROGEN OXIDE
SYMBOL : N2O

OPERATING CONDITIONS : PULSED AND CONTINUOUS DISCHARGE IN A MIXTURE OF N2O+N2+HE. TYPICAL MIXING RATIO 1:3.5:40 AT A TOTAL PRESSURE OF 13 TORR.
ALSO OPTICAL PUMPING WITH A HBR LASER AT 4.465 MICROMETER [302].
VERY PRECISE FREQUENCY MEASUREMENTS HAVE BEEN REPORTED AT [385].

| | WAVELENGTH IN VACUUM [MICROMETER] | | TRANSITION |
|---|---|---|---|
| N2O | 10.3456 | [269] | 001-100 R(35) |
| N2O | 10.3532 | [269] | 001-100 R(34) |
| N2O | 10.3609 | [269] | 001-100 R(33) |
| N2O | 10.3687 | [269] | 001-100 R(32) |
| N2O | 10.3765 | [269] | 001-100 R(31) |
| N2O | 10.3843 | [269] | 001-100 R(30) |
| N2O | 10.3922 | [269] | 001-100 R(29) |
| N2O | 10.4001 | [269] | 001-100 R(28) |
| N2O | 10.4081 | [269] | 001-100 R(27) |
| N2O | 10.4161 | [269] | 001-100 R(26) |
| N2O | 10.4242 | [269] | 001-100 R(25) |
| N2O | 10.4323 | [269] | 001-100 R(24) |
| N2O | 10.4405 | [269] | 001-100 R(23) |
| N2O | 10.4487 | [269] | 001-100 R(22) |
| N2O | 10.4570 | [269] | 001-100 R(21) |
| N2O | 10.4653 | [269,234] | 001-100 R(20) |
| N2O | 10.4737 | [269,234] | 001-100 R(19) |
| N2O | 10.4821 | [269,234] | 001-100 R(18) |
| N2O | 10.4906 | [269,234] | 001-100 R(17) |
| N2O | 10.4991 | [269,234] | 001-100 R(16) |
| N2O | 10.5077 | [269,234] | 001-100 R(15) |
| N2O | 10.5163 | [269,234] | 001-100 R(14) |
| N2O | 10.5250 | [269,234] | 001-100 R(13) |
| N2O | 10.5337 | [269,234] | 001-100 R(12) |
| N2O | 10.5425 | [269,234] | 001-100 R(11) |
| N2O | 10.5513 | [269] | 001-100 R(10) |
| N2O | 10.5602 | [269] | 001-100 R(09) |
| N2O | 10.5692 | [269] | 001-100 R(08) |
| N2O | 10.5781 | [269] | 001-100 R(07) |
| N2O | 10.5872 | [269] | 001-100 R(06) |
| N2O | 10.5963 | [269] | 001-100 R(05) |
| N2O | 10.6054 | [269] | 001-100 R(04) |
| N2O | 10.6146 | [269] | 001-100 R(03) |
| N2O | 10.6239 | [269] | 001-100 R(02) |
| N2O | 10.6332 | [269] | 001-100 R(01) |
| N2O | 10.6426 | [270] | 001-100 R(00) |
| N2O | 10.6614 | [270] | 001-100 P(01) |
| N2O | 10.6710 | [269] | 001-100 P(02) |

| | | WAVELENGTH IN VACUUM [MICROMETER] | | TRANSITION |
|---|---|---|---|---|
| N2O | | 10.6806 | [269] | 001-100 P(03) |
| N2O | | 10.6903 | [269] | 001-100 P(04) |
| N2O | | 10.6999 | [269] | 001-100 P(05) |
| N2O | | 10.7097 | [269] | 001-100 P(06) |
| N2O | | 10.7195 | [269] | 001-100 P(07) |
| N2O | | 10.7294 | [269] | 001-100 P(08) |
| N2O | | 10.7393 | [269] | 001-100 P(09) |
| N2O | | 10.7493 | [269] | 001-100 P(10) |
| N2O | | 10.7593 | [269] | 001-100 P(11) |
| N2O | | 10.7694 | [269,235) | 001-100 P(12) |
| N2O | | 10.7796 | [269,235] | 001-100 P(13) |
| N2O | | 10.7898 | [269,235] | 001-100 P(14) |
| N2O | | 10.8000 | [269,235] | 001-100 P(15) |
| N2O | | 10.8104 | [269,235] | 001-100 P(16) |
| N2O | | 10.8208 | [269,235] | 001-100 P(17) |
| N2O | | 10.8312 | [269,235] | 001-100 P(18) |
| N2O | | 10.8418 | [269,235] | 001-100 P(19) |
| N2O | | 10.8523 | [269,235,236] | 001-100 P(20) |
| N2O | | 10.8629 | [269,235,236] | 001-100 P(21) |
| N2O | | 10.8736 | [269,235,236] | 001-100 P(22) |
| N2O | | 10.8844 | [269,235,236] | 001-100 P(23) |
| N2O | | 10.8952 | [269,235,336] | 001-100 P(24) |
| N2O | | 10.9061 | [269,235,236] | 001-100 P(25) |
| N2O | | 10.9170 | [269,235,236] | 001-100 P(26) |
| N2O | | 10.9280 | [269,235,236] | 001-100 P(27) |
| N2O | | 10.9390 | [269,235,236] | 001-100 P(28) |
| N2O | | 10.9501 | [269,235] | 001-100 P(29) |
| N2O | | 10.9613 | [269,235] | 001-100 P(30) |
| N2O | | 10.9726 | [269,235] | 001-100 P(31) |
| N2O | | 10.9839 | [269,235] | 001-100 P(32) |
| N2O | | 10.9953 | [269,235] | 001-100 P(33) |
| N2O | | 11.0067 | [269,235] | 001-100 P(34) |
| N2O | | 11.0182 | [269,235] | 001-100 P(35) |
| N2O | | 11.0298 | [269,235] | 001-100 P(36) |
| N2O | | 11.0415 | [269,235] | 001-100 P(37) |
| | | | | |
| N2O | @ | 10.40331 | [478] | 002-101 R(31) |
| N2O | | 10.41107 | [478] | 002-101 R(30) |
| N2O | | 10.41889 | [478] | 002-101 R(29) |
| N2O | | 10.42676 | [478] | 002-101 R(28) |
| N2O | | 10.43468 | [478] | 002-101 R(27) |
| N2O | | 10.44265 | [478] | 002-101 R(26) |
| N2O | | 10.45067 | [478] | 002-101 R(25) |
| N2O | | 10.45874 | [478] | 002-101 R(24) |
| N2O | | 10.46686 | [478] | 002-101 R(23) |
| N2O | | 10.47503 | [478] | 002-101 R(22) |
| N2O | | 10.48325 | [478] | 002-101 R(21) |
| N2O | | 10.49157 | [478] | 002-101 R(20) |
| N2O | | 10.49985 | [478] | 002-101 R(19) |
| N2O | | 10.50823 | [478] | 002-101 R(18) |
| N2O | | 10.51666 | [478] | 002-101 R(17) |
| N2O | | 10.52513 | [478] | 002-101 R(16) |
| N2O | | 10.53367 | [478] | 002-101 R(15) |
| N2O | | 10.54225 | [478] | 002-101 R(14) |
| N2O | | 10.55089 | [478] | 002-101 R(13) |
| N2O | @ | 10.55958 | [478] | 002-101 R(12) |
| N2O | | 10.56832 | [478] | 002-101 R(11) |

| | WAVELENGTH IN VACUUM [MICROMETER] | | TRANSITION |
|---|---|---|---|
| N2O | 10.57712 | [478] | 002-101 R(10) |
| N2O | 10.58596 | [478] | 002-101 R(09) |
| N2O | 10.59486 | [478] | 002-101 R(08) |
| N2O | 10.60382 | [478] | 002-101 R(07) |
| N2O | 10.61282 | [478] | 002-101 R(06) |
| N2O | 10.62189 | [478] | 002-101 R(05) |
| N2O | @ 10.63100 | [478] | 002-101 R(04) |
| N2O | @ 10.72517 | [478] | 002-101 P(05) |
| N2O | 10.73489 | [478] | 002-101 P(06) |
| N2O | 10.74468 | [478] | 002-101 P(07) |
| N2O | 10.75450 | [478] | 002-101 P(08) |
| N2O | 10.76439 | [478] | 002-101 P(09) |
| N2O | 10.77434 | [478] | 002-101 P(10) |
| N2O | 10.78435 | [478] | 002-101 P(11) |
| N2O | 10.79441 | [478] | 002-101 P(12) |
| N2O | 10.80453 | [478] | 002-101 P(13) |
| N2O | 10.81471 | [478] | 002-101 P(14) |
| N2O | 10.82495 | [478] | 002-101 P(15) |
| N2O | 10.83524 | [478] | 002-101 P(16) |
| N2O | 10.84560 | [478] | 002-101 P(17) |
| N2O | 10.85601 | [478] | 002-101 P(18) |
| N2O | 10.86648 | [478] | 002-101 P(19) |
| N2O | 10.87701 | [478] | 002-101 P(20) |
| N2O | 10.88760 | [478] | 002-101 P(21) |
| N2O | 10.89825 | [478] | 002-101 P(22) |
| N2O | 10.90896 | [478] | 002-101 P(23) |
| N2O | 10.91973 | [478] | 002-101 P(24) |
| N2O | 10.93056 | [478] | 002-101 P(25) |
| N2O | 10.94145 | [478] | 002-101 P(26) |
| N2O | 10.95241 | [478] | 002-101 P(27) |
| N2O | 10.96342 | [478] | 002-101 P(28) |
| N2O | 10.97450 | [478] | 002-101 P(29) |
| N2O | 10.98564 | [478] | 002-101 P(30) |
| N2O | @ 10.99684 | [478] | 002-101 P(31) |

NUMBER OF LINES IN DINITROGEN OXIDE 123

ACTIVE MEDIUM : CARBON OXYSULFIDE
SYMBOL : OCS

OPERATING CONDITIONS : PULSED DISCHARGE IN OCS OR OCS+N2, OCS+HE, OCS+CO, AND OCS+CO+HE MIXTURES.
OPTICAL PUMPING WITH PULSED CO2-LASER [397].

| | WAVELENGTH IN VACUUM [MICROMETER] | | TRANSITION |
|---|---|---|---|
| OCS | 3.428 | [275] | ? |
| OCS | 8.2388 | [227] | 001-100 R(26) |
| OCS | 8.2416 | [227] | 001-100 R(25) |
| OCS | 8.2439 | [227] | 001-100 R(24) |
| OCS | 8.2518 | [227] | 001-100 R(21) |
| OCS | 8.2543 | [227] | 001-100 R(20) |

| | WAVELENGTH IN VACUUM [MICROMETER] | | TRANSITION |
|---|---|---|---|
| OCS | 8.2571 | [227] | 001-100 R(19) |
| OCS | 8.2595 | [227] | 001-100 R(18) |
| OCS | 8.2623 | [227] | 001-100 R(17) |
| OCS | 8.2645 | [227] | 001-100 R(16) |
| OCS | 8.2673 | [227] | 001-100 R(15) |
| OCS | 8.3625 | [227] | 001-100 P(18) |
| OCS | 8.3654 | [227] | 001-100 P(19) |
| OCS | 8.3685 | [227] | 001-100 P(20) |
| OCS | 8.3715 | [227] | 001-100 P(21) |
| OCS | 8.3746 | [227] | 001-100 P(22) |
| OCS | 8.3779 | [227] | 001-100 P(23) |
| OCS | 8.3809 | [227] | 001-100 P(24) |
| OCS | 8.3839 | [227] | 001-100 P(25) |
| OCS | 8.3870 | [227] | 001-100 P(26) |
| OCS | 8.3900 | [227] | 001-100 P(27) |
| OCS | 8.3930 | [227] | 001-100 P(28) |
| OCS | 8.3962 | [227] | 001-100 P(29) |
| OCS | 8.3999 | [227] | 001-100 P(30) |
| OCS | 8.4024 | [227] | 001-100 P(31) |
| OCS | 8.4055 | [227] | 001-100 P(32) |
| OCS | 8.4085 | [227] | 001-100 P(33) |
| OCS | 8.4117 | [227] | 001-100 P(34) |
| OCS | 8.4146 | [227] | 001-100 P(35) |
| OCS | 8.4178 | [227] | 001-100 P(36) |
| OCS | 8.4213 | [227] | 001-100 P(37) |
| OCS | 8.4243 | [227] | 001-100 P(38) |
| OCS | 18,983 | [397] | ? |
| OCS | 19,057 | [397] | ? |
| OCS | 123 | [267] | ? |
| OCS | 132 | [267] | ? |

NUMBER OF LINES IN CARBON OXYSULFIDE 36

ACTIVE MEDIUM : CARBON DISULFIDE
SYMBOL : CS2

OPERATING CONDITIONS : EMISSION IN A FLOWING MIXTURE OF 0.1 TORR CS2 AND 2 TORR N2. PRIOR TO MIXING THE GASES, NITROGEN IS EXCITED IN A CONTINUOUS DISCHARGE. PULSED EMISSION IN AN ELECTRON-BEAM-STABILIZED TRANSVERSE DISCHARGE AT ATMOSPHERIC PRESSURE [377].

| | WAVELENGTH IN VACUUM [MICROMETER] | | TRANSITION |
|---|---|---|---|
| CS2 | 11.482 | [228,268,377] | ?001-100 P(28) |
| CS2 | > 11.489 | [228,268,377] | ?001-100 P(30) |
| CS2 | 11.596 | [228,268] | ?001-100 P(32) |
| CS2 | > 11.503 | [228,268,377] | ?001-100 P(34) |
| CS2 | > 11.510 | [228,268,377] | ?001-100 P(36) |
| CS2 | 11.517 | [228,268] | ?001-100 P(38) |
| CS2 | 11.524 | [228,268] | ?001-100 P(40) |
| CS2 | 11.531 | [228,268,377] | ?001-100 P(42) |

| WAVELENGTH IN VACUUM [MICROMETER] | | | | TRANSITION |
|---|---|---|---|---|
| CS2 | > | 11.538 | [228,268,377] | ?001-100 P(44) |
| CS2 | > | 11.545 | [228,268,377] | ?001-100 P(46) |
| CS2 | > | 11.960 | [377] | 001-100 R(40) C[13]S[32]2 |
| CS2 | > | 11.965 | [377] | 001-100 R(38) C[13]S[32]2 |
| CS2 | > | 11.986 | [377] | 001-100 R(30) C[13]S[32]2 |
| CS2 | > | 12.217 | [377] | ?001-100 P(40) C[13]S[32]2 |
| CS2 | > | 12.241 | [377] | ?001-100 P(46) C[13]S[32]2 |
| CS2 | > | 12.249 | [377] | ?001-100 P(48) C[13]S[32]2 |

NUMBER OF LINES IN CARBON DISULFIDE 16

---

ACTIVE MEDIUM : HYDROGEN CYANIDE
SYMBOL : HCN

OPERATING CONDITIONS : PULSED DISCHARGE WITH HIGH CURRENT DENSITY IN DIFFERENT GAS MIXTURES, FOR EXAMPLE IN CH4+NH3, CH4+N2, CH3CN, (CH3)2NH, OR HCN. MANY LINES (ESPECIALLY AT 337 MICROMETER) ALSO CONTINUOUSLY, 100 MILLIWATT OUTPUT POWER IN THE WAVEGUIDE RESONATOR [368]. PEAK POWERS UP TO 1 KW [200].

| WAVELENGTH IN VACUUM [MICROMETER] | | | | WAVELENGTH IN VACUUM [MICROMETER] | | | |
|---|---|---|---|---|---|---|---|
| HCN | ? | 81.554 | [213] | HCN | ? | 201.059 | [213,256] |
| HCN | ? | 96.401 | [213] | HCN | ? | 211.001 | [213,256] |
| HCN | ? | 98.693 | [213] | HCN | ? | 222.949 | [213,256] |
| HCN | ? | 101.257 | [213] | HCN | | 284 | [207] |
| HCN15 | ? | 110.240 | [213] | HCN | | 309.7140 | [207,213,255] |
| HCN | ? | 112.066 | [213] | HCN | | 310.8870 | [207,208,255] |
| HCN15 | ? | 113.311 | [213] | HCN | | 335.1831 | [207,255] |
| HCN | ? | 116.132 | [213] | HCN | > | 336.5578 | [213,207,208] |
| HCN | | 126.164 | [213,255,256] | HCN | | 372.5283 | [213,207,255] |
| HCN | | 128.629 | [213,255,256] | HCN | ? | 538.2 | [213] |
| HCN | | 130.838 | [213,255,256] | HCN | ? | 545.4 | [213] |
| HCN | | 134.932 | [213,255,256] | HCN | ? | 676 | [213] |
| HCN15 | ? | 138.768 | [213] | HCN | ? | 773.5 | [213] |
| HCN15 | ? | 165.150 | [213] | | | | |

NUMBER OF LINES IN HYDROGEN CYANIDE 27

---

ACTIVE MEDIUM : DEUTERIUM CYANIDE
SYMBOL : DCN

OPERATING CONDITIONS : PULSED (SOME LINES ALSO CONTINUOUS) DISCHARGE IN D2+BRCN OR CD4+ND3.

| | WAVELENGTH IN AIR [MICROMETER] | |
|---|---|---|
| DCN | 181.789 | [213,257] |
| DCN | 189.9490 | [213,257] |
| DCN | 190.0080 | [255] |
| DCN | 194.7027 | [213,257] |
| DCN | 194.7644 | [255] |
| DCN | 204.3872 | [213,257] |

NUMBER OF LINES IN DEUTERIUM CYANIDE 6

ACTIVE MEDIUM : WATER
SYMBOL : H2O

OPERATING CONDITIONS : PULSED DISCHARGE IN H2O VAPOR OF 0.4-1 TORR PRESSURE. SOME LINES ALSO CONTINUOUS [201]. TUNING TO SINGLE LINES IN A DISPERSIVE RESONATOR [169].

| | | WAVELENGTH IN AIR [MICROMETER] | |
|---|---|---|---|
| H2O | ? | 2.28 | [381] |
| H2O | ? | 4.77 | [381] |
| H2O | | 7.093 | [473] |
| H2O | | 7.204 | [473] |
| H2O | | 7.265 | [473] |
| H2O | | 7.297 | [473] |
| H2O | | 7.390 | [473] |
| H2O | | 7.425 | [473] |
| H2O | | 7.543 | [473] |
| H2O | | 7.590 | [473] |
| H2O | | 7.7069 | [170] |
| H2O | | 7.709 | [473] |
| H2O | | 7.7097 | [108,381] |
| H2O | | 7.740 | [473] |
| H2O | | 7.4568 | [170,381] |
| H2O | | 7.5945 | [108,170,381] |
| H2O | | 9.3938 | [108] |
| H2O | | 9.4747 | [108] |
| H2O | | 9.5674 | [108] |

| | | WAVELENGTH IN VACUUM [MICROMETER] | |
|---|---|---|---|
| H2O | ? | 11.83 | [381] |
| H2O | ? | 11.96 | [381] |
| H2O | | 16.932 | [201,381] |
| H2O | | 23.13 | [169,381] |
| H2O | | 23.365 | [201,381] |
| H2O | ? | 24.966 | [381] |
| H2O[18] | | 25.162 | [381] |
| H2O[18] | | 26.595 | [381] |
| H2O | | 26.660 | [201,381] |
| H2O | > | 27.970755 | [201,381,382] |
| H2O | | 28.054 | [201,381] |
| H2O | | 28.270 | [201,381] |
| H2O[18] | | 28.295 | [381] |
| H2O | | 28.356 | [201,381] |
| H2O | ? | 28.451 | [381] |
| H2O | | 32.924 | [201,381] |
| H2O[18] | | 33.308 | [381] |
| H2O | > | 33.329 | [201,381] |
| H2O | ? | 34.60 | [169,381] |
| H2O | ? | 35.017 | [201,381] |
| H2O[18] | ? | 35.383 | [381] |
| H2O | ? | 35.633 | [201,381] |
| H2O | ? | 36.606 | [201,381] |
| H2O | ? | 37.648 | [201,381] |
| H2O | ? | 38.086 | [201,381] |
| H2O | ? | 39.695 | [201,381] |
| H2O | ? | 40.45 | [381] |
| H2O | ? | 40.638 | [201,381] |
| H2O | ? | 42.51 | [169,381] |
| H2O | ? | 45.517 | [201,381] |
| H2O | ? | 45.91 | [381] |
| H2O | | 47.244 | [169,201,381] |
| H2O | | 47.39 | [169,381] |
| H2O | | 47.468 | [201,381] |
| H2O | | 47.687 | [201,381] |
| H2O | | 48.19 | [169,381] |
| H2O[18] | | 48.366 | [381] |
| H2O | ? | 48.676 | [201,381] |
| H2O[18] | | 48.765 | [381] |
| H2O | | 49.06 | [169,381] |
| H2O[18] | | 49.430 | [381] |
| H2O | | 53.910 | [201,381] |

| WAVELENGTH IN VACUUM [MICROMETER] | | | WAVELENGTH IN VACUUM [MICROMETER] | | |
|---|---|---|---|---|---|
| H2O | ? 55.000 | [381] | H2O | ? 85.564 | [381] |
| H2O | 55.088 | [201,381] | H2O | ? 86.301 | [381] |
| H2O[18] | 56.129 | [381] | H2O | ? 86.471 | [381] |
| H2O | 57.659 | [201,381] | H2O | ? 87.323 | [381] |
| H2O | ? 57.799 | [381] | H2O | ? 87.469 | [381] |
| H2O | ? 66.800 | [381] | H2O | 89.772 | [201,381] |
| H2O | ? 66.903 | [381] | H2O | 115.32 | [201,381] |
| H2O | ? 67.169 | [201,381] | H2O | 118.59104 | [201,381,382] |
| H2O | ? 72.856 | [381] | H2O | 120.08 | [201,381] |
| H2O | 73.401 | [201,381] | H2O | 220.230 | [203,204,381] |
| H2O | 78.443329 | [201,381,382] | H2O | 350.20 | [331] |
| H2O | 79.091010 | [201,381,382] | | | |

NUMBER OF LINES IN WATER 84

ACTIVE MEDIUM : HEAVY WATER
SYMBOL : D2O

OPERATING CONDITIONS : PULSED DISCHARGE IN D2O VAPOR OF 0.2-1 TORR PRESSURE. SOME LINES ALSO IN CONTINUOUS OPERATION [173,201]. OPTICAL PUMPING WITH SINGLE CO2-LASER LINES [436,437,438].

| WAVELENGTH IN VACUUM [MICROMETER] | | | WAVELENGTH IN VACUUM [MICROMETER] | | |
|---|---|---|---|---|---|
| D2O | ? 26.36 | [171,381] | D2O | 73.337 | [201] |
| D2O | ? 33.896 | [201,381] | D2O | ? 74.341 | [381] |
| D2O | 35.081 | [201,381] | D2O | ? 74.526 | [201,381] |
| D2O | 36.096 | [171,381] | D2O | 76.305 | [201,381] |
| D2O | 36.324 | [201,381] | D2O | 78.16 | [171,381] |
| D2O | 36.526 | [201,381] | D2O | ? 83.730 | [171,381] |
| D2O | 37.788 | [201,381] | D2O | 84.111 | [201,381] |
| D2O | 37.864 | [171,381] | D2O | > 84.278897 | [173,381,382] |
| D2O | 39.53 | [171,381] | D2O | ? 99.00 | [171,381] |
| D2O | 40.994 | [201,381] | D2O | ?103.33 | [171,381] |
| D2O | ? 41.79 | [171,381] | D2O | >107.72019 | [203,381,382] |
| D2O | ? 48.80 | [171,381] | D2O | ?107.91 | [381] |
| D2O | ? 50.71 | [171,381] | D2O | ?108.88 | [171,381] |
| D2O | ? 54.73 | [171,381] | D2O | ?110.49 | [171,381] |
| D2O | ? 56.830 | [201,381] | D2O | ?111.74 | [171,381] |
| D2O | ? 61.182 | [381] | D2O | 170.08 | [381] |
| D2O | 71.944 | [201,381] | D2O | 171.67 | [203,173,381] |
| D2O | ? 72.427 | [201,381] | D2O | 218.5 | [172] |
| D2O | > 72.747780 | [201,381,382] | | | |

| WAVELENGTH IN VACUUM [MICROMETER] | | | PUMP TRANSITION |
|---|---|---|---|
| D2O | 50,5 | [436,439] | 001-020 P(32) CO2 |
| D2O | 66 | [436,439] | 001-020 P(32) CO2 |
| D2O | 83 | [436,439] | 001-020 P(32) CO2 |

| | WAVELENGTH IN VACUUM [MICROMETER] | | PUMP TRANSITION |
|---|---|---|---|
| D2O | 94 | [436,439] | 001-020 R(12) CO2 |
| D2O | 94.62 | [539] | 002-021 R(17) CO2 |
| D2O | 98 | [500] | 001-020 R(32) CO2 |
| D2O | 99 | [500] | 001-020 R(30) CO2 |
| D2O | 112.58 | [539] | 002-021 R(17) CO2 |
| D2O | 114 | [436,439] | 001-020 R(12) CO2 |
| D2O | 116 | [437] | 001-020 R(32) CO2 |
| D2O | 119 | [437] | 001-020 P(32) CO2 |
| D2O | 142 | [437] | 001-020 R(12) CO2 |
| D2O | 239 | [500] | 001-020 R(22) CO2 |
| D2O | 263 | [437] | 001-020 R(34) CO2 |
| D2O | 276 | [500] | 001-020 R(22) CO2 |
| D2O | 358.5 | [439,500] | 001-020 R(22) CO2 |
| D2O | 385 | [436,437] | 001-020 R(22) CO2 |

NUMBER OF LINES IN HEAVY WATER 54

ACTIVE MEDIUM : SULFUR DIOXIDE
SYMBOL : SO2

OPERATING CONDITIONS : PULSED DISCHARGE IN A MIXTURE OF SO2+HE. TYPICAL PRESSURE 0.4 TORR SO2, 4 TORR HE. SOME LINES ALSO CONTINUOUS.

| | WAVELENGTH IN VACUUM [MICROMETER] | | | WAVELENGTH IN VACUUM [MICROMETER] | |
|---|---|---|---|---|---|
| SO2 | 139.83 | [272] | SO2 | 151.08 | [272] |
| SO2 | > 140.82 | [272] | SO2 | 151.35 | [272,273] |
| SO2 | > 141.06 | [272,273] | SO2 | 192.60 | [272,273] |
| SO2 | 142.00 | [272] | SO2 | 206.53 | [272] |
| SO2 | 149.94 | [272] | SO2 | 215.27 | [272,273,274] |

NUMBER OF LINES IN SULFUR DIOXIDE 10

ACTIVE MEDIUM : HYDROGEN SULFIDE
SYMBOL : H2S

OPERATING CONDITIONS : PULSED DISCHARGE IN H2S OF 0.15 TORR PRESSURE BECAUSE OF DISSOCIATION, AFTER EACH PULSE THE GAS HAS TO BE EXCHANGED. ISOTOPIC SPECIES SHOW SUPERFLUORESCENT EMISSION AFTER OPTICAL PUMPING WITH CO2-LASER [556].

| | WAVELENGTH IN VACUUM [MICROMETER] | | | WAVELENGTH IN VACUUM [MICROMETER] | |
|---|---|---|---|---|---|
| H2S | 33.30 | [267,274] | H2S | 96.4 | [267,274] |
| H2S | 37.600 | [267,274] | H2S | 103.3 | [267,274] |
| H2S | 48.70 | [267,274] | H2S | 108.6 | [267,274] |
| H2S | 52.307 | [267,274] | H2S | 116.6 | [267,274] |
| H2S | 55.612 | [267,274] | H2S | 126.2 | [267,274] |
| H2S | 60.224 | [267,274] | H2S | 129.1 | [267,274] |
| H2S | > 61.413 | [267,274] | H2S | 130.60 | [267,274] |
| H2S | 62.6 | [267,274] | H2S | 135.3 | [267,274] |
| H2S | 73.54 | [267,274] | H2S | 140.6 | [267,274] |
| H2S | 81.05 | [267,274] | H2S | 162.4 | [267,274] |
| H2S | 83.45 | [267,274] | H2S | 192.9 | [267,274] |
| H2S | > 87.580 | [267,274] | H2S | > 225.3 | [267,274] |
| H2S | 92.0 | [267,274] | | | |

| | WAVELENGTH IN VACUUM [MICROMETER] | | PUMP TRANSITION |
|---|---|---|---|
| D2S | 90.5 | [556] | 001-100 R(18) CO2 |
| D2S | 135.3 | [556] | 001-100 P(32) CO2 |
| HDS | 183.2 | [556] | 001-100 R(12) CO2 |

NUMBER OF LINES IN HYDROGEN SULFIDE 26

ACTIVE MEDIUM : BORON TRICHLORIDE
SYMBOL : BCL3

OPERATING CONDITIONS : CW OPERATION BY ADDING BCL3 VAPOR TO THE GAS MIXTURE OF A CO2 LASER.

| | WAVELENGTH IN AIR [MICROMETER] | | | WAVELENGTH IN AIR [MICROMETER] | |
|---|---|---|---|---|---|
| BCL3 | 18.3 | [370] | BCL3 | 20.2 | [370] |
| BCL3 | 18.6 | [370] | BCL3 | 20.6 | [370] |
| BCL3 | 19.1 | [370] | BCL3 | 22.4 | [370] |
| BCL3 | 19.4 | [370] | BCL3 | 23.0 | [370] |

NUMBER OF LINES IN BORON TRICHLORIDE 8

ACTIVE MEDIUM : SULFUR HEXAFLUORIDE
SYMBOL : SF6

OPERATING CONDITIONS : PULSED EXCITATION BY TWO PHOTON ABSORPTION OF CO2-LASER LINES.
(P(14) AND P(16) LINES OF 10.6 MICRON - BAND).

| WAVELENGTH IN AIR [MICROMETER] | | | WAVELENGTH IN AIR [MICROMETER] | | |
|---|---|---|---|---|---|
| SF6 | 15.9005 | [375] | | | |

NUMBER OF LINES IN SULFUR HEXAFLUORIDE 1

ACTIVE MEDIUM : AMMONIA
SYMBOL : NH3

OPERATING CONDITIONS : PULSED DISCHARGE WITH HIGH CURRENT DENSITY IN NH3 GAS OF 0.5-1 TORR PRESSURE [240,459]. OPTICAL PUMPING WITH A N2O-LASER (P(13) LINE AT 10.78 MICROMETER)[241], OR WITH CO2 LASER AND STARK SHIFTING THE ABSORPTION LINE OF NH3 [321]. OPTICAL PUMPING WITH DIFFERENT CO2-LASER LINES, PEAK POWER 0.5 MW [515],MEAN POWER 20 W [576]. TWO PHOTON PUMPING WITH CO2-LASER [464]. OPTICAL PUMPING WITH HF-LASER [549].

| WAVELENGTH IN AIR [MICROMETER] | | | WAVELENGTH IN AIR [MICROMETER] | | |
|---|---|---|---|---|---|
| NH3 | 14.78 | [240] | NH3 | 23.86 | [240] |
| NH3 | 15.04 | [240] | NH3 | 24.918 | [229,230] |
| NH3 | 15.08 | [240] | NH3 | 25.12 | [240] |
| NH3 | 15.41 | [240] | NH3 | 26.282 | [229,230] |
| NH3 | 15.47 | [240] | NH3 | 30.69 | [240] |
| NH3 | 16.21 | [240] | NH3 | 31.47 | [240] |
| NH3 | 21.471 | [229,230] | NH3 | 31.951 | [229,230] |
| NH3 | 22.542 | [229,230] | NH3 | 32.13 | [240] |
| NH3 | 22.563 | [229,230] | | | |
| NH3 | 22.71 | [240] | N[15]H3 | 15.08 | [459] |
| NH3 | 23.675 | [22 ,230] | N[15]H3 | 15.91 | [459] |

| WAVELENGTH IN AIR [MICROMETER] | | | PUMP TRANSITION |
|---|---|---|---|
| NH3 | 6.27 | [464] | 001-100 P(34),P(18) CO2 |
| NH3 | 6.69 | [464] | 001-100 P(34),P(18) CO2 |
| NH3 | 9.3 | [446] | 001-020 R(30) CO2 |
| NH3 | 9.6 | [446] | 001-020 R(30) CO2 |
| NH3 | 9.7 | [446] | 001-020 R(30) CO2 |
| NH3 | 9.9 | [446] | 001-020 R(30) CO2 |
| NH3 | 10.2 | [446] | 001-020 R(30) CO2 |
| NH3 | 10.5 | [446] | 001-020 R(30) CO2 |
| NH3 | 10.6 | [446] | 001-020 R(30) CO2 |
| NH3 | 10.7 | [446] | 001-020 R(30) CO2 |
| NH3 | 11.0 | [446] | 001-020 R(30) CO2 |
| NH3 | 11.446 | [446] | 001-020 R(30) CO2 |
| NH3 | 11.459 | [465,543] | 001-100 R(14) CO2 |
| NH3 | > 11.526 | [446,543] | 001-020 R(30) CO2 |
| NH3 | 11.5547 | [605] | 100R(14)CO2,020P(14)N[13]O2 |
| NH3 | 11.721 | [446] | 001-020 R(30) CO2 |
| NH3 | 11.80 | [465] | 001-100 R(06) CO2 |
| NH3 | 11.811 | [461] | 001-020 R(16) CO2 |

| | | WAVELENGTH IN AIR [MICROMETER] | | PUMP TRANSITION |
|---|---|---|---|---|
| NH3 | | 11.994 | [446] | 001-020 R(30) CO2 |
| NH3 | | 12.010 | [446] | 001-020 R(30) CO2 |
| HH3 | | 12.078 | [543] | 001-020 R(16) CO2 |
| HH3 | | 12.078 | [543] | 001-020 R(30) CO2 |
| NH3 | | 12.0791 | [461,465,515] | 001-020 R(30) CO2 |
| NH3 | | 12.1143 | [464,605] | 001-100 P(34),P(16) CO2 |
| NH3 | | 12.1558 | [605] | 020P(08)C[13]O2,020P(24)CO2 |
| NH3 | | 12.1846 | [605] | 020P(34)CO2,100P(24)CO2 |
| HH3 | | 12.245 | [543] | 001-020 R(16) CO2 |
| HH3 | | 12.245 | [543] | 001-020 R(30) CO2 |
| NH3 | | 12.251 | [446] | 001-020 R(30) CO2 |
| NH3 | | 12.266 | [446] | 001-020 R(30) CO2 |
| HH3 | | 12.280 | [543] | 001-100 P(32) CO2 |
| NH3 | | 12.286 | [446,543] | 001-020 R(30) CO2 |
| NH3 | > | 12.316 | [446] | 001-020 R(30) CO2 |
| NH3 | | 12.348 | [446] | 001-020 R(30) CO2 |
| HH3 | | 12.520 | [543] | 001-020 R(30) CO2 |
| NH3 | > | 12.526 | [446] | 001-020 R(30) CO2 |
| NH3 | > | 12.541 | [446] | 001-020 R(30) CO2 |
| NH3 | | 12.566 | [446] | 001-020 R(30) CO2 |
| NH3 | | 12.591 | [446] | 001-020 R(30) CO2 |
| NH3 | > | 12.631 | [446] | 001-020 R(30) CO2 |
| NH3 | > | 12.689 | [446] | 001-020 R(30) CO2 |
| NH3 | > | 12.8115 | [515,446,461] | 001-020 R(30) CO2 |
| NH3 | | 12.812 | [465,543] | 001-020 R(16) CO2 |
| NH3 | > | 12.851 | [446] | 001-020 R(30) CO2 |
| NH3 | | 12.876 | [446] | 001-020 R(30) CO2 |
| NH3 | > | 12.921 | [446] | 001-020 R(30) CO2 |
| NH3 | | 13.031 | [446] | 001-020 R(30) CO2 |
| NH3 | | 13.114 | [446] | 001-020 R(30) CO2 |
| NH3 | | 13.124 | [446] | 001-020 R(30) CO2 |
| NH3 | | 13.145 | [446] | 001-020 R(30) CO2 |
| NH3 | | 13.176 | [446] | 001-020 R(30) CO2 |
| NH3 | | 13.269 | [446] | 001-020 R(30) CO2 |
| NH3 | | 13.218 | [446] | 001-020 R(30) CO2 |
| NH3 | | 13.331 | [446] | 001-020 R(30) CO2 |
| NH3 | | 13.411 | [446] | 001-020 R(30) CO2 |
| NH3 | | 13.576 | [446] | 001-020 R(30) CO2 |
| NH3 | | 13.7261 | [464,605] | 001-100 P(34),P(16) CO2 |
| NH3 | | 13.821 | [446] | 001-020 R(30) CO2 |
| NH3 | | 15.8782 | [464,573,605] | 001-100 P(34),P(16) CO2 |
| NH3 | | 15.9452 | [464,573,605] | 001-100 P(34),P(16) CO2 |
| NH3 | | 16.9250 | [464,605] | 001-100 P(34),P(16) CO2 |
| NH3 | | 19.5497 | [464,605] | 001-100 P(34),P(16) CO2 |
| NH3 | | 25.4744 | [605] | 001-100 P(26) CO2 |
| NH3 | | 25.8839 | [605] | 001-020 P(10) CO2 |
| NH3 | | 26.1046 | [464,605] | 001-100 P(34),P(16) CO2 |
| NH3 | | 26.4416 | [543,605] | 001-020 P(24) CO2 |
| NH3 | | 26.7068 | [605] | 020P(34)CO2,100P(24)CO2 |
| NH3 | | 27.8437 | [543,605] | 001-100 P(20) CO2 |
| NH3 | | 34.2248 | [332,543,605] | 001-100 P(12) CO2 |
| NH3 | | 35.1573 | [332,605] | 001-100 P(14) CO2 |
| NH3 | | 35.5011 | [464,605] | 001-100 P(34),P(16) CO2 |
| NH3 | | 36.02 | [543] | 001-100 P(20) CO2 |
| NH3 | | 36.1686 | [332,543,605] | 001-020 P(24) CO2 |
| NH3 | | 49.0356 | [605] | 001-020 P(24) CO2 |
| NH3 | | 54.45 | [598] | 001-100 P(14) N2O |

| WAVELENGTH IN AIR [MICROMETER] | | | PUMP TRANSITION |
|---|---|---|---|
| NH3 | 56.8631 | [332,543,605] | 001-100 P(20) CO2 |
| NH3 | 58.01 | [332,316,543] | 001-100 R(04) CO2 |
| NH3 | 63.25 | [543] | 001-020 R(16) CO2 |
| NH3 | 64.5 | [598] | 001-020 R(30) CO2 |
| NH3 | 64.7274 | [332,543,605] | 001-020 P(24) CO2 |
| NH3 | 67.19 | [543] | 001-020 R(16) CO2 |
| NH3 | 67.24 | [332,598] | 001-020 R(30) CO2 |
| NH3 | 72.6 | [316] | 001-100 R(36) CO2 |
| NH3 | 72.76 | [543] | 001-020 R(16) CO2 |
| NH3 | 74.15 | [332,543] | 001-100 R(36) CO2 |
| NH3 | 78.28 | [598] | 001-100 P(06) N2O |
| NH3 | 81.53 | [241] | 001-100 P(13) N2O |
| NH3 | 83.60 | [543] | 001-100 R(36) CO2 |
| NH3 | 83.65 | [332,543] | 001-020 R(30) CO2 |
| NH3 | 84.64 | [543] | 001-100 P(32) CO2 |
| NH3 | 87.1 | [539] | 002-021 P(17) CO2 |
| NH3 | 88.05 | [572,598] | 001-100 R(06) CO2 |
| NH3 | 88.20 | [321] | 001-100 R(06) CO2 |
| NH3 | 88.90 | [332,543] | 001-020 R(30) CO2 |
| NH3 | 90.50 | [332,543] | 001-020 R(16) CO2 |
| NH3 | 90.93 | [572] | 001-020 R(16) CO2 |
| NH3 | 92.87 | [572,598] | 001-100 P(30) CO2 |
| NH3 | 112.98 | [572] | 001-020 P(10) CO2 |
| NH3 | 114.29 | [321,572] | 001-020 P(12) CO2 |
| NH3 | 119.02 | [332,543] | 001-100 R(04) CO2 |
| NH3 | 147.04 | [317] | 001-020 R(30) CO2 |
| NH3 | 147.2 | [543] | 001-020 R(16) CO2 |
| NH3 | 147.15 | [332,543,598] | 001-020 R(30) CO2 |
| NH3 | 151.49 | [321,332,543] | 001-100 P(32) CO2 |
| NH3 | 151.5 | [598] | 001-100 P(07) N2O |
| NH3 | 155.17 | [332] | 001-100 R(06) CO2 |
| NH3 | 155.28 | [572] | 001-020 P(36) CO2 |
| NH3 | 216.44 | [332,543] | 001-100 P(06) CO2 |
| NH3 | 223.91 | [572] | 001-100 P(32) CO2 |
| NH3 | 225.39 | [332] | 001-100 P(04) CO2 |
| NH3 | 225.07 | [598] | 001-100 R(23) N2O |
| NH3 | 256.61 | [332,543] | 001-100 R(14) CO2 |
| NH3 | 263.43 | [241] | 001-100 P(13) N2O |
| NH3 | 263.43 | [598] | 001-100 R(23) N2O |
| NH3 | 280.5 | [316,543] | 001-100 R(08) CO2 |
| NH3 | 281.35 | [332] | 001-100 R(14) CO2 |
| NH3 | 281.48 | [332,316] | 001-100 R(08) CO2 |
| NH3 | 290.4 | [598] | 001-020 R(30) CO2 |
| NH3 | 290.9 | [539] | 002-021 P(17) CO2 |
| NH3 | 291.2 | [316] | 001-100 R(06) CO2 |
| NH3 | 291.35 | [332] | 001-100 P(32) CO2 |
| NH3 | 291.95 | [332,316] | 001-100 R(06) CO2 |
| NH3 | 301.2 | [598] | 001-020 R(06) CO2 |
| NH3 | 306.28 | [572] | 001-020 P(12) CO2 |
| NH3 | 311.75 | [332] | 001-100 R(04) CO2 |
| NH3 | 388 | [332] | 001-100 R(30) CO2 |
| NH3 | 404.69 | [598] | 001-100 P(08) N2O |
| | | | |
| N[15]H3 | 14.3 | [549] | 0- 2 P(06) HF |
| N[15]H3 | 14.8 | [549] | 0- 2 P(06) HF |
| N[15]H3 | 15.2 | [549] | 0- 2 P(06) HF |
| N[15]H3 | 15.7 | [549] | 0- 2 P(06) HF |

| WAVELENGTH IN AIR [MICROMETER] | | | PUMP TRANSITION |
|---|---|---|---|
| N[15]H3 | 16.0 | [549] | 0- 2 P(06) HF |
| N[15]H3 | 17.8 | [549] | 0- 2 P(06) HF |
| N[15]H3 | 89.68 | [543] | 001-020 R(10) CO2 |
| N[15]H3 | 375.9 | [514] | 001-100 R(42) CO2 |
| N[15]H3 | 218.9 | [514] | 002-101 P(35) CO2 |
| N[15]H3 | 111.9 | [514] | 002-101 P(31) CO2 |

NUMBER OF LINES IN AMMONIA 158

ACTIVE MEDIUM : PHOSPHINE
SYMBOL : PH3

OPERATING CONDITIONS : OPTICAL PUMPING WITH SINGLE CO2-LASER LINES IN PULSED OPERATION.

| WAVELENGTH IN AIR [MICROMETER] | | | PUMP TRANSITION |
|---|---|---|---|
| PH3 | ? 77.58 | [517] | 001-020 R(22) CO2 |
| PH3 | 83.77 | [517] | 001-100 R(34) CO2 |
| PH3 | 89.76 | [517] | 001-020 R(24) CO2 |
| PH3 | 89.80 | [517] | 001-020 R(32) CO2 |
| PH3 | 90.26 | [517] | 001-100 R(34) CO2 |
| PH3 | 97.19 | [517] | 001-020 R(24) CO2 |
| PH3 | 97.30 | [517] | 001-020 R(32) CO2 |
| PH3 | 102.62 | [517] | 001-020 R(18) CO2 |
| PH3 | ?104.4 | [517] | 001-020 R(14) CO2 |
| PH3 | 106.04 | [517] | 001-020 R(24) CO2 |
| PH3 | 106.05 | [517] | 001-020 R(32) CO2 |
| PH3 | 106.09 | [517] | 001-020 R(10) CO2 |
| PH3 | 106.23 | [517] | 001-020 R(18) CO2 |
| PH3 | ?109.7 | [517] | 001-020 P(24) CO2 |
| PH3 | 116.88 | [517] | 001-020 R(18) CO2 |
| PH3 | 116.88 | [517] | 001-020 R(10) CO2 |
| PH3 | 117.01 | [517] | 001-020 R(14) CO2 |
| PH3 | 121.45 | [517] | 001-020 P(18) CO2 |
| PH3 | 129.78 | [517] | 001-100 R(30) CO2 |
| PH3 | 129.98 | [517] | 001-020 R(18) CO2 |
| PH3 | 129.98 | [517] | 001-020 R(10) CO2 |
| PH3 | 130.14 | [517] | 001-020 R(14) CO2 |
| PH3 | 135.95 | [517] | 001-020 P(12) CO2 |
| PH3 | 136.71 | [517] | 001-020 P(10) CO2 |
| PH3 | 140.85 | [517] | 001-100 R(34) CO2 |
| PH3 | 145.88 | [517] | 001-020 P(14) CO2 |
| PH3 | 146.07 | [517] | 001-100 R(30) CO2 |
| PH3 | 146.07 | [517] | 001-100 R(32) CO2 |
| PH3 | 146.34 | [517] | 001-020 R(10) CO2 |
| PH3 | 155.07 | [517] | 001-020 P(12) CO2 |
| PH3 | 156.34 | [517] | 001-020 P(10) CO2 |
| PH3 | 166.73 | [517] | 001-020 P(22) CO2 |
| PH3 | 166.79 | [517] | 001-020 P(14) CO2 |
| PH3 | 166.84 | [517] | 001-100 R(34) CO2 |
| PH3 | 166.87 | [517] | 001-100 R(32) CO2 |

| WAVELENGTH IN AIR [MICROMETER] | | | PUMP TRANSITION |
|---|---|---|---|
| PH3 | 180.54 | [517] | 001-020 P(12) CO2 |
| PH3 | ?182 | [517] | 001-020 R(22) CO2 |
| PH3 | 186.25 | [517] | 001-020 P(06) CO2 |
| PH3 | 187.56 | [517] | 001-100 P(20) CO2 |
| PH3 | 194.47 | [517] | 001-020 P(22) CO2 |
| PH3 | 194.70 | [517] | 001-020 P(14) CO2 |
| PH3 | 194.89 | [517] | 001-100 R(32) CO2 |
| PH3 | 195.18 | [517] | 001-020 P(24) CO2 |
| PH3 | 223.07 | [517] | 001-020 P(06) CO2 |

NUMBER OF LINES IN PHOSPHINE 44

ACTIVE MEDIUM : NITROSYL CHLORIDE
SYMBOL : NOCL

OPERATING CONDITIONS : OPTICAL PUMPING WITH SINGLE CO2-LASER LINES.

| WAVELENGTH IN AIR [MICROMETER] | | | PUMP TRANSITION |
|---|---|---|---|
| NOCL | 16.4 | [468] | 001-100 P(26) CO2 |
| NOCL | 16.52 | [468] | 001-100 P(28) CO2 |
| NOCL | 16.57 | [468] | 001-100 P(30) CO2 |
| NOCL | 16.69 | [468] | 001-100 P(34) CO2 |
| NOCL | 16.7 | [468] | 001-100 P(36) CO2 |
| NOCL | 16.7 | [468] | 001-100 P(38) CO2 |
| NOCL | 16.75 | [468] | 001-100 P(40) CO2 |
| NOCL | 16.86 | [468] | 001-100 P(42) CO2 |
| NOCL | 16.9 | [468] | 001-100 P(44) CO2 |
| NOCL | 16.99 | [468] | 001-100 P(34) CO2 |

NUMBER OF LINES IN NITROSYL CHLORIDE 10

ACTIVE MEDIUM : CARBON TETRAFLUORO METHANE
SYMBOL : CF4

OPERATING CONDITIONS : OPTICAL PUMPING WITH SINGLE CO2-LASER LINES IN PULSED OPERATION. PULSE REPETITION FREQUENCY 100 HZ WITH AVERAGE POWER OF 0.2W [588]. SPECTROSCOPIC DATA AND ASSIGNMENTS [587,589].

| WAVELENGTH IN AIR [MICROMETER] | | | PUMP TRANSITION |
|---|---|---|---|
| CF4 | 15.33 | [587,468] | 001-020 P(10) CO2 |
| CF4 | 15.41 | [587,468] | 001-020 P(08) CO2 |
| CF4 | 15.49 | [587,468] | 001-020 P(06) CO2 |
| CF4 | 15.49 | [587] | 001-020 R(24) CO2 |

| | WAVELENGTH IN AIR [MICROMETER] | | PUMP TRANSITION |
|---|---|---|---|
| CF4 | 15.50 | [587] | 001-020 P(32) C[12]O[18]2 |
| CF4 | 15.55 | [587,468] | 001-020 P(04) CO2 |
| CF4 | 15.55 | [587,468] | 001-020 R(20) CO2 |
| CF4 | 15.56 | [587,468] | 001-020 P(04) CO2 |
| CF4 | 15.56 | [587] | 001-020 P(30) C[12]O[18]2 |
| CF4 | 15.58 | [587] | 001-020 R(18) CO2 |
| CF4 | 15.60 | [587] | 001-020 R(06) CO2 |
| CF4 | 15.61 | [587,468] | 001-020 R(16) CO2 |
| CF4 | 15.62 | [587] | 001-020 R(06) CO2 |
| CF4 | 15.62 | [587] | 001-020 P(20) C[12]O[18]2 |
| CF4 | 15.70 | [587] | 001-020 R(10) CO2 |
| CF4 | 15.74 | [587] | 001-020 P(26) C[12]O[18]2 |
| CF4 | @ 15.76 | [587] | 001-020 R(04) CO2 |
| CF4 | @ 15.77 | [587] | 001-020 P(22) C[12]O[18]2 |
| CF4 | 15.84 | [587] | 001-020 R(10) CO2 |
| CF4 | 15.84 | [587,468] | 001-020 P(12) CO2 |
| CF4 | ? 15.85 | [587] | 001-020 P(04) CO2 |
| CF4 | 15.85 | [587] | 001-020 R(06) CO2 |
| CF4 | 15.85 | [587,468] | 001-020 R(18) CO2 |
| CF4 | 15.85 | [587,468] | 001-020 R(22) CO2 |
| CF4 | 15.91 | [587] | 001-020 P(24) C[12]O[18]2 |
| CF4 | 15.94 | [587] | 001-020 P(28) C[12]O[18]2 |
| CF4 | 16.00 | [587] | 001-020 P(20) C[12]O[18]2 |
| CF4 | 16.03 | [587] | 001-020 P(06) CO2 |
| CF4 | @ 16.07 | [587] | 001-020 P(08) CO2 |
| CF4 | 16.10 | [587] | 001-020 P(18) C[12]O[18]2 |
| CF4 | 16.12 | [587] | 001-020 P(10) CO2 |
| CF4 | 16.18 | [587] | 001-020 R(10) CO2 |
| CF4 | 16.20 | [587] | 001-020 P(14) CO2 |
| CF4 | 16.24 | [587] | 001-020 R(12) CO2 |
| CF4 | 16.24 | [587] | 001-020 P(28) C[12]O[18]2 |
| CF4 | 16.26 | [587,468] | 001-020 R(12) CO2 |
| CF4 | 16.27 | [587] | 001-020 R(08) CO2 |
| CF4 | 16.31 | [587] | 001-020 R(14) CO2 |
| CF4 | 16.35 | [587] | 001-020 R(14) CO2 |
| CF4 | 16.40 | [587,468] | 001-020 P(04) CO2 |
| CF4 | 16.85 | [587] | 001-020 P(14) CO2 |

NUMBER OF LINES IN CARBON TETRAFLUORO METHANE 38

ACTIVE MEDIUM : TRIFLUOROBROMO METHANE
SYMBOL : CF3BR

OPERATING CONDITIONS : OPTICAL PUMPING WITH SINGLE CO2-LASER LINES.

| | WAVELENGTH IN AIR [MICROMETER] | | PUMP TRANSITION |
|---|---|---|---|
| CFBR | 823.4 | [599] | 001-020 R(08) CO2 |
| CFBR | 685.2 | [599] | 001-020 R(10) CO2 |

NUMBER OF LINES IN TRIFLUOROBROMO METHANE 2

ACTIVE MEDIUM : TRIFLUOROIODO METHANE
SYMBOL : CF3I

OPERATING CONDITIONS : OPTICAL PUMPING WITH SINGLE CO2-LASER LINES.

| WAVELENGTH IN AIR [MICROMETER] | | | PUMP TRANSITION |
|---|---|---|---|
| CF3I | 13.63 | [468] | 001-020 P(30) CO2 |
| CF3I | 13.57 | [468] | 001-020 P(34) CO2 |
| CF3I | 13.54 | [468] | 001-020 P(36) CO2 |

NUMBER OF LINES IN TRIFLUOROIODO METHANE 3

ACTIVE MEDIUM : ACETYLENE
SYMBOL : C2H2

OPERATING CONDITIONS : PULSED DISCHARGE IN A FLOWING MIXTURE OF C2H2+H2+HE. RATIO 1:2:20.
ALSO E-BEAM-CONTROLLED DISCHARGE EXCITATION OF CO AND SUBSEQUENT ENERGY TRANSFER TO C2H2 [456].
OPTICAL PUMPING WITH SINGLE CO2-LASER LINES.

| WAVELENGTH IN VACUUM [MICROMETER] | | | WAVELENGTH IN VACUUM [MICROMETER] | | |
|---|---|---|---|---|---|
| C2H2 | 8.0334 | [456] | C2H2 | 8.0360 | [258] |
| C2H2 | 8.0340 | [258] | C2H2 | 8.0409 | [258,456] |
| C2H2 | 8.0347 | [456] | C2H2 | 8.0442 | [258,456] |
| C2H2 | 8.0356 | [258,456] | | | |

| WAVELENGTH IN VACUUM [MICROMETER] | | | PUMP TRANSITION |
|---|---|---|---|
| C2D2 | 20.01 | [552] | 001-020 R(12) CO2 |
| C2D2 | 17.77 | [552] | 001-020 R(12) CO2 |
| C2D2 | 19.67 | [552] | 001-020 R(20) CO2 |
| C2D2 | 17.61 | [552] | 001-020 R(20) CO2 |
| C2D2 | 18.96 | [552] | 001-020 P(24) CO2 |
| C2D2 | 18.65 | [552] | 001-020 P(24) CO2 |
| C2D2 | 19.21 | [552] | 001-020 P(36) CO2 |
| C2D2 | 17.45 | [552] | 001-020 R(14) CO2 |
| C2D2 | 18.67 | [552] | 001-020 R(14) CO2 |
| C2D2 | 20.44 | [552] | 001-020 R(14) CO2 |
| C2D2 | 19.27 | [552] | 001-020 P(38) CO2 |
| C2D2 | 19.03 | [552] | 001-020 P(38) CO2 |
| C2D2 | 18.79 | [552] | 001-020 P(38) CO2 |
| C2D2 | 18.79 | [552] | 001-020 P(26) CO2 |
| C2D2 | 19.13 | [552] | 001-020 P(26) CO2 |

NUMBER OF LINES IN ACETYLENE 22

ACTIVE MEDIUM : ETHYLENE
SYMBOL : C2H4

OPERATING CONDITIONS : OPTICAL PUMPING WITH SINGLE CO2-LASER LINES OFF RESONANCE.

| WAVELENGTH IN AIR [MICROMETER] | | | PUMP TRANSITION |
|---|---|---|---|
| C2H4 | 10.53 | [465] | 001-100 R(10) CO2 |
| C2H4 | 10.98 | [465] | 001-100 R(16) CO2 |

NUMBER OF LINES IN ETHYLENE 2

---

ACTIVE MEDIUM : FORMALDEHYDE
SYMBOL : H2CO

OPERATING CONDITIONS : PULSED DISCHARGE IN A CLOSED DISCHARGE TUBE. INNER DIAMETER 270 MM.
TYPICAL PRESSURE 0.05-0.4 TORR.
OPTICAL PUMPING WITH SINGLE CO2-LASER LINES.

| WAVELENGTH IN VACUUM [MICROMETER] | | | WAVELENGTH IN VACUUM [MICROMETER] | | |
|---|---|---|---|---|---|
| H2CO | 101.9 | [301,412] | H2CO | 157.6 | [412] |
| H2CO | 119.6 | [301,412] | H2CO | 159.5 | [301,412] |
| H2CO | 122.8 | [301,412] | H2CO | 163.8 | [412] |
| H2CO | 125.9 | [412] | H2CO | 170.2 | [412] |
| H2CO | 155.1 | [412] | H2CO | 184.4 | [412] |

| WAVELENGTH IN VACUUM [MICROMETER] | | | PUMP TRANSITION |
|---|---|---|---|
| HDCO | 195 | [488] | 001-020 R(26) CO2 |
| HDCO | 196 | [488] | 001-020 P(08) CO2 |
| D2CO | 233 | [488] | 001-020 R(14) CO2 |
| D2CO | 245 | [488] | 001-020 R(24) CO2 |
| D2CO | 279 | [488] | 001-100 P(08) CO2 |
| D2CO | 733.5739 | [488] | 001-020 P(32) CO2 |
| D2CO | 752.6807 | [488] | 001-020 R(32) CO2 |

NUMBER OF LINES IN FORMALDEHYDE 17

---

ACTIVE MEDIUM : TRIOXANE
SYMBOL : [H2CO]3 = C3H6O3

OPERATING CONDITIONS : OPTICAL PUMPING WITH SINGLE CO2-LASER LINES IN CONTINUOUS OPERATION.

| WAVELENGTH IN VACUUM [MICROMETER] | | | PUMP TRANSITION |
|---|---|---|---|
| C3H6O3 | 384 | [488] | 001-020 R(30) CO2 |
| C3H6O3 | 433 | [488] | 001-100 P(44) CO2 |
| C3H6O3 | 460 | [488] | 001-020 R(22) CO2 |
| C3H6O3 | 512 | [488] | 001-100 P(40) CO2 |
| C3H6O3 | 619 | [488] | 001-100 R(22) CO2 |
| C3H6O3 | 680 | [488] | 001-100 P(34) CO2 |
| C3H6O3 | 696 | [488] | 001-020 R(16) CO2 |
| C3H6O3 | 712 | [488] | 001-020 R(32) CO2 |
| C3H6O3 | 750 | [488] | 001-100 P(18) CO2 |
| C3H6O3 | 815 | [488] | 001-020 P(32) CO2 |
| C3H6O3 | 890 | [488] | 001-020 R(26) CO2 |
| C3H6O3 | 891 | [488] | 001-020 R(20) CO2 |
| C3H6O3 | 948.9247 | [488] | 001-020 R(24) CO2 |

NUMBER OF LINES IN TRIOXANE 13

ACTIVE MEDIUM : FORMIC ACID
SYMBOL : HCOOH

OPERATING CONDITIONS : OPTICAL PUMPING WITH SINGLE CO2-LASER LINES. ASSIGNMENTS OF SOME LINES [575].

| WAVELENGTH IN VACUUM [MICROMETER] | | | PUMP TRANSITION |
|---|---|---|---|
| HCOOH | 229.39 | [317] | 001-020 R(32) CO2 |
| HCOOH | 254.60 | [317,334,548] | 001-020 P(20) CO2 |
| HCOOH | 278.61 | [317,334,548] | 001-020 P(30) CO2 |
| HCOOH | 302.08 | [317,334] | 001-020 P(08) CO2 |
| HCOOH | 302.2781 | [548] | 001-020 R(04) CO2 |
| HCOOH | 309.23 | [317,334] | 001-020 R(04) CO2 |
| HCOOH | 311.45 | [317,334,548] | 001-100 R(22) CO2 |
| HCOOH | 319.48 | [317] | 001-100 R(22) CO2 |
| HCOOH | 334.62 | [317] | 001-020 P(18) CO2 |
| HCOOH | 334.91 | [317] | 001-020 R(14) CO2 |
| HCOOH | 336 | [548] | 001-020 R(12) CO2 |
| HCOOH | 336.3 | [334] | 001-020 R(12) CO2 |
| HCOOH | 342.74 | [317] | 001-020 R(14) CO2 |
| HCOOH | 359.81 | [317] | 001-020 R(34) CO2 |
| HCOOH | 368 | [335] | 001-020 R(18) CO2 |
| HCOOH | 388 | [335] | 001-020 R(16) CO2 |
| HCOOH | 392 | [335] | 001-020 R(18) CO2 |
| HCOOH | 393.6311 | [548] | 001-020 R(18) CO2 |
| HCOOH | 393.6311 | [317,334,383] | 001-020 R(18) CO2 |
| HCOOH | 394.2 | [334] | 001-020 R(16) CO2 |
| HCOOH | 396 | [548] | 001-020 R(42) CO2 |
| HCOOH | 401 | [335] | 001-020 R(16) CO2 |
| HCOOH | 403 | [335] | 001-020 R(18) CO2 |
| HCOOH | 404.1 | [317] | 001-100 R(42) CO2 |
| HCOOH | 405.5848 | [317,548] | 001-020 R(18) CO2 |
| HCOOH | 405.75 | [317,548] | 001-020 P(26) CO2 |
| HCOOH | 406.0 | [334] | 001-020 R(16) CO2 |

| | WAVELENGTH IN VACUUM [MICROMETER] | | PUMP TRANSITION |
|---|---|---|---|
| HCOOH | 413 | [335] | 001-020 R(16) CO2 |
| HCOOH | 414 | [335] | 001-020 R(22) CO2 |
| HCOOH | 418.51 | [317] | 001-020 R(24) CO2 |
| HCOOH | 418.6 | [548] | 001-020 R(22) CO2 |
| HCOOH | 419.55 | [317] | 001-020 R(22) CO2 |
| HCOOH | 420.0 | [334] | 001-020 R(18) CO2 |
| HCOOH | 420.26 | [317] | 001-020 R(08) CO2 |
| HCOOH | 421 | [317] | 001-020 R(18) CO2 |
| HCOOH | 421.0 | [334,548] | 001-020 R(06) CO2 |
| HCOOH | 428 | [335] | 001-020 R(18) CO2 |
| HCOOH | 432.1093 | [548] | 001-020 R(22) CO2 |
| HCOOH | 432.6313 | [317,383] | 001-020 R(20) CO2 |
| HCOOH | 432.6325 | [548] | 001-020 R(20) CO2 |
| HCOOH | 433.10 | [317,334,346] | 001-020 R(22) CO2 |
| HCOOH | 435 | [335] | 001-020 P(16) CO2 |
| HCOOH | 437.70 | [317,334] | 001-020 P(16) CO2 |
| HCOOH | 438 | [548] | ?001-020 P(16) CO2 |
| HCOOH | 441 | [335] | 001-020 R(18) CO2 |
| HCOOH | 445.21 | [317] | 001-100 P(14) CO2 |
| HCOOH | 445.81 | [317] | 001-020 R(20) CO2 |
| HCOOH | 445.8971 | [548] | 001-020 R(20) CO2 |
| HCOOH | 446.5054 | [548] | 001-020 R(22) CO2 |
| HCOOH | 446.75 | [317] | 001-020 R(16) CO2 |
| HCOOH | 446.8730 | [548] | 001-020 R(16) CO2 |
| HCOOH | 447.0 | [334] | 001-020 R(14) CO2 |
| HCOOH | 447.58 | [317] | 001-020 R(22) CO2 |
| HCOOH | 458.43 | [317] | 001-020 R(36) CO2 |
| HCOOH | 458.5229 | [548] | 001-020 R(38) CO2 |
| HCOOH | 458.6 | [334] | 001-020 R(34) CO2 |
| HCOOH | 460.51 | [317] | 001-020 R(10) CO2 |
| HCOOH | 492 | [317] | 001-020 P(42) CO2 |
| HCOOH | 493.28 | [317] | 001-020 P(14) CO2 |
| HCOOH | 496 | [335] | 001-020 R(18) CO2 |
| HCOOH | 512.88 | [317,334,335] | 001-020 R(28) CO2 |
| HCOOH | 513.2 | [334] | 001-020 R(24) CO2 |
| HCOOH | 515.1690 | [548] | 001-020 P(16) CO2 |
| HCOOH | 518.83 | [317] | 001-020 P(16) CO2 |
| HCOOH | 530 | [335] | 001-020 R(26) CO2 |
| HCOOH | 530 | [335] | 001-020 R(28) CO2 |
| HCOOH | 533.6773 | [548] | 001-020 P(16) CO2 |
| HCOOH | 533.6773 | [548] | 001-020 R(28) CO2 |
| HCOOH | 534.5 | [334] | 001-020 P(18) CO2 |
| HCOOH | 534.8 | [334] | 001-020 R(24) CO2 |
| HCOOH | 577 | [335] | 001-020 P(38) CO2 |
| HCOOH | 580.3872 | [548] | 001-020 R(22) CO2 |
| HCOOH | 580.52 | [317] | 001-020 P(38) CO2 |
| HCOOH | 582.0 | [334,548] | 001-020 P(38) CO2 |
| HCOOH | 669.5308 | [548] | 001-020 R(30) CO2 |
| HCOOH | 670.0 | [334] | 001-020 R(26) CO2 |
| HCOOH | 742.5723 | [548] | 001-020 R(40) CO2 |
| HCOOH | 743.0 | [334] | 001-020 R(36) CO2 |
| HCOOH | 744.0503 | [548] | 001-020 R(24) CO2 |
| HCOOH | 745.0 | [334] | 001-020 R(20) CO2 |
| HCOOH | 761 | [346,548] | 001-020 R(24) CO2 |
| HCOOH | 785 | [346] | 001-020 R(10) CO2 |
| HCOOH | 786.1617 | [548] | 001-020 R(40) CO2 |

| WAVELENGTH IN VACUUM [MICROMETER] | | | PUMP TRANSITION |
|---|---|---|---|
| HCOOD | 462 | [548] | 001-100 P(16) CO2 |
| HCOOD | 925 | [548] | 001-100 R(32) CO2 |
| DCOOD | 305 | [548] | 001-100 R(24) CO2 |
| DCOOD | 349 | [548] | 001-100 R(40) CO2 |
| DCOOD | 381 | [548] | 001-100 R(12) CO2 |
| DCOOD | 492 | [548] | 001-100 P(08) CO2 |
| DCOOD | 526 | [548] | 001-100 P(34) CO2 |
| DCOOD | 569 | [548] | 001-100 R(26) CO2 |
| DCOOD | 790 | [548] | 001-100 R(20) CO2 |
| DCOOD | 937 | [548] | 001-020 P(16) CO2 |
| H13COOH | 260 | [548] | 001-020 P(16) CO2 |
| H13COOH | 313 | [548] | 001-020 P(06) CO2 |
| H13COOH | 448.5335 | [548] | 001-020 R(22) CO2 |
| H13COOH | 480 | [548] | 001-100 R(46) CO2 |
| H13COOH | 788.9192 | [548] | 001-020 P(12) CO2 |
| H13COOH | 1030.3782 | [548] | 001-020 R(30) CO2 |

NUMBER OF LINES IN FORMIC ACID 99

ACTIVE MEDIUM : METHYLENE FLUORIDE
SYMBOL : CH2F2

OPERATING CONDITIONS : OPTICAL PUMPING WITH SINGLE CO2-LASER LINES IN CW OPERATION. HIGH CONVERSION EFFICIENCY (30%). OUTPUT POWER UP TO 33 MILLIWATTS [591].

| WAVELENGTH IN AIR [MICROMETER] | | | PUMP TRANSITION |
|---|---|---|---|
| CH2F2 | 95.5 | [590,526,591] | 001-020 R(12) CO2 |
| CH2F2 | 105.5 | [590] | 001-020 P(16) CO2 |
| CH2F2 | 109.3 | [590,591] | 001-020 P(24) CO2 |
| CH2F2 | > 117.7 | [590,526,591] | 001-020 R(20) CO2 |
| CH2F2 | 121.7 | [526] | 001-020 R(22) CO2 |
| CH2F2 | > 122.4 | [590,526,591] | 001-020 R(22) CO2 |
| CH2F2 | 122.4 | [590] | 001-020 P(08) CO2 |
| CH2F2 | > 134.0 | [590] | 001-020 P(22) CO2 |
| CH2F2 | 135.3 | [590,591] | 001-020 P(24) CO2 |
| CH2F2 | > 158.5 | [590,591] | 001-020 P(10) CO2 |
| CH2F2 | 158.9 | [590] | 001-020 P(20) CO2 |
| CH2F2 | 165.8 | [526] | 001-020 R(22) CO2 |
| CH2F2 | 165.9 | [526] | 001-020 R(20) CO2 |
| CH2F2 | > 166.6 | [590,526] | 001-020 R(22) CO2 |
| CH2F2 | > 166.6 | [590,526,591] | 001-020 R(20) CO2 |
| CH2F2 | > 184.3 | [590,591] | 001-020 R(32) CO2 |
| CH2F2 | > 191.8 | [590] | 001-020 P(22) CO2 |
| CH2F2 | 193.9 | [590,526] | 001-020 R(22) CO2 |
| CH2F2 | 194.5 | [590,526,591] | 001-020 R(12) CO2 |
| CH2F2 | 202.5 | [590,526,591] | 001-020 R(06) CO2 |
| CH2F2 | > 214.5 | [590,591] | 001-020 R(34) CO2 |

| WAVELENGTH IN AIR [MICROMETER] | | | PUMP TRANSITION |
|---|---|---|---|
| CH2F2 | 227.6 | [590] | 001-020 P(18) CO2 |
| CH2F2 | > 230.1 | [590] | 001-020 R(42) CO2 |
| CH2F2 | > 235.7 | [590,591] | 001-020 R(32) CO2 |
| CH2F2 | > 236.5 | [590,526,591] | 001-020 R(06) CO2 |
| CH2F2 | 255.9 | [590,591] | 001-020 P(24) CO2 |
| CH2F2 | 261.7 | [590,591] | 001-020 P(38) CO2 |
| CH2F2 | 270.0 | [590,526] | 001-020 R(22) CO2 |
| CH2F2 | > 272.2 | [590,591] | 001-020 P(10) CO2 |
| CH2F2 | > 287.7 | [590,591] | 001-020 R(34) CO2 |
| CH2F2 | > 289.4 | [590] | 001-020 P(04) CO2 |
| CH2F2 | 293.9 | [590] | 001-020 P(20) CO2 |
| CH2F2 | 298.2 | [590] | 001-020 R(36) CO2 |
| CH2F2 | 326.5 | [590,526,591] | 001-020 R(14) CO2 |
| CH2F2 | 355.2 | [590] | 001-020 P(08) CO2 |
| CH2F2 | 381.8 | [590] | 001-020 R(36) CO2 |
| CH2F2 | > 382.9 | [590,591] | 001-020 P(10) CO2 |
| CH2F2 | > 394.7 | [590] | 001-020 P(06) CO2 |
| CH2F2 | 418.1 | [590,526,591] | 001-020 R(12) CO2 |
| CH2F2 | 432.4 | [526] | 001-020 R(06) CO2 |
| CH2F2 | · 434.9 | [590,526] | 001-020 R(06) CO2 |
| CH2F2 | 464.5 | [590] | 001-020 P(06) CO2 |
| CH2F2 | 503.6 | [590] | 001-020 R(06) CO2 |
| CH2F2 | 511.3 | [590,526] | 001-020 R(28) CO2 |
| CH2F2 | 540.8 | [590] | 001-020 R(42) CO2 |
| CH2F2 | 567.5 | [590] | 001-020 R(28) CO2 |
| CH2F2 | 588.1 | [590] | 001-020 R(46) CO2 |
| CH2F2 | 642.5 | [590] | 001-020 R(44) CO2 |
| CH2F2 | 657.2 | [590] | 001-020 P(10) CO2 |
| CH2F2 | 725.1 | [590] | 001-020 P(04) CO2 |
| CH2F2 | 1448.1 | [590] | 001-020 R(44) CO2 |

NUMBER OF LINES IN METHYLENE FLUORIDE 51

ACTIVE MEDIUM : METHYLENE CHLORIDE
SYMBOL : CH2CL2

OPERATING CONDITIONS : OPTICAL PUMPING WITH SINGLE CO2-LASER LINES IN CONTINUOUS OPERATION.

| WAVELENGTH IN VACUUM [MICROMETER] | | | PUMP TRANSITION |
|---|---|---|---|
| CD2CL2 | 249 | [525] | 001-100 R(16) CO2 |
| CD2CL2 | 254 | [525] | 001-100 R(36) CO2 |
| CH2CL2 | 258 | [341] | 001-100 P(26) CO2 |
| CD2CL2 | 342 | [525] | 001-100 P(16) CO2 |
| CD2CL2 | 469 | [525] | 001-100 R(04) CO2 |
| CD2CL2 | 520 | [525] | 001-100 R(12) CO2 |
| CD2CL2 | 631 | [525] | 001-100 R(18) CO2 |
| CD2CL2 | 829 | [525] | 001-100 P(06) CO2 |

NUMBER OF LINES IN METHYLENE CHLORIDE 8

ACTIVE MEDIUM : METHYL FLUORIDE
SYMBOL : CH3F

OPERATING CONDITIONS : OPTICAL PUMPING WITH SINGLE CO2-LASER LINES IN PULSED OPERATION. THE LINES AT 496 MICROMETER ALSO CONTINUOUSLY [372]. PEAK POWER 600 KW [344]. THE LINE AT 1.2 MM IN C[13]H3F. EMISSION WITH SMALL LINE WIDTH [333,348]. CW OPERATION IN A STARK-TUNED WAVEGUIDE RESONATOR [593].

| WAVELENGTH IN VACUUM [MICROMETER] | | | PUMP TRANSITION |
|---|---|---|---|
| CH3F | 9.75 | [533] | TWO PHOTON PUMPING |
| CH3F | 190.3 | [316] | 001-020 P(30) CO2 |
| CH3F | 192.78 | [271] | 001-100 R(32) CO2 |
| CH3F | 195.0 | [316] | 001-020 R(42) CO2 |
| CH3F | 196 | [611] | 001-020 R(10) C[12]O[18]2 |
| CH3F | 199.14 | [271] | 001-100 R(32) CO2 |
| CH3F | 200.3 | [316] | 001-020 R(40) CO2 |
| CH3F | 215.3 | [316] | 001-020 R(36) CO2 |
| CH3F | 251.91 | [271] | 001-100 R(34) CO2 |
| CH3F | 372.68 | [271] | 001-020 P(50) CO2 |
| CH3F | 397.51 | [271] | 001-020 P(50) CO2 |
| CH3F | 419 | [333] | 001-020 P(20) CO2 |
| CH3F | 451.903 | [239] | 001-020 P(20) CO2 |
| CH3F | 451.924 | [239] | 001-020 P(20) CO2 |
| CH3F | 494 | [611] | 001-020 P(22) C[12]O[18]2 |
| CH3F | > 496.072 | [239] | 001-020 P(20) CO2 |
| CH3F | > 496.1009 | [239,383] | 001-020 P(20) CO2 |
| CH3F | 541.113 | [239] | 001-020 P(20) CO2 |
| CH3F | 541.147 | [239] | 001-020 P(20) CO2 |
| CH3F | 595 | [333] | 001-020 P(20) CO2 |
| CH3F | 1221.79 | [271,340] | 001-020 P(32) CO2 |

NUMBER OF LINES IN METHYL FLUORIDE 21

ACTIVE MEDIUM : METHYL CHLORIDE
SYMBOL : CH3CL

OPERATING CONDITIONS : OPTICAL PUMPING WITH SINGLE CO2-LASER LINES.

ASSIGNMENT OF LINES [535].

| WAVELENGTH IN VACUUM [MICROMETER] | | | PUMP TRANSITION |
|---|---|---|---|
| CH3CL | 227.15 | [369] | 001-020 P(48) CO2 |
| CH3CL | 236.25 | [369] | 001-020 R(02) CO2 |
| CH3CL | 240.98 | [369] | 001-100 P(10) CO2 |
| CH3CL | 250.4 | [316] | 001-020 P(30) CO2 |
| CH3CL | 254 | [341] | 001-100 P(10) CO2 |
| CH3CL | 261.03 | [369] | 001-100 P(34) CO2 |
| CH3CL | 271.29 | [369] | 001-100 P(20) CO2 |
| CH3CL | 273.7 | [316] | 001-020 P(12) CO2 |

| WAVELENGTH IN VACUUM [MICROMETER] | | | PUMP TRANSITION |
|---|---|---|---|
| CH3CL | 275.00 | [369] | 001-020 R(14) CO2 |
| CH3CL | 275.09 | [369] | 001-020 R(36) CO2 |
| CH3CL | 281.67 | [369] | 001-020 R(14) CO2 |
| CH3CL | 286.79 | [369] | 001-100 R(34) CO2 |
| CH3CL | 307.65 | [369] | 011-110 P(19) CO2 |
| CH3CL | 333.96 | [369] | 001-020 P(42) CO2 |
| CH3CL | 349.34 | [369] | 001-100 R(18) CO2 |
| CH3CL | 354 | [341] | 001-100 R(18) CO2 |
| CH3CL | 364.5 | [316] | 001-020 R(16) CO2 |
| CH3CL | 378.57 | [369] | 001-020 R(16) CO2 |
| CH3CL | 397.6 | [316] | 001-100 R(24) CO2 |
| CH3CL | 461.20 | [369] | 001-020 R(42) CO2 |
| CH3CL | 511.90 | [369] | 001-100 R(52) CO2 |
| CH3CL | 568.81 | [369] | 001-100 R(26) CO2 |
| CH3CL | 870.80 | [369] | 001-020 P(52) CO2 |
| CH3CL | 943.97 | [369] | 001-020 R(12) CO2 |
| CH3CL | 958.25 | [369] | 001-020 P(38) CO2 |
| CH3CL | 968 | [341] | 001-020 R(12) CO2 |
| CH3CL | 1886.87 | [369] | 001-020 P(26) CO2 |
| CD3CL | 224 | [416] | 001-020 R(28) CO2 |
| CD3CL | 245 | [416] | 001-020 P(32) CO2 |
| CD3CL | 246 | [416] | 001-100 R(14) CO2 |
| CD3CL | 249 | [416] | 001-020 P(18) CO2 |
| CD3CL | 288 | [416] | 001-100 R(18) CO2 |
| CD3CL | 288 | [416] | 001-020 P(16) CO2 |
| CD3CL | 291.27 | [416] | 001-020 P(24) CO2 |
| CD3CL | 318 | [416] | 001-100 R(28) CO2 |
| CD3CL | 383.28 | [416] | 001-020 R(34) CO2 |
| CD3CL | 443.26 | [416] | 001-020 P(10) CO2 |
| CD3CL | 449.79 | [416] | 001-100 R(20) CO2 |
| CD3CL | 464.76 | [416] | 001-100 R(20) CO2 |
| CD3CL | 480.31 | [416] | 001-020 P(36) CO2 |
| CD3CL | 519.30 | [416] | 001-020 P(36) CO2 |
| CD3CL | 698.55 | [416] | 001-020 P(06) CO2 |
| CD3CL | 735.12 | [416] | 001-020 P(06) CO2 |
| CD3CL | 883.59 | [416] | 001-020 P(34) CO2 |
| CD3CL | 1239.47 | [416] | 001-020 P(12) CO2 |
| CD3CL | 1990.75 | [416] | 001-020 P(14) CO2 |

NUMBER OF LINES IN METHYL CHLORIDE 46

---

ACTIVE MEDIUM : METHYL BROMIDE
SYMBOL : CH3BR

OPERATING CONDITIONS : OPTICAL PUMPING WITH SINGLE CO2-LASER LINES.

| | WAVELENGTH IN AIR [MICROMETER] | | PUMP TRANSITION |
|---|---|---|---|
| CH3BR | 245.04 | [369] | 001-020 P(28) CO2 |
| CH3BR | 264.05 | [369] | 001-100 R(10) CO2 |
| CH3BR | 279.81 | [369] | 001-100 R(52) CO2 |
| CH3BR | 294.28 | [369] | 001-100 R(28) CO2 |
| CH3BR | 311.07 | [369] | 001-100 R(12) CO2 |
| CH3BR | 311.10 | [369] | 001-100 R(20) CO2 |
| CH3BR | 311.20 | [369] | 001-100 P(40) CO2 |
| CH3BR | 311.21 | [369] | 001-100 R(50) CO2 |
| CH3BR | 332.86 | [369] | 001-100 R(06) CO2 |
| CH3BR | 333.15 | [369] | 001-100 P(08) CO2 |
| CH3BR | 352.75 | [369] | 001-020 P(18) CO2 |
| CH3BR | 380.02 | [369] | 001-100 R(18) CO2 |
| CH3BR | 407.72 | [369] | 001-020 P(28) CO2 |
| CH3BR | 414.98 | [369] | 001-100 R(02) CO2 |
| CH3BR | 418.31 | [369] | 001-100 P(26) CO2 |
| CH3BR | 422.78 | [369] | 001-100 R(26) CO2 |
| CH3BR | 508.48 | [369] | 001-100 R(42) CO2 |
| CH3BR | 531.06 | [369] | 001-100 P(24) CO2 |
| CH3BR | 545.21 | [369] | 001-100 P(38) CO2 |
| CH3BR | 545.39 | [369] | 001-100 R(32) CO2 |
| CH3BR | 564.68 | [369] | 001-100 P(38) CO2 |
| CH3BR | 585.72 | [369] | 001-020 P(40) CO2 |
| CH3BR | 631.93 | [369] | 001-100 P(16) CO2 |
| CH3BR | 632.00 | [369] | 001-100 P(22) CO2 |
| CH3BR | 658.53 | [369] | 001-020 P(56) CO2 |
| CH3BR | 660.70 | [369] | 001-100 R(20) CO2 |
| CH3BR | 715.40 | [369] | 001-100 R(14) CO2 |
| CH3BR | 749.29 | [369] | 001-100 P(14) CO2 |
| CH3BR | 749.36 | [369] | 001-100 R(14) CO2 |
| CH3BR | 831.13 | [369] | 001-100 P(28) CO2 |
| CH3BR | 925.52 | [369] | 001-100 R(46) CO2 |
| CH3BR | 990.15 | [369] | 001-100 P(10) CO2 |
| CH3BR | 1310.38 | [369] | 001-100 R(04) CO2 |
| CH3BR | 1572.64 | [369] | 001-100 P(04) CO2 |
| CH3BR | 1965.34 | [369] | 001-100 P(28) CO2 |

NUMBER OF LINES IN METHYL BROMIDE 35

ACTIVE MEDIUM : METHYL IODIDE
SYMBOL : CH3I

OPERATING CONDITIONS : OPTICAL PUMPING WITH SINGLE CO2-LASER LINES.

| | WAVELENGTH IN VACUUM [MICROMETER] | | PUMP TRANSITION |
|---|---|---|---|
| CH3I | 377.45 | [416,369] | 001-020 R(16) CO2 |
| CH3I | 390.53 | [369,416] | 001-100 P(42) CO2 |
| CH3I | 392.48 | [369] | 001-020 R(14) CO2 |
| CH3I | 447.1424 | [416,369,383] | 001-100 P(18) CO2 |
| CH3I | 457.25 | [369] | 001-100 P(18) CO2 |
| CH3I | 459.18 | [369] | 001-100 P(08) CO2 |

| WAVELENGTH IN VACUUM [MICROMETER] | | | PUMP TRANSITION |
|---|---|---|---|
| CH3I | 477.67 | [369] | 001-020 P(26) CO2 |
| CH3I | 508.37 | [369] | 001-020 P(34) CO2 |
| CH3I | 517.33 | [369] | 001-100 P(14) CO2 |
| CH3I | 525.32 | [369] | 001-020 P(04) CO2 |
| CH3I | 529.28 | [369] | 001-100 P(36) CO2 |
| CH3I | 542.99 | [369,416] | 001-100 P(26) CO2 |
| CH3I | 576.17 | [369,416] | 001-100 P(16) CO2 |
| CH3I | 578.90 | [369] | 001-100 R(34) CO2 |
| CH3I | 583.87 | [369] | 001-020 P(04) CO2 |
| CH3I | 639.73 | [369] | 001-020 P(06) CO2 |
| CH3I | 670.99 | [369] | 001-100 P(28) CO2 |
| CH3I | 719.30 | [369,416] | 001-100 P(22) CO2 |
| CH3I | 964 | [416] | 001-100 P(22) CO2 |
| CH3I | 1063.29 | [369,416] | 001-100 P(38) CO2 |
| CH3I | 1253.738 | [416,369,383] | 001-100 P(32) CO2 |
| CD3I | 272 | [416] | 001-020 P(12) CO2 |
| CD3I | 301 | [416] | 001-020 R(26) CO2 |
| CD3I | 390 | [416] | 001-020 P(26) CO2 |
| CD3I | 433.1038 | [416] | 001-020 P(28) CO2 |
| CD3I | 444.3862 | [416] | 001-020 R(32) CO2 |
| CD3I | 460.5619 | [416] | 001-020 R(12) CO2 |
| CD3I | 487.2260 | [416] | 001-020 P(10) CO2 |
| CD3I | 490.3909 | [416] | 001-020 R(22) CO2 |
| CD3I | 523.4061 | [416] | 001-100 P(38) CO2 |
| CD3I | 540 | [416] | 001-020 R(06) CO2 |
| CD3I | 556.6755 | [416] | 001-100 P(36) CO2 |
| CD3I | 569.4773 | [416] | 001-100 P(36) CO2 |
| CD3I | 599.5499 | [416] | 001-100 R(22) CO2 |
| CD3I | 614.1098 | [416] | 001-100 R(22) CO2 |
| CD3I | 640 | [416] | 001-100 R(18) CO2 |
| CD3I | 644 | [416] | 001-100 P(16) CO2 |
| CD3I | 660.5822 | [416] | 001-100 P(46) CO2 |
| CD3I | 667.2322 | [416] | 001-100 P(10) CO2 |
| CD3I | 670.0940 | [416] | 001-100 R(08) CO2 |
| CD3I | 670.1143 | [416] | 001-100 R(08) CO2 |
| CD3I | 691.1292 | [416] | 001-020 R(20) CO2 |
| CD3I | 730.3234 | [416] | 001-020 R(28) CO2 |
| CD3I | 734.2624 | [416] | 001-020 P(22) CO2 |
| CD3I | 745 | [416] | 001-100 P(08) CO2 |
| CD3I | 786.48 | [416] | 001-100 P(12) CO2 |
| CD3I | 895 | [416] | 001-100 P(30) CO2 |
| CD3I | 918.6101 | [416] | 001-020 R(28) CO2 |
| CD3I | 953.6799 | [416] | 001-020 R(28) CO2 |
| CD3I | 981.7094 | [416] | 001-100 P(22) CO2 |
| CD3I | 1005.3476 | [416] | 001-100 P(34) CO2 |
| CD3I | 1099.5441 | [416] | 001-100 P(22) CO2 |
| CD3I | 1549.5048 | [416] | 001-020 R(10) CO2 |
| C13D3I | 690 | [340] | 001-100 P(10) CO2 |
| C13D3I | 806 | [340] | 001-100 P(12) CO2 |

NUMBER OF LINES IN METHYL IODIDE 55

ACTIVE MEDIUM : METHYL ALCOHOL
SYMBOL : CH3OH

OPERATING CONDITIONS : OPTICAL PUMPING WITH SINGLE CO2-LASER LINES IN CONTINUOUS OPERATION. ASSIGNMENTS OF CH3OH LINES AT [458,495]. AT 118 MICRONS 400 MILLIWATTS OUTPUT [463].
CH3OH* = C[13]H3OH.

| | WAVELENGTH IN VACUUM [MICROMETER] | | PUMP TRANSITION |
|---|---|---|---|
| CH3OH | 37.5 | [347] | 001-020 P(32) CO2 |
| CH3OH | 40.2 | [347] | 001-020 P(34) CO2 |
| CH3OH | 42.18 | [347,555] | 001-020 P(32) CO2 |
| CH3OH | 43.4 | [347] | 001-020 P(34) CO2 |
| CH3OH | 43.47 | [555] | 001-100 R(34) CO2 |
| CH3OH | 55.39 | [555] | 001-020 P(40) CO2 |
| CH3OH | 58.1 | [349] | 001-020 P(14) CO2 |
| CH3OH | 60.25 | [555] | 001-020 P(40) CO2 |
| CH3OH | 65.1 | [349] | 001-020 R(18) CO2 |
| CH3OH | 65.6 | [347] | 001-020 P(34) CO2 |
| CH3OH | 69.70 | [555] | 001-100 R(16) CO2 |
| CH3OH | 70.511716 | [239,347,382] | 001-020 P(34) CO2 |
| CH3OH | 73.30 | [555] | 001-020 P(40) CO2 |
| CH3OH | 77.92 | [555] | 001-100 R(16) CO2 |
| CH3OH | 80.3 | [347] | 001-020 P(34) CO2 |
| CH3OH | 80.6 | [523] | 002-021 P(21) CO2 |
| CH3OH | 85.59 | [555] | 001-020 P(40) CO2 |
| CH3OH | 92.60 | [555] | 001-020 P(24) CO2 |
| CH3OH | 92.69 | [555] | 001-100 R(34) CO2 |
| CH3OH | 96.522394 | [317,346,382] | 001-020 R(10) CO2 |
| CH3OH | 97.48 | [555] | 001-100 R(40) CO2 |
| CH3OH | 117.95 | [555] | 001-020 P(14) CO2 |
| CH3OH | 118.83409 | [239,347,382] | 001-020 P(36) CO2 |
| CH3OH | 129.5497 | [414,555] | 001-100 R(34) CO2 |
| CH3OH | 133.1196 | [555] | 001-020 P(24) CO2 |
| CH3OH | 151.35 | [317] | 001-020 R(26) CO2 |
| CH3OH | 159.2 | [523] | 002-021 P(31) CO2 |
| CH3OH | 162 | [346] | 001-100 R(38) CO2 |
| CH3OH | 163.03353 | [317,382] | 001-100 R(34) CO2 |
| CH3OH | 163.9 | [316] | 001-020 P(12) CO2 |
| CH3OH | 164 | [559] | 001-100 R(38) CO2 |
| CH3OH | 164.3 | [239] | 001-020 P(16) CO2 |
| CH3OH | 164.5076 | [555] | 001-020 P(14) CO2 |
| CH3OH | 164.77 | [555] | 001-020 P(24) CO2 |
| CH3OH | 164.7832 | [414,555] | 001-020 R(10) CO2 |
| CH3OH | 170.57638 | [239,347,382] | 001-020 P(36) CO2 |
| CH3OH | 171.3 | [523] | 002-021 P(21) CO2 |
| CH3OH | 185.5 | [239] | 001-020 P(34) CO2 |
| CH3OH | 186.03 | [317,555] | 001-020 R(18) CO2 |
| CH3OH | 190.3209 | [239,383] | 001-020 P(34) CO2 |
| CH3OH | 191.2 | [317] | 001-100 R(04) CO2 |
| CH3OH | 191.57 | [317] | 001-100 R(10) CO2 |
| CH3OH | 191.58 | [555] | 001-100 R(10) CO2 |
| CH3OH | 191.63 | [317] | 001-020 R(16) CO2 |
| CH3OH | 193.2 | [239,347] | 001-020 P(38) CO2 |
| CH3OH | 194.01 | [555] | 001-020 R(14) CO2 |
| CH3OH | 198.8 | [239,347] | 001-020 P(38) CO2 |

| | WAVELENGTH IN VACUUM [MICROMETER] | | PUMP TRANSITION |
|---|---|---|---|
| CH3OH | 202.4 | [239] | 001-020 P(36) CO2 |
| CH3OH | 205.3 | [316] | 001-020 P(18) CO2 |
| CH3OH | 206.90 | [555] | 001-020 P(12) CO2 |
| CH3OH | 209.89 | [555] | 001-020 R(14) CO2 |
| CH3OH | 211.25 | [555] | 001-020 P(12) CO2 |
| CH3OH | 214.35 | [555] | 001-020 P(10) CO2 |
| CH3OH | 218.22 | [555] | 001-020 P(10) CO2 |
| CH3OH | 223.5 | [239] | 001-020 P(16) CO2 |
| CH3OH | 232.85 | [555] | 001-020 R(22) CO2 |
| CH3OH | 232.93906 | [317,346,382] | 001-020 R(08) CO2 |
| CH3OH | 237.6 | [239] | 001-020 P(34) CO2 |
| CH3OH | 242.4727 | [555] | 001-100 R(34) CO2 |
| CH3OH | 242.79 | [317] | 001-100 R(32) CO2 |
| CH3OH | 246 | [559] | 001-100 R(38) CO2 |
| CH3OH | 250.78129 | [346,382,555] | 001-100 R(34) CO2 |
| CH3OH | 251.13983 | [382] | 001-100 R(38) CO2 |
| CH3OH | 251.56 | [555] | 001-020 R(18) CO2 |
| CH3OH | 253.6 | [239] | 001-020 P(34) CO2 |
| CH3OH | 254.1 | [239,346] | 001-020 P(34) CO2 |
| CH3OH | 263.7 | [239] | 001-020 P(34) CO2 |
| CH3OH | 264.6 | [239] | 001-020 P(34) CO2 |
| CH3OH | 267.4432 | [555] | 001-100 R(34) CO2 |
| CH3OH | 278.8 | [239] | 001-020 P(38) CO2 |
| CH3OH | 280.96 | [555] | 001-020 R(18) CO2 |
| CH3OH | 290.62 | [555] | 001-020 P(12) CO2 |
| CH3OH | 292.2 | [239,346] | 001-020 P(38) CO2 |
| CH3OH | 292.5 | [239] | 001-020 P(34) CO2 |
| CH3OH | 293.78 | [555] | 001-100 R(10) CO2 |
| CH3OH | 301.9943 | [555] | 001-020 P(14) CO2 |
| CH3OH | 369.11368 | [239,382] | 001-020 P(16) CO2 |
| CH3OH | 386.20 | [555] | 001-020 P(14) CO2 |
| CH3OH | 390.1 | [523] | 002-021 P(13) CO2 |
| CH3OH | 392.06871 | [239,347,382] | 001-020 P(36) CO2 |
| CH3OH | 416.5224 | [555] | 001-020 P(14) CO2 |
| CH3OH | 417.8 | [239] | 001-020 P(36) CO2 |
| CH3OH | 451.9 | [349] | 001-020 P(12) CO2 |
| CH3OH | 469.02330 | [382] | 001-100 R(38) CO2 |
| CH3OH | 470 | [559] | 001-100 R(38) CO2 |
| CH3OH | 471 | [346] | 001-100 R(38) CO2 |
| CH3OH | 486.1 | [523] | 002-021 P(13) CO2 |
| CH3OH | 570.56864 | [239,347,382] | 001-020 P(16) CO2 |
| CH3OH | 603.06 | [555] | 001-020 P(24) CO2 |
| CH3OH | 614.92 | [555] | 001-020 P(24) CO2 |
| CH3OH | 627.34 | [317] | 001-020 P(16) CO2 |
| CH3OH | 694.17 | [555] | 001-020 P(24) CO2 |
| CH3OH | 695 | [555] | 001-100 R(16) CO2 |
| CH3OH | 699.42258 | [239,347,382] | 001-020 P(34) CO2 |
| | | | |
| CH3OH* | 85.31729 | [617] | 001-020 P(22) CO2 |
| CH3OH* | 86.11179 | [617] | 001-020 P(10) CO2 |
| CH3OH* | 103.48081 | [617] | 001-020 P(22) CO2 |
| CH3OH* | 115.82324 | [617] | 001-100 R(16) CO2 |
| CH3OH* | 118.01314 | [617] | 001-020 P(22) CO2 |
| CH3OH* | 146.09739 | [617] | 001-020 P(10) CO2 |
| CH3OH* | 148.59041 | [617] | 001-100 R(16) CO2 |
| CH3OH* | 149.27226 | [617] | 001-020 P(22) CO2 |
| CH3OH* | 152.07568 | [617] | 001-100 R(16) CO2 |

| WAVELENGTH IN VACUUM [MICROMETER] | | | PUMP TRANSITION |
|---|---|---|---|
| CH3OH* | 157.92848 | [617] | 001-020 P(12) CO2 |
| CH3OH* | 203.63578 | [617] | 001-100 R(16) CO2 |
| CH3OH* | 208.41205 | [617] | 001-020 P(10) CO2 |
| CH3OH* | 238.52268 | [617] | 001-020 P(12) CO2 |
| CH3OH* | 268.57203 | [617] | 001-100 R(16) CO2 |
| CH3OH* | 280.21826 | [617] | 001-100 R(16) CO2 |
| CH3OH* | 280.23974 | [617] | 001-100 R(16) CO2 |
| CH3OH* | 332.6033 | [617] | 001-100 R(16) CO2 |
| CH3OH* | 338.9637 | [617] | 001-020 P(22) CO2 |
| CH3OH* | 461.3848 | [617] | 001-020 P(12) CO2 |
| CD3OH | 34.8 | [493] | 001-100 P(22) CO2 |
| CD3OH | 37.6 | [493] | 001-100 R(34) CO2 |
| CD3OH | 40.1 | [493] | 001-100 P(22) CO2 |
| CD3OH | 41.5 | [493] | 001-100 R(08) CO2 |
| CD3OH | 41.8 | [493] | 001-100 R(18) CO2 |
| CD3OH | 43.9 | [493] | 001-100 R(18) CO2 |
| CD3OH | 49.8 | [550] | 001-020 R(28) CO2 |
| CD3OH | 52.9 | [550] | 001-020 R(34) CO2 |
| CD3OH | 60.8 | [550] | 001-020 R(34) CO2 |
| CD3OH | 71.0 | [493] | 001-100 R(08) CO2 |
| CD3OH | 76.1 | [493] | 001-100 P(32) CO2 |
| CD3OH | 81.2 | [493] | 001-100 R(16) CO2 |
| CD3OH | 86.4 | [493] | 001-100 R(16) CO2 |
| CD3OH | 102.6 | [493] | 001-100 R(34) CO2 |
| CD3OH | 112.3 | [493] | 001-100 R(34) CO2 |
| CD3OH | 128.7 | [493] | 001-100 R(34) CO2 |
| CD3OH | 144.0 | [493] | 001-100 P(18) CO2 |
| CD3OH | 158 | [550] | 001-020 R(28) CO2 |
| CD3OH | 179 | [550] | 001-020 R(14) CO2 |
| CD3OH | 182.4 | [493] | 001-100 R(34) CO2 |
| CD3OH | 184 | [494] | 001-020 R(08) CO2 |
| CD3OH | 191.9 | [493] | 001-100 R(34) CO2 |
| CD3OH | 201 | [550] | 001-020 P(40) CO2 |
| CD3OH | 219.9 | [493] | 001-100 R(18) CO2 |
| CD3OH | 222 | [494] | 001-020 P(06) CO2 |
| CD3OH | 223 | [550] | 001-020 P(08) CO2 |
| CD3OH | 232 | [494] | 001-020 R(34) CO2 |
| CD3OH | 236 | [550] | 001-020 R(14) CO2 |
| CD3OH | 238.3 | [493] | 001-100 P(24) CO2 |
| CD3OH | 253.2 | [493] | 001-100 R(36) CO2 |
| CD3OH | 258.7 | [493,494] | 001-100 P(22) CO2 |
| CD3OH | 266 | [494] | 001-020 P(20) CO2 |
| CD3OH | 266.2 | [493,494] | 001-100 R(34) CO2 |
| CD3OH | 267 | [494] | 001-100 R(14) CO2 |
| CD3OH | 268 | [494] | 001-020 P(14) CO2 |
| CD3OH | 276 | [494] | 001-100 P(28) CO2 |
| CD3OH | 277 | [494] | 001-020 R(26) CO2 |
| CD3OH | 278 | [494] | 001-100 R(24] CO2 |
| CD3OH | 285 | [494] | 001-020 P(40) CO2 |
| CD3OH | 286.6 | [493,494] | 001-100 P(24) CO2 |
| CD3OH | 287.4 | [493,494] | 001-100 P(18) CO2 |
| CD3OH | 290.0 | [493] | 001-100 P(18) CO2 |
| CD3OH | 297 | [494] | 001-100 R(34) CO2 |
| CD3OH | 297 | [494] | 001-020 R(18) CO2 |
| CD3OH | 299 | [494] | 001-020 R(06) CO2 |
| CD3OH | 309 | [494] | 001-100 P(20) CO2 |

| | WAVELENGTH IN VACUUM [MICROMETER] | | PUMP TRANSITION |
|---|---|---|---|
| CD3OH | 310 | [494] | 001-100 R(28) CO2 |
| CD3OH | 321 | [494] | 001-020 R(16) CO2 |
| CD3OH | 336 | [494] | 001-020 R(36) CO2 |
| CD3OH | 336 | [494] | 001-100 R(30) CO2 |
| CD3OH | 346 | [550] | 001-020 R(14) CO2 |
| CD3OH | 350 | [494] | 001-020 P(32) CO2 |
| CD3OH | 351 | [494] | 001-100 R(30) CO2 |
| CD3OH | 351 | [550,494] | 001-020 P(32) CO2 |
| CD3OH | 352 | [494] | 001-020 R(28) CO2 |
| CD3OH | 352 | [494] | 001-020 R(14) CO2 |
| CD3OH | 353 | [494] | ? |
| CD3OH | 370 | [494] | 001-020 R(26) CO2 |
| CD3OH | 370 | [494] | 001-020 P(28) CO2 |
| CD3OH | 385 | [494] | 001-020 P(30) CO2 |
| CD3OH | 386 | [494] | 001-020 P(16) CO2 |
| CD3OH | 398 | [494] | 001-100 R(28) CO2 |
| CD3OH | 407 | [494] | 001-020 R(44) CO2 |
| CD3OH | 409 | [494] | 001-020 R(34) CO2 |
| CD3OH | 410 | [494] | 001-020 P(32) CO2 |
| CD3OH | 412 | [494] | 001-100 R(12) CO2 |
| CD3OH | 419.0 | [493,494] | 001-100 R(36) CO2 |
| CD3OH | 421 | [494] | 001-100 R(32) CO2 |
| CD3OH | 422 | [494] | 001-020 P(20) CO2 |
| CD3OH | 435 | [550] | 001-020 P(28) CO2 |
| CD3OH | 455 | [494] | 001-020 P(18) CO2 |
| CD3OH | 472 | [494] | 001-020 R(16) CO2 |
| CD3OH | 480 | [494] | 001-020 P(16) CO2 |
| CD3OH | 483 | [494] | 001-100 P(26) CO2 |
| CD3OH | 483 | [494] | 001-020 R(22) CO2 |
| CD3OH | 495 | [494] | 001-100 R(18) CO2 |
| CD3OH | 498.0 | [493] | 001-100 R(34) CO2 |
| CD3OH | 508 | [494] | 001-020 P(08) CO2 |
| CD3OH | 517 | [494] | 001-100 P(42) CO2 |
| CD3OH | 551 | [494] | 001-020 P(22) CO2 |
| CD3OH | 553 | [494] | 001-100 R(08) CO2 |
| CD3OH | 553 | [494] | 001-020 R(14) CO2 |
| CD3OH | 554 | [494] | 001-020 R(32) CO2 |
| CD3OH | 583 | [550,494] | 001-020 R(28) CO2 |
| CD3OH | 583 | [494] | 001-020 R(22) CO2 |
| CD3OH | 599 | [494] | 001-100 R(16) CO2 |
| CD3OH | 646 | [494] | 001-100 R(36) CO2 |
| CD3OH | 648 | [494] | 001-100 R(08) CO2 |
| CD3OH | 680 | [494] | 001-020 P(06) CO2 |
| CD3OH | 685 | [494] | 001-020 R(34) CO2 |
| CD3OH | 695 | [494] | 001-020 P(10) CO2 |
| CD3OH | 702 | [494] | 001-020 P(24) CO2 |
| CD3OH | 703 | [494] | 001-100 R(36) CO2 |
| CD3OH | 711 | [494] | 001-020 P(08) CO2 |
| CD3OH | 722 | [494] | 001-100 P(20) CO2 |
| CD3OH | 745 | [494] | 001-020 R(26) CO2 |
| CD3OH | 760 | [494] | 001-100 P(18) CO2 |
| CD3OH | 774 | [494] | 001-020 P(30) CO2 |
| CD3OH | 862.0 | [493,494] | 001-100 R(18) CO2 |
| CD3OH | 968 | [494] | 001-020 R(20) CO2 |
| CD3OH | 1100 | [494] | 001-020 P(12) CO2 |
| CD3OH | 1146 | [494] | 001-020 P(24) CO2 |
| CD3OH | 1290 | [494] | 001-100 R(20) CO2 |

| WAVELENGTH IN VACUUM [MICROMETER] | | | PUMP TRANSITION |
|---|---|---|---|
| CH3OD | 46.7 | [607] | 001-020 R(08) CO2 |
| CH3OD | 57 | [414] | 001-020 R(08) CO2 |
| CH3OD | 69.5 | [580] | 001-020 R(06) CO2 |
| CH3OD | 70.3 | [580] | 001-020 R(16) CO2 |
| CH3OD | 100.8 | [580] | 001-020 P(26) CO2 |
| CH3OD | 104 | [414] | 001-020 P(30) CO2 |
| CH3OD | 110 | [414] | 001-100 R(44) CO2 |
| CH3OD | 117 | [414] | 001-020 P(26) CO2 |
| CH3OD | 134 | [597] | 001-020 P(10) CO2 |
| CH3OD | 134.7 | [607] | 001-020 P(06) CO2 |
| CH3OD | 134.8 | [580] | 001-020 P(10) CO2 |
| CH3OD | 136 | [597] | 001-020 P(24) CO2 |
| CH3OD | 145.66171 | [414,395] | 001-020 P(30) CO2 |
| CH3OD | 212.8 | [580] | 001-020 R(04) CO2 |
| CH3OD | 215.37244 | [414,395] | 001-020 R(14) CO2 |
| CH3OD | 225 | [580] | 001-020 R(06) CO2 |
| CH3OD | 229.1 | [607] | 001-020 P(06) CO2 |
| CH3OD | 238 | [580] | 001-020 R(14) CO2 |
| CH3OD | 294.81097 | [414,395] | 001-020 R(08) CO2 |
| CH3OD | 305.72610 | [414,395] | 001-020 R(08) CO2 |
| CH3OD | 330.1 | [607] | 001-020 R(04) CO2 |
| CH3OD | 417.1 | [607] | 001-020 P(06) CO2 |
| | | | |
| CD3OD | 35 | [597] | 001-100 R(28) CO2 |
| CD3OD | 41 | [413,597] | 001-100 R(24) CO2 |
| CD3OD | 78 | [597] | 001-100 P(10) CO2 |
| CD3OD | 119 | [597] | 001-100 R(26) CO2 |
| CD3OD | 150 | [597] | 001-100 R(30) CO2 |
| CD3OD | 165 | [597] | 001-100 R(22) CO2 |
| CD3OD | 184 | [413] | 001-100 R(10) CO2 |
| CD3OD | 229 | [413,597] | 001-100 R(10) CO2 |
| CD3OD | 255 | [413,597] | 001-100 R(36) CO2 |
| CD3OD | 312 | [413] | 001-100 R(24) CO2 |
| CD3OD | 339 | [413,597] | 001-100 R(04) CO2 |
| CD3OD | 299 | [413,597] | 001-100 R(24) CO2 |
| CD3OD | 312 | [597] | 001-100 R(10) CO2 |
| CD3OD | 354 | [597] | 001-100 R(16) CO2 |
| CD3OD | 406 | [413] | 001-100 R(12) CO2 |
| CD3OD | 414 | [597] | 001-100 R(12) CO2 |
| CD3OD | 495 | [597] | 001-100 R(24) CO2 |
| CD3OD | 869 | [340] | 001-100 R(18) CO2 |
| | | | |
| CH2DOH | 109 | [525] | 001-020 P(12) CO2 |
| CH2DOH | 125 | [525] | 001-100 P(30) CO2 |
| CH2DOH | 151 | [525] | 001-100 R(34) CO2 |
| CH2DOH | 164 | [525] | 001-020 R(18) CO2 |
| CH2DOH | 167 | [525] | 001-020 P(18) CO2 |
| CH2DOH | 171 | [525] | 001-020 P(12) CO2 |
| CH2DOH | 207 | [525] | 001-020 P(14) CO2 |
| CH2DOH | 238 | [525] | 001-100 P(18) CO2 |
| CH2DOH | 250 | [525] | 001-100 P(34) CO2 |
| CH2DOH | 272 | [525] | 001-020 R(24) CO2 |
| CH2DOH | 295 | [525] | 001-100 R(34) CO2 |
| CH2DOH | 296 | [525] | 001-020 P(10) CO2 |
| CH2DOH | 308 | [525] | 001-020 P(14) CO2 |
| CH2DOH | 322 | [525] | 001-020 P(12) CO2 |

| WAVELENGTH IN VACUUM [MICROMETER] | | | PUMP TRANSITION |
|---|---|---|---|
| CH2DOH | 363 | [525] | 001-100 R(16) CO2 |
| CH2DOH | 374 | [525] | 001-100 P(46) CO2 |
| CH2DOH | 396 | [525] | 001-020 P(18) CO2 |
| CH2DOH | 468 | [525] | 001-020 P(26) CO2 |
| CH2DOH | 616 | [525] | 001-020 P(26) CO2 |
| CHD2OH | 165 | [525] | 001-020 R(18) CO2 |
| CHD2OH | 168 | [525] | 001-100 R(38) CO2 |
| CHD2OH | 179 | [525] | 001-100 R(16) CO2 |
| CHD2OH | 238 | [525] | 001-100 P(18) CO2 |
| CHD2OH | 260 | [525] | 001-100 R(20) CO2 |
| CHD2OH | 346 | [525] | 001-020 P(20) CO2 |
| CHD2OH | 355 | [525] | 001-100 P(18) CO2 |
| CHD2OH | 363 | [525] | 001-100 R(16) CO2 |
| CHD2OH | 426 | [525] | 001-100 R(38) CO2 |
| CHD2OH | 483 | [525] | 001-020 P(06) CO2 |
| CHD2OH | 513 | [525] | 001-020 P(30) CO2 |

NUMBER OF LINES IN METHYL ALCOHOL 286

ACTIVE MEDIUM : METHYL CYANIDE
SYMBOL : CH3CN

OPERATING CONDITIONS : OPTICAL PUMPING WITH SINGLE CO2-LASER LINES. CONTINUOUS OPERATION IN THE WAVEGUIDE RESONATOR [346].

| WAVELENGTH IN VACUUM [MICROMETER] | | | PUMP TRANSITION |
|---|---|---|---|
| CH3CN | 281.18 | [369] | 001-020 P(34) CO2 |
| CH3CN | 281.96 | [369] | 001-020 P(50) CO2 |
| CH3CN | 286.88 | [369] | 001-020 P(50) CO2 |
| CH3CN | 303.54 | [271] | 001-100 P(10) CO2 |
| CH3CN | 346.32 | [369] | 001-020 P(16) CO2 |
| CH3CN | 372.87 | [271] | 001-100 P(20) CO2 |
| CH3CN | 380.71 | [271] | 001-100 P(16) CO2 |
| CH3CN | 386.41 | [369] | 001-020 P(46) CO2 |
| CH3CN | 387.31 | [346,369] | 001-020 R(12) CO2 |
| CH3CN | 388.39 | [369] | 001-020 P(22) CO2 |
| CH3CN | 422.14 | [271] | 001-100 P(24) CO2 |
| CH3CN | 427.04 | [369] | 001-020 P(26) CO2 |
| CH3CN | 430.55 | [271] | 001-100 P(18) CO2 |
| CH3CN | 441.15 | [369] | 001-020 R(16) CO2 |
| CH3CN | 453.41 | [346,369] | 001-020 R(16) CO2 |
| CH3CN | 466.25 | [369] | 001-020 R(16) CO2 |
| CH3CN | 480.01 | [369] | 001-020 R(16) CO2 |
| CH3CN | 494.74 | [346,369] | 001-020 P(06) CO2 |
| CH3CN | 510.16 | [369] | 001-020 P(06) CO2 |
| CH3CN | 561.41 | [369] | 001-020 R(08) CO2 |
| CH3CN | 652.68 | [346,369] | 001-020 P(30) CO2 |
| CH3CN | 704.53 | [369] | 001-020 R(34) CO2 |
| CH3CN | 713.72 | [271] | 001-100 P(32) CO2 |

| WAVELENGTH IN VACUUM [MICROMETER] | | | PUMP TRANSITION |
|---|---|---|---|
| CH3CN | 741.62 | [369] | 001-020 R(08) CO2 |
| CH3CN | 854.41 | [369] | 001-020 P(16) CO2 |
| CH3CN | 1014.89 | [369] | 001-020 R(14) CO2 |
| CH3CN | 1016.33 | [369] | 001-020 P(08) CO2 |
| CH3CN | 1086.69 | [369] | 001-020 P(40) CO2 |
| CH3CN | 1146.83 | [369] | 001-020 P(10) CO2 |
| CH3CN | 1351.78 | [369] | 001-020 R(20) CO2 |
| CH3CN | 1814.37 | [271] | 001-100 P(46) CO2 |

NUMBER OF LINES IN METHYL CYANIDE 31

ACTIVE MEDIUM : METHYL ISOCYANIDE
SYMBOL : CH3NC

OPERATING CONDITIONS : OPTICAL PUMPING WITH SINGLE CO2-LASER LINES IN CW OPERATION.

| WAVELENGTH IN AIR [MICROMETER] | | | PUMP TRANSITION |
|---|---|---|---|
| CH3NC | 280 | [611] | 001-100 R(24) CO2 |
| CH3NC | 286 | [611] | 001-100 R(18) CO2 |
| CH3NC | 404 | [611] | 001-100 R(14) CO2 |

NUMBER OF LINES IN METHYL ISOCYANIDE 3

ACTIVE MEDIUM : METHYL ACETYLENE
SYMBOL : CH3CCH

OPERATING CONDITIONS : OPTICAL PUMPING WITH SINGLE CO2-LASER LINES.

| WAVELENGTH IN AIR [MICROMETER] | | | PUMP TRANSITION |
|---|---|---|---|
| CH3CCH | 427.89 | [271] | 001-100 P(10) CO2 |
| CH3CCH | 428.87 | [369] | 001-020 R(38) CO2 |
| CH3CCH | 488.88 | [271] | 001-100 P(12) CO2 |
| CH3CCH | 516.77 | [369] | 001-020 R(12) CO2 |
| CH3CCH | 531.08 | [369] | 001-020 P(06) CO2 |
| CH3CCH | 563.13 | [271] | 001-100 P(24) CO2 |
| CH3CCH | 566.44 | [369] | 001-020 P(18) CO2 |
| CH3CCH | 583.77 | [369] | 001-020 P(20) CO2 |
| CH3CCH | 647.69 | [271] | 001-100 P(14) CO2 |
| CH3CCH | 649.59 | [271] | 001-100 P(34) CO2 |
| CH3CCH | 675.29 | [369] | 001-020 P(40) CO2 |
| CH3CCH | 757.41 | [271] | 001-100 P(10) CO2 |
| CH3CCH | 798.55 | [271] | 001-100 P(20) CO2 |
| CH3CCH | 1097.11 | [369] | 001-020 P(08) CO2 |

| WAVELENGTH IN AIR [MICROMETER] | | | PUMP TRANSITION |
|---|---|---|---|
| CH3CCH | 1174.87 | [271] | 001-100 P(44) CO2 |

NUMBER OF LINES IN METHYL ACETYLENE 15

ACTIVE MEDIUM : METHYL AMINE
SYMBOL : CH3NH2

OPERATING CONDITIONS : OPTICAL PUMPING WITH SINGLE CO2-LASER LINES.

| WAVELENGTH IN AIR [MICROMETER] | | | PUMP TRANSITION |
|---|---|---|---|
| CH3NH2 | 99.5 | [335] | 001-020 R(14) CO2 |
| CH3NH2 | 104 | [335] | 001-020 P(28) CO2 |
| CH3NH2 | 115.5 | [335] | 001-020 P(44) CO2 |
| CH3NH2 | 118 | [335] | 001-020 P(08) CO2 |
| CH3NH2 | 126 | [335] | 001-100 R(06) CO2 |
| CH3NH2 | 134 | [335] | 001-020 R(14) CO2 |
| CH3NH2 | 134 | [335] | 001-020 R(18) CO2 |
| CH3NH2 | 139 | [335] | 001-020 R(14) CO2 |
| CH3NH2 | 141 | [335] | 001-100 R(22) CO2 |
| CH3NH2 | 143 | [335] | 001-020 R(14) CO2 |
| CH3NH2 | 147 | [335] | 001-020 P(24) CO2 |
| CH3NH2 | 148.5 | [312] | 001-020 P(24) CO2 |
| CH3NH2 | 153 | [335] | 001-020 P(08) CO2 |
| CH3NH2 | 159 | [335] | 001-020 P(24) CO2 |
| CH3NH2 | 164 | [335] | 001-020 R(18) CO2 |
| CH3NH2 | 166 | [335] | 001-020 P(32) CO2 |
| CH3NH2 | 166 | [335] | 001-020 R(22) CO2 |
| CH3NH2 | 176 | [335] | 001-100 R(32) CO2 |
| CH3NH2 | 175 | [335] | 001-100 R(06) CO2 |
| CH3NH2 | 177 | [335] | 001-020 R(12) CO2 |
| CH3NH2 | 180 | [346] | 001-020 P(46) CO2 |
| CH3NH2 | 183 | [335] | 001-020 R(14) CO2 |
| CH3NH2 | 194 | [335] | 001-020 R(08) CO2 |
| CH3NH2 | 198 | [335] | 001-020 R(20) CO2 |
| CH3NH2 | 198.0 | [312] | 001-020 P(24) CO2 |
| CH3NH2 | 201 | [335] | 001-020 R(12) CO2 |
| CH3NH2 | 208 | [335] | 001-020 R(12) CO2 |
| CH3NH2 | 218.0 | [312,335] | 001-020 P(24) CO2 |
| CH3NH2 | 219 | [335] | 001-020 P(32) CO2 |
| CH3NH2 | 243 | [335] | 001-020 P(24) CO2 |
| CH3NH2 | 251.0 | [312] | 001-020 P(04) CO2 |
| CH3NH2 | 267 | [335] | 001-020 P(40) CO2 |
| CH3NH2 | 268 | [335] | 001-020 R(12) CO2 |
| CH3NH2 | 288 | [346] | 001-020 R(04) CO2 |
| CH3NH2 | 314 | [346] | 001-020 R(04) CO2 |
| CH3NH2 | 347 | [335] | 001-100 R(20) CO2 |

NUMBER OF LINES IN METHYL AMINE 36

ACTIVE MEDIUM : ETHYLENE DIFLUORIDE
SYMBOL : CH2CF2

OPERATING CONDITIONS : OPTICAL PUMPING WITH SINGLE CO2-LASER LINES.

ASSIGNEMENT OF LINES [402].

| WAVELENGTH IN AIR [MICROMETER] | | | PUMP TRANSITION |
|---|---|---|---|
| CH2CF2 | 288.5 | [312] | 001-100 P(12) CO2 |
| CH2CF2 | 375.0 | [312] | 001-100 P(12) CO2 |
| CH2CF2 | 415 | [347] | 001-100 P(14) CO2 |
| CH2CF2 | 458.0 | [312] | 001-100 P(30) CO2 |
| CH2CF2 | 464.3 | [312] | 001-100 R(20) CO2 |
| CH2CF2 | 532 | [346] | 001-100 P(16) CO2 |
| CH2CF2 | 554.4 | [312,347] | 001-100 P(14) CO2 |
| CH2CF2 | 568 | [347] | 001-100 P(24) CO2 |
| CH2CF2 | 663.3 | [312,347] | 001-100 P(24) CO2 |
| CH2CF2 | 764.1 | [607] | 001-100 P(10) CO2 |
| CH2CF2 | 884 | [347] | 001-100 P(12) CO2 |
| CH2CF2 | 890.0 | [312,347] | 001-100 P(22) CO2 |
| CH2CF2 | 890.1 | [312] | 001-100 P(22) CO2 |
| CH2CF2 | 990.0 | [312,347] | 001-100 P(22) CO2 |
| CH2CF2 | 1020 | [347] | 001-100 P(14) CO2 |

NUMBER OF LINES IN ETHYLENE DIFLUORIDE 15

ACTIVE MEDIUM : ETHYLENE GLYCOL
SYMBOL : C2H4[OH]2 = C2H6O2

OPERATING CONDITIONS : OPTICAL PUMPING WITH SINGLE CO2-LASER LINES IN PULSED OPERATION.

| WAVELENGTH IN AIR [MICROMETER] | | | PUMP TRANSITION |
|---|---|---|---|
| C2H6O2 | 62.5 | [335] | 001-100 R(16) CO2 |
| C2H6O2 | 69.1 | [335] | 001-100 R(16) CO2 |
| C2H6O2 | 70.1 | [335] | 001-020 P(34) CO2 |
| C2H6O2 | 75.2 | [335] | 001-020 P(32) CO2 |
| C2H6O2 | 77.4 | [335] | 001-100 R(16) CO2 |
| C2H6O2 | 90.8 | [335] | 001-020 P(32) CO2 |
| C2H6O2 | 95.8 | [335] | 001-020 R(10) CO2 |
| C2H6O2 | 109.1 | [335] | 001-020 R(16) CO2 |
| C2H6O2 | 117.1 | [335] | 001-020 P(14) CO2 |
| C2H6O2 | 118 | [335] | 001-100 P(30) CO2 |
| C2H6O2 | 118.9 | [335] | 001-020 P(34) CO2 |
| C2H6O2 | 125.8 | [335] | 001-020 P(34) CO2 |
| C2H6O2 | 132 | [335] | 001-020 P(24) CO2 |
| C2H6O2 | 132 | [335] | 001-020 P(36) CO2 |
| C2H6O2 | 135 | [335] | 001-020 P(36) CO2 |
| C2H6O2 | 164 | [335] | 001-020 P(14) CO2 |
| C2H6O2 | 164 | [335] | 001-020 R(10) CO2 |
| C2H6O2 | 164 | [335] | 001-020 P(16) CO2 |

| WAVELENGTH IN AIR [MICROMETER] | | | PUMP TRANSITION |
|---|---|---|---|
| C2H6O2 | 169 | [335] | 001-020 P(36) CO2 |
| C2H6O2 | 171 | [335] | 001-020 R(08) CO2 |
| C2H6O2 | 185 | [335] | 001-020 P(34) CO2 |
| C2H6O2 | 185 | [335] | 001-020 R(18) CO2 |
| C2H6O2 | 189 | [335] | 001-020 P(34) CO2 |
| C2H6O2 | 189 | [335] | 001-020 P(36) CO2 |
| C2H6O2 | 192 | [335] | 001-020 P(38) CO2 |
| C2H6O2 | 197 | [335] | 001-020 P(38) CO2 |
| C2H6O2 | 200 | [335] | 001-020 P(36) CO2 |
| C2H6O2 | 231 | [335] | 001-020 R(10) CO2 |
| C2H6O2 | 240 | [335] | 001-020 R(10) CO2 |
| C2H6O2 | 250 | [335] | 001-020 R(18) CO2 |
| C2H6O2 | 252 | [335] | 001-020 P(34) CO2 |
| C2H6O2 | 262 | [335] | 001-020 P(34) CO2 |
| C2H6O2 | 277 | [335] | 001-020 P(38) CO2 |
| C2H6O2 | 288 | [335] | 001-020 P(12) CO2 |
| C2H6O2 | 290 | [335] | 001-020 P(38) CO2 |
| C2H6O2 | 299 | [335] | 001-020 P(34) CO2 |
| C2H6O2 | 344 | [335] | 001-020 P(22) CO2 |
| C2H6O2 | 358 | [335] | 001-020 P(34) CO2 |
| C2H6O2 | 388 | [335] | 001-020 P(36) CO2 |
| C2H6O2 | 415 | [335] | 001-020 P(14) CO2 |
| C2H6O2 | 696 | [335] | 001-020 P(34) CO2 |

NUMBER OF LINES IN ETHYLENE GLYCOL 41

ACTIVE MEDIUM : ETHYL MONOFLUORIDE

SYMBOL : CH3CH2F = C2H5F

OPERATING CONDITIONS : OPTICAL PUMPING WITH SINGLE CO2-LASER LINES. CONTINUOUS OPERATION IN WAVEGUIDE RESONATOR [346].

| WAVELENGTH IN VACUUM [MICROMETER] | | | PUMP TRANSITION |
|---|---|---|---|
| C2H5F | 206.60 | [317] | 001-100 P(36) CO2 |
| C2H5F | 217.1 | [317] | 001-020 R(14) CO2 |
| C2H5F | 226.9 | [317] | 001-100 P(40) CO2 |
| C2H5F | 264.7 | [317,346] | 001-020 P(18) CO2 |
| C2H5F | 282.3 | [317] | 001-020 R(12) CO2 |
| C2H5F | 330.2 | [317] | 001-020 R(22) CO2 |
| C2H5F | 336.7 | [317,346] | 001-020 R(16) CO2 |
| C2H5F | 362.1 | [317] | 001-020 R(18) CO2 |
| C2H5F | 376.0 | [317] | 001-020 R(14) CO2 |
| C2H5F | 378.0 | [317] | 001-020 R(32) CO2 |
| C2H5F | 404 | [317] | 001-020 P(34) CO2 |
| C2H5F | 404 | [346] | 001-020 R(30) CO2 |
| C2H5F | 405.0 | [346] | 001-020 P(34) CO2 |
| C2H5F | 405.50 | [317] | 001-020 R(30) CO2 |
| C2H5F | 462.92 | [317] | 001-020 P(32) CO2 |
| C2H5F | 486 | [346] | 001-020 R(24) CO2 |
| C2H5F | 502.2 | [317,346] | 001-020 R(24) CO2 |

| WAVELENGTH IN VACUUM [MICROMETER] | | | PUMP TRANSITION |
|---|---|---|---|
| C2H5F | 519 | [346] | 001-020 R(04) CO2 |
| C2H5F | 540.9 | [317] | 001-020 P(38) CO2 |
| C2H5F | 593.32 | [317,346] | 001-020 P(36) CO2 |
| C2H5F | 620.4 | [317,346] | 001-020 P(22) CO2 |
| C2H5F | 851.9 | [317,346] | 001-020 P(30) CO2 |
| C2H5F | 1013 | [346] | 001-020 P(28) CO2 |
| C2H5F | 1069 | [346] | 001-020 R(10) CO2 |
| C2H5F | 1546 | [346] | 001-020 P(10) CO2 |

NUMBER OF LINES IN ETHYL MONOFLUORIDE 25

ACTIVE MEDIUM : ETHYL DIFLUORIDE
SYMBOL : CH3CHF2 = C2H4F2

OPERATING CONDITIONS : OPTICAL PUMPING WITH SINGLE CO2-LASER LINES IN CONTINUOUS OPERATION.

| WAVELENGTH IN VACUUM [MICROMETER] | | | PUMP TRANSITION |
|---|---|---|---|
| C2H4F2 | 458 | [347] | 001-100 P(20) CO2 |
| C2H4F2 | 464 | [316] | 001-100 P(20) CO2 |
| C2H4F2 | 533 | [347] | 001-100 P(20) CO2 |
| C2H4F2 | 755 | [316] | 001-100 P(14) CO2 |

NUMBER OF LINES IN ETHYL DIFLUORIDE 4

ACTIVE MEDIUM : ETHYL TRIFLUORIDE
SYMBOL : CH3CF3 = C2H3F3

OPERATING CONDITIONS : OPTICAL PUMPING WITH SINGLE CO2-LASER LINES.

| WAVELENGTH IN VACUUM [MICROMETER] | | | PUMP TRANSITION |
|---|---|---|---|
| C2H3F3 | 379 | [316] | 001-100 P(32) CO2 |

NUMBER OF LINES IN ETHYL TRIFLUORIDE 1

ACTIVE MEDIUM : ETHYL CHLORIDE
SYMBOL : C2H5CL

OPERATING CONDITIONS : OPTICAL PUMPING WITH SINGLE CO2-LASER LINES IN CONTINUOUS OPERATION.

| WAVELENGTH IN VACUUM [MICROMETER] | | PUMP TRANSITION |
|---|---|---|
| C2H5CL 900 | [341] | 001-100 R(30) CO2 |
| C2H5CL 1350 | [341] | 001-100 R(30) CO2 |
| C2H5CL 1400 | [341] | 001-100 R(38) CO2 |
| C2H5CL 1720 | [341] | 001-100 R(28) CO2 |

NUMBER OF LINES IN ETHYL CHLORIDE 4

ACTIVE MEDIUM : ETHYL ALCOHOL
SYMBOL : C2H5OH

OPERATING CONDITIONS : OPTICAL PUMPING WITH SINGLE CO2-LASER LINES IN CONTINUOUS OPERATION.

| WAVELENGTH IN VACUUM [MICROMETER] | | PUMP TRANSITION |
|---|---|---|
| C2H5OH 396 | [341] | 001-020 P(32) CO2 |

NUMBER OF LINES IN ETHYL ALCOHOL 1

ACTIVE MEDIUM : VINYL CHLORIDE
SYMBOL : C2H3CL

OPERATING CONDITIONS : OPTICAL PUMPING WITH SINGLE CO2-LASER LINES. CONTINUOUS OPERATION IN THE WAVEGUIDE RESONATOR [346].

| WAVELENGTH IN VACUUM [MICROMETER] | | PUMP TRANSITION |
|---|---|---|
| C2H3CL 368.0 | [239] | 001-100 P(22) CO2 |
| C2H3CL 424 | [346] | 001-100 R(28) CO2 |
| C2H3CL 445 | [346] | 001-100 R(18) CO2 |
| C2H3CL 487 | [346] | 001-020 P(10) CO2 |
| C2H3CL 507.7 | [239] | 001-100 P(22) CO2 |
| C2H3CL 519 | [346] | 001-100 P(34) CO2 |
| C2H3CL 532 | [346] | 001-020 P(16) CO2 |
| C2H3CL 538 | [346] | 001-100 R(04) CO2 |
| C2H3CL 574 | [346] | 001-100 P(16) CO2 |
| C2H3CL 603 | [346] | 001-100 P(38) CO2 |
| C2H3CL 634.4 | [239] | 001-020 P(20) CO2 |
| C2H3CL 638 | [346] | 001-100 P(06) CO2 |
| C2H3CL 699 | [346] | 001-020 P(22) CO2 |
| C2H3CL 707 | [346] | 001-020 P(18) CO2 |
| C2H3CL 828 | [346] | 001-020 P(24) CO2 |
| C2H3CL 935 | [346] | 001-100 P(46) CO2 |
| C2H3CL 995 | [346] | 001-100 R(26) CO2 |
| C2H3CL 1041 | [346] | 001-100 R(36) CO2 |

NUMBER OF LINES IN VINYL CHLORIDE 18

ACTIVE MEDIUM : VINYL BROMIDE
SYMBOL : C2H3BR

OPERATING CONDITIONS : OPTICAL PUMPING WITH SINGLE CO2-LASER LINES. CW OPERATION.

| WAVELENGTH IN AIR [MICROMETER] | | | PUMP TRANSITION |
|---|---|---|---|
| C2H3BR | 283 | [441] | 001-100 R(20) CO2 |
| C2H3BR | 356 | [441] | 001-100 R(20) CO2 |
| C2H3BR | 370 | [441] | 001-100 P(28) CO2 |
| C2H3BR | 396 | [441] | 001-020 P(42) CO2 |
| C2H3BR | 411 | [441] | 001-100 R(26) CO2 |
| C2H3BR | 416 | [441] | 001-100 R(22) CO2 |
| C2H3BR | 419 | [441] | 001-100 R(32) CO2 |
| C2H3BR | 424 | [441] | 001-100 P(20) CO2 |
| C2H3BR | 427 | [441] | 001-100 R(24) CO2 |
| C2H3BR | 438.5 | [441] | 001-100 P(28) CO2 |
| C2H3BR | 443.5 | [441] | 001-100 P(24) CO2 |
| C2H3BR | 445 | [441] | 001-100 P(22) CO2 |
| C2H3BR | 482.96 | [441] | 001-100 P(26) CO2 |
| C2H3BR | 490.08 | [441] | 001-100 P(16) CO2 |
| C2H3BR | 506 | [441] | 001-100 R(38) CO2 |
| C2H3BR | 528.49 | [441] | 001-100 R(40) CO2 |
| C2H3BR | 553.69 | [441] | 001-100 P(40) CO2 |
| C2H3BR | 594.72 | [441] | 001-100 P(32) CO2 |
| C2H3BR | 618.44 | [441] | 001-100 R(30) CO2 |
| C2H3BR | 624.09 | [441] | 001-100 R(18) CO2 |
| C2H3BR | 635.35 | [441] | 001-100 R(26) CO2 |
| C2H3BR | 646 | [441] | 001-100 R(26) CO2 |
| C2H3BR | >649.42 | [441] | 001-100 P(18) CO2 |
| C2H3BR | >680.54 | [441] | 001-100 R(16) CO2 |
| C2H3BR | 693.13 | [441] | 001-100 R(16) CO2 |
| C2H3BR | 707.22 | [441] | 001-100 R(24) CO2 |
| C2H3BR | 712 | [441] | 001-100 R(10) CO2 |
| C2H3BR | 724.13 | [441] | 001-100 P(14) CO2 |
| C2H3BR | 741.11 | [441] | 001-100 P(20) CO2 |
| C2H3BR | 780.13 | [441] | 001-100 R(14) CO2 |
| C2H3BR | 784.26 | [441] | 001-100 P(24) CO2 |
| C2H3BR | 826.94 | [441] | 001-100 P(22) CO2 |
| C2H3BR | 853.43 | [441] | 001-100 P(10) CO2 |
| C2H3BR | 900.13 | [441] | 001-100 R(18) CO2 |
| C2H3BR | 943.22 | [441] | 001-020 P(28) CO2 |
| C2H3BR | 936.15 | [441] | 001-100 R(32) CO2 |
| C2H3BR | >963.48 | [441] | 001-100 P(10) CO2 |
| C2H3BR | 985.65 | [441] | 001-100 R(02) CO2 |
| C2H3BR | 989.19 | [441] | 001-100 P(16) CO2 |
| C2H3BR | 990.63 | [441] | 001-100 R(04) CO2 |
| C2H3BR | 1247.59 | [441] | 001-100 R(12) CO2 |
| C2H3BR | 1383.88 | [441] | 001-100 P(24) CO2 |
| C2H3BR | 1394.06 | [441] | 001-100 R(20) CO2 |
| C2H3BR | 1614.88 | [441] | 001-100 P(26) CO2 |
| C2H3BR | 1899.889 | [441] | 001-100 P(20) CO2 |

NUMBER OF LINES IN VINYL BROMIDE 45

ACTIVE MEDIUM : VINYL CYANIDE
SYMBOL : C2H3CN

OPERATING CONDITIONS : OPTICAL PUMPING WITH SINGLE CO2-LASER LINES. CONTINUOUS OPERATION IN THE WAVEGUIDE RESONATOR [346].

| WAVELENGTH IN AIR [MICROMETER] | | | PUMP TRANSITION |
|---|---|---|---|
| C2H3CN | 270.6 | [312] | 001-100 P(26) CO2 |
| C2H3CN | 489 | [346] | 001-100 P(08) CO2 |
| C2H3CN | 503 | [346] | 001-020 R(12) CO2 |
| C2H3CN | 550.0 | [312] | 001-100 P(14) CO2 |
| C2H3CN | 574.4 | [312] | 001-100 R(16) CO2 |
| C2H3CN | 578 | [346] | 001-100 R(14) CO2 |
| C2H3CN | 584.0 | [312] | 001-100 P(12) CO2 |
| C2H3CN | 586.6 | [312] | 001-100 P(20) CO2 |
| C2H3CN | 623 | [346] | 001-100 R(12) CO2 |
| C2H3CN | 631 | [346] | 001-100 R(06) CO2 |
| C2H3CN | 722 | [346] | 001-100 P(42) CO2 |
| C2H3CN | 738 | [346] | 001-100 P(16) CO2 |
| C2H3CN | 775 | [346] | 001-100 R(42) CO2 |
| C2H3CN | 793 | [346] | 001-100 R(40) CO2 |
| C2H3CN | 828 | [346] | 001-100 R(18) CO2 |
| C2H3CN | 910 | [346] | 001-100 R(12) CO2 |
| C2H3CN | 940 | [346] | 001-100 P(28) CO2 |
| C2H3CN | 1156 | [346] | 001-100 P(26) CO2 |
| C2H3CN | 1184 | [346] | 001-100 R(38) CO2 |

NUMBER OF LINES IN VINYL CYANIDE 19

---

ACTIVE MEDIUM : DIMETHYL ETHER
SYMBOL : CH3OCH3 = C2H6O

OPERATING CONDITIONS : OPTICAL PUMPING WITH SINGLE CO2-LASER LINES IN PULSED OPERATION.

| WAVELENGTH IN AIR [MICROMETER] | | | PUMP TRANSITION |
|---|---|---|---|
| C2H6O | 375 | [335] | 001-100 P(20) CO2 |
| C2H6O | 461 | [335] | 001-100 P(34) CO2 |
| C2H6O | 480 | [335] | 001-100 P(34) CO2 |
| C2H6O | 492 | [335] | 001-100 P(34) CO2 |
| C2H6O | 495 | [335] | 001-100 P(12) CO2 |
| C2H6O | 520 | [335] | 001-100 P(12) CO2 |

NUMBER OF LINES IN DIMETHYL ETHER 6

ACTIVE MEDIUM : PROPYNAL
SYMBOL : C3H2O

OPERATING CONDITIONS : OPTICAL PUMPING WITH SINGLE CO2-LASER LINES IN CW OPERATION.

| WAVELENGTH IN AIR [MICROMETER] | | | PUMP TRANSITION |
|---|---|---|---|
| C3H2O | 148 | [611] | 001-100 P(18) CO2 |
| C3H2O | 156 | [611] | 001-100 P(22) CO2 |
| C3H2O | 336 | [611] | 001-100 P(26) CO2 |
| C3H2O | 516 | [611] | 001-100 P(14) CO2 |

NUMBER OF LINES IN PROPYNAL 4

ACTIVE MEDIUM : CHLORINE DIOXIDE
SYMBOL : CLO2

OPERATING CONDITIONS : OPTICAL PUMPING WITH SINGLE CO2-LASER LINES IN CW OPERATION.

| WAVELENGTH IN AIR [MICROMETER] | | | PUMP TRANSITION |
|---|---|---|---|
| CLO2 | ? 112 | [611] | 001-020 P(12) C[12]O[18]2 |
| CLO2 | 176 | [611] | 001-020 R(14) C[12]O[18]2 |
| CLO2 | 196 | [611] | 001-020 R(30) C[12]O[18]2 |
| CLO2 | 204 | [611] | 001-020 R(16) C[12]O[18]2 |
| CLO2 | 216 | [611] | 001-020 R(38) C[12]O[18]2 |

NUMBER OF LINES IN CHLORINE DIOXIDE 5

ACTIVE MEDIUM : FLUORO ETHYNE
SYMBOL : HCCF

OPERATING CONDITIONS : OPTICAL PUMPING WITH SINGLE CO2-LASER LINES IN CW OPERATION.

| WAVELENGTH IN AIR [MICROMETER] | | | PUMP TRANSITION |
|---|---|---|---|
| HCCF | 509 | [611] | 001-020 P(12) C[12]O[18]2 |
| HCCF | 1028 | [611] | 001-020 R(18) CO2 |

NUMBER OF LINES IN FLUORO ETHYNE 2

ACTIVE MEDIUM : FLUORINE CYANIDE
SYMBOL : FCN

OPERATING CONDITIONS : OPTICAL PUMPING WITH SINGLE $CO_2$-LASER LINES IN CW OPERATION.

| WAVELENGTH IN AIR [MICROMETER] | | | PUMP TRANSITION |
|---|---|---|---|
| FCN | 308 | [611] | 001-020 R(22) C[12]O[18]2 |

NUMBER OF LINES IN FLUORINE CYANIDE 1

# Table of Laser Lines Arranged by Wavelength

| WAVELENGTH [MICROMETER] | ACTIVE MEDIUM | PAGE |
|---|---|---|
| 0.109816 | H2 P | 32 |
| 0.110205 | H2 P | 32 |
| 0.111336 | D2 | 34 |
| 0.111515 | H2 P | 32 |
| 0.111894 | H2 P | 32 |
| 0.113770 | D2 | 34 |
| 0.113864 | HD | 36 |
| 0.114154 | HD | 36 |
| 0.114462 | H2 P | 32 |
| 0.114757 | D2 | 34 |
| 0.114862 | H2 P | 32 |
| 0.115198 | HD | 36 |
| 0.115650 | D2 | 34 |
| 0.115840 | D2 | 35 |
| 0.115976 | H2 | 32 |
| 0.116003 | H2 P | 32 |
| 0.116136 | H2 | 32 |
| 0.116390 | H2 P | 32 |
| 0.116617 | H2 | 32 |
| 0.117436 | H2 | 32 |
| 0.117456 | H2 P | 32 |
| 0.117586 | H2 | 32 |
| 0.117806 | HD | 36 |
| 0.117830 | H2 P | 32 |
| 0.118050 | H2 | 32 |
| 0.118811 | D2 | 34 |
| 0.118936 | H2 | 32 |
| 0.118995 | HD | 36 |
| 0.119005 | D2 | 35 |
| 0.119281 | HD | 36 |
| 0.119753 | D2 | 34 |
| 0.119940 | D2 | 35 |
| 0.120103 | HD | 36 |
| 0.120497 | H2 | 32 |
| 0.120536 | H2 P | 32 |
| 0.120640 | D2 | 35 |
| 0.120668 | H2 | 32 |
| 0.120821 | D2 | 35 |
| 0.120929 | H2 P | 32 |
| 0.121125 | HD | 36 |
| 0.121734 | H2 | 32 |
| 0.121767 | H2 P | 32 |
| 0.121900 | H2 | 32 |
| 0.121946 | H2 P | 33 |
| 0.122143 | H2 P | 32 |
| 0.122358 | H2 | 32 |
| 0.122800 | D2 | 35 |
| 0.122837 | HD | 36 |
| 0.122874 | H2 P | 32 |
| 0.123004 | H2 | 32 |
| 0.123230 | H2 P | 32 |
| 0.123556 | D2 | 35 |
| 0.123833 | H2 P | 32 |
| 0.123956 | H2 | 32 |
| 0.124167 | H2 P | 33 |
| 0.124239 | D2 | 35 |
| 0.124412 | D2 | 35 |
| 0.124567 | HD | 36 |
| 0.124620 | H2 P | 33 |
| 0.124831 | D2 | 35 |
| 0.124997 | D2 | 35 |
| 0.125202 | H2 P | 33 |
| 0.125276 | HD | 36 |
| 0.125329 | D2 | 35 |
| 0.1261 | AR2 | 30 |
| 0.126839 | H2 P | 33 |
| 0.130334 | HD | 36 |
| 0.130363 | D2 | 35 |
| 0.133856 | H2 P | 33 |
| 0.134226 | H2 | 33 |
| 0.134590 | D2 | 35 |
| 0.135507 | HD | 36 |
| 0.135984 | H2 P | 33 |
| 0.136799 | H2 P | 33 |
| 0.138879 | D2 | 35 |
| 0.139895 | H2 P | 33 |
| 0.140264 | H2 | 33 |
| 0.140728 | H2 P | 33 |
| 0.140770 | HD | 36 |
| 0.143217 | D2 | 35 |
| 0.143262 | H2 P | 33 |
| 0.143622 | H2 | 33 |
| 0.143757 | H2 P | 33 |
| 0.144049 | H2 | 33 |
| 0.144061 | H2 P | 33 |
| 0.1457 | KR2 | 30 |
| 0.146017 | H2 P | 33 |
| 0.146383 | H2 | 33 |
| 0.146411 | H2 P | 33 |
| 0.146841 | H2 P | 33 |
| 0.148652 | H2 P | 33 |
| 0.148843 | HD | 36 |

| WAVELENGTH [MICROMETER] | ACTIVE MEDIUM | PAGE |
|---|---|---|
| 0.149171 | H2 P | 33 |
| 0.14942 | H2 | 33 |
| 0.149522 | H2 | 33 |
| 0.151359 | HD | 36 |
| 0.151570 | H2 P | 33 |
| 0.151867 | H2 | 33 |
| 0.151994 | H2 P | 33 |
| 0.152325 | H2 | 33 |
| 0.152989 | HD | 36 |
| 0.153494 | H2 P | 33 |
| 0.154493 | H2 P | 33 |
| 0.15482 | C3+ | 19 |
| 0.155010 | H2 P | 33 |
| 0.15508 | C3+ | 19 |
| 0.155345 | H2 | 33 |
| 0.156201 | HD | 36 |
| 0.15655 | H2 | 33 |
| 0.156629 | H2 P | 34 |
| 0.156644 | H2 P | 34 |
| 0.15671 | F2 | 48 |
| 0.156725 | H2 | 33 |
| 0.156753 | H2 P | 34 |
| 0.157136 | HD | 36 |
| 0.157199 | H2 | 33 |
| 0.157242 | HD | 36 |
| 0.157267 | HD | 36 |
| 0.15743 | H2 | 33 |
| 0.157434 | H2 P | 34 |
| 0.15748 | F2 | 48 |
| 0.157585 | D2 | 35 |
| 0.15759 | F2 | 48 |
| 0.157739 | H2 | 33 |
| 0.157771 | H2 P | 34 |
| 0.157919 | H2 | 33 |
| 0.157998 | H2 | 33 |
| 0.157998 | H2 P | 34 |
| 0.158008 | HD | 36 |
| 0.158077 | H2 | 33 |
| 0.158085 | HD | 36 |
| 0.158110 | H2 P | 34 |
| 0.158140 | H2 P | 34 |
| 0.158185 | HD | 36 |
| 0.158253 | HD | 36 |
| 0.158305 | HD | 36 |
| 0.158634 | D2 | 35 |
| 0.158642 | D2 | 35 |
| 0.158675 | D2 | 35 |
| 0.158714 | D2 | 35 |
| 0.158720 | D2 | 35 |
| 0.158899 | H2 P | 34 |
| 0.15890 | D2 | 35 |
| 0.158983 | D2 | 35 |
| 0.159130 | D2 | 35 |
| 0.159131 | H2 | 33 |
| 0.159137 | D2 | 35 |
| 0.159226 | D2 | 35 |
| 0.15923 | D2 | 35 |

| WAVELENGTH [MICROMETER] | ACTIVE MEDIUM | PAGE |
|---|---|---|
| 0.159257 | D2 | 35 |
| 0.159340 | H2 | 33 |
| 0.159340 | H2 P | 34 |
| 0.159378 | HD | 36 |
| 0.159524 | HD | 36 |
| 0.159606 | H2 | 33 |
| 0.159713 | HD | 36 |
| 0.159926 | H2 P | 34 |
| 0.160044 | D2 | 35 |
| 0.160086 | D2 | 35 |
| 0.160210 | D2 | 35 |
| 0.160233 | HD | 36 |
| 0.160236 | H2 P | 34 |
| 0.160354 | D2 | 35 |
| 0.160365 | HD | 36 |
| 0.160448 | H2 | 33 |
| 0.160465 | HD | 36 |
| 0.160496 | HD | 36 |
| 0.160569 | HD | 36 |
| 0.160578 | D2 | 35 |
| 0.160594 | H2 P | 34 |
| 0.160623 | H2 P | 34 |
| 0.160623 | H2 | 33 |
| 0.16063 | D2 | 35 |
| 0.160647 | HD | 36 |
| 0.160648 | HD | 36 |
| 0.160650 | D2 | 35 |
| 0.160674 | HD | 36 |
| 0.160681 | D2 | 35 |
| 0.160692 | HD | 37 |
| 0.160747 | HD | 37 |
| 0.160751 | H2 | 33 |
| 0.160769 | D2 | 35 |
| 0.160794 | HD | 37 |
| 0.160827 | HD | 37 |
| 0.160829 | H2 P | 34 |
| 0.160829 | H2 | 33 |
| 0.160839 | H2 | 33 |
| 0.160844 | H2 P | 34 |
| 0.160848 | D2 | 35 |
| 0.160893 | HD | 37 |
| 0.160902 | H2 | 33 |
| 0.160955 | D2 | 35 |
| 0.160961 | H2 P | 34 |
| 0.161005 | HD | 37 |
| 0.161019 | H2 P | 34 |
| 0.161033 | H2 P | 34 |
| 0.161033 | H2 | 33 |
| 0.161075 | D2 | 35 |
| 0.161080 | D2 | 35 |
| 0.161131 | HD | 37 |
| 0.161147 | D2 | 35 |
| 0.161165 | D2 | 35 |
| 0.161165 | H2 P | 34 |
| 0.161166 | H2 | 33 |
| 0.161171 | D2 | 35 |
| 0.161198 | D2 | 35 |

| WAVELENGTH [MICROMETER] | ACTIVE MEDIUM | PAGE |
|---|---|---|
| 0.161236 | D2 | 35 |
| 0.161251 | D2 | 35 |
| 0.161257 | D2 | 35 |
| 0.161318 | D2 | 35 |
| 0.161318 | H2 P | 34 |
| 0.161319 | H2 | 33 |
| 0.161320 | D2 | 35 |
| 0.161324 | D2 | 35 |
| 0.161412 | D2 | 35 |
| 0.16148 | H2 | 33 |
| 0.161485 | H2 P | 34 |
| 0.16165 | H2 | 33 |
| 0.161658 | D2 | 35 |
| 0.1722 | XE2 | 30 |
| 0.1750 | ARCL | 31 |
| 0.175641 | KR3+ | 5 |
| 0.181085 | CO | 67 |
| 0.183243 | KR4+ | 5 |
| 0.184340 | AR3+ | 4 |
| 0.184343 | AR4+ | 4 |
| 0.187831 | CO | 67 |
| 0.189784 | CO | 67 |
| 0.1933 | ARF | 31 |
| 0.195006 | CO | 67 |
| 0.195027 | KR3+ | 5 |
| 0.196808 | KR3+ | 5 |
| 0.197013 | CO | 67 |
| 0.2018424 | NE3+ | 1 |
| 0.2022186 | NE3+ | 1 |
| 0.2051082 | KR3+ | 6 |
| 0.2065 | KRBR | 31 |
| 0.2065304 | NE3+ | 1 |
| 0.2113982 | AR3+ | 4 |
| 0.2177705 | NE2+ | 1 |
| 0.2180858 | NE2+ | 1 |
| 0.2191916 | KR3+ | 6 |
| 0.2229 | KRCL | 31 |
| 0.2232442 | XE3+ | 7 |
| 0.22434 | AG+ | 12 |
| 0.2248840 | AR3+ | 4 |
| 0.2254638 | KR3+ | 6 |
| 0.22657 | NE4+ | 1 |
| 0.22774 | AG+ | 12 |
| 0.2285793 | NE3+ | 1 |
| 0.2315357 | XE3+ | 7 |
| 0.2338478 | KR3+ | 6 |
| 0.2357980 | NE3+ | 1 |
| 0.2362465 | BR3+? | 27 |
| 0.2373200 | NE3+ | 1 |
| 0.2417843 | KR3+ | 6 |
| 0.2428 | AU | 13 |
| 0.2473398 | NE2+ | 1 |
| 0.247718 | XE2+ | 7 |
| 0.2481 | KRF | 31 |
| 0.2484 | KRF | 31 |
| 0.2485 | KRF | 31 |
| 0.24858 | CU+ | 11 |
| 0.2495 | KRF | 31 |
| 0.25063 | CU+ | 11 |
| 0.2513298 | AR3+ | 4 |
| 0.2526664 | XE3+ | 7 |
| 0.25337 | AU+ | 13 |
| 0.2580 | CL2 | 49 |
| 0.2581246 | BR3+ | 27 |
| 0.25905 | CU+ | 11 |
| 0.25988 | CU+ | 11 |
| 0.2609982 | NE2+ | 1 |
| 0.26134 | NE2+ | 1 |
| 0.26165 | AU+ | 13 |
| 0.2621377 | AR3+ | 4 |
| 0.2624882 | AR3+ | 4 |
| 0.2632686 | CL2+ | 27 |
| 0.2638964 | S2+ | 24 |
| 0.2640 | O4+ | 23 |
| 0.2649357 | KR3+ | 6 |
| 0.2664398 | KR3+ | 6 |
| 0.2676 | AU | 13 |
| 0.2677918 | NE2+ | 1 |
| 0.2678690 | NE2+ | 1 |
| 0.2691939 | XE2+ | 7 |
| 0.27032 | CU+ | 11 |
| 0.2741380 | KR3+ | 6 |
| 0.2753884 | AR2+ | 4 |
| 0.275958 | F2+ | 26 |
| 0.2777634 | NE2+ | 1 |
| 0.278139 | O4+ | 23 |
| 0.2787619 | BR2+ | 27 |
| 0.2818 | XEBR | 31 |
| 0.28225 | AU+ | 13 |
| 0.282612 | F3+ | 26 |
| 0.28470 | AU+ | 13 |
| 0.285 | CLF | 53 |
| 0.2855374 | AR2+ | 4 |
| 0.28633 | AU+ | 13 |
| 0.2866726 | NE2+ | 1 |
| 0.2884216 | AR2+ | 4 |
| 0.28882 | AU+ | 13 |
| 0.28933 | AU+ | 13 |
| 0.2912924 | AR3+ | 4 |
| 0.2915 | BR2 | 49 |
| 0.29182 | AU+ | 13 |
| 0.2926227 | AR3+ | 4 |
| 0.29594 | AU+ | 13 |
| 0.29837 | XE3+ | 7 |
| 0.298378 | O2+ | 23 |
| 0.3002642 | AR2+ | 4 |
| 0.302405 | AR2+ | 4 |
| 0.304713 | O2+ | 23 |
| 0.3049704 | KR2+ | 6 |
| 0.305484 | AR2+ | 4 |
| 0.306345 | O3+ | 23 |
| 0.30792 | XECL | 31 |
| 0.3079738 | XE2+ | 7 |
| 0.30816 | XECL | 31 |

| WAVELENGTH [MICROMETER] | ACTIVE MEDIUM | PAGE |
|---|---|---|
| 0.3121501 | F2+ | 26 |
| 0.3122 | AU | 13 |
| 0.3124363 | KR2+ | 6 |
| 0.315756 | N2 | 38 |
| 0.315778 | N2 | 38 |
| 0.315798 | N2 | 38 |
| 0.315803 | N2 | 38 |
| 0.315816 | N2 | 38 |
| 0.315827 | N2 | 38 |
| 0.315832 | N2 | 38 |
| 0.315844 | N2 | 38 |
| 0.315853 | N2 | 38 |
| 0.315861 | N2 | 38 |
| 0.315870 | N2 | 38 |
| 0.315874 | N2 | 38 |
| 0.315883 | N2 | 38 |
| 0.315891 | N2 | 38 |
| 0.315900 | N2 | 38 |
| 0.315911 | N2 | 38 |
| 0.315919 | N2 | 38 |
| 0.317413 | F2+ | 26 |
| 0.31807 | AG+ | 12 |
| 0.3191424 | CL2+ | 27 |
| 0.320276 | F+ | 26 |
| 0.3239512 | KR2+ | 6 |
| 0.3246922 | XE3+ | 7 |
| 0.3250 | CD+ | 16 |
| 0.3305957 | XE3+ | 7 |
| 0.330599 | XE2+ | 7 |
| 0.3319745 | NE+ | 1 |
| 0.3323745 | NE+ | 1 |
| 0.3324859 | S2+ | 24 |
| 0.332717 | NE+ | 1 |
| 0.332923 | NE+ | 1 |
| 0.3330869 | XE3+ | 7 |
| 0.333114 | NE2+ | 1 |
| 0.333613 | AR2+ | 4 |
| 0.334472 | AR2+ | 4 |
| 0.3345446 | NE+ | 1 |
| 0.334769 | P3+ | 21 |
| 0.334974 | XE3+ | 7 |
| 0.335849 | AR2+ | 4 |
| 0.3364909 | N2 | 38 |
| 0.3365425 | N2 | 38 |
| 0.3365476 | N2 | 38 |
| 0.3366913 | N2 | 38 |
| 0.336734 | N2+ | 21 |
| 0.3369541 | N2 | 38 |
| 0.3369552 | N2 | 38 |
| 0.3369769 | N2 | 38 |
| 0.3369823 | N2 | 39 |
| 0.3369835 | N2 | 39 |
| 0.3369907 | N2 | 39 |
| 0.3370027 | N2 | 39 |
| 0.3370075 | N2 | 39 |
| 0.3370081 | N2 | 39 |
| 0.3370137 | N2 | 39 |
| 0.3370174 | N2 | 39 |
| 0.3370288 | N2 | 39 |
| 0.3370295 | N2 | 39 |
| 0.3370312 | N2 | 39 |
| 0.3370381 | N2 | 39 |
| 0.3370438 | N2 | 39 |
| 0.3370466 | N2 | 39 |
| 0.3370474 | N2 | 39 |
| 0.3370555 | N2 | 39 |
| 0.3370562 | N2 | 39 |
| 0.3370608 | N2 | 39 |
| 0.3370619 | N2 | 39 |
| 0.3370665 | N2 | 39 |
| 0.3370677 | N2 | 39 |
| 0.3370714 | N2 | 39 |
| 0.3370726 | N2 | 39 |
| 0.3370749 | N2 | 39 |
| 0.3370758 | N2 | 39 |
| 0.3370782 | N2 | 39 |
| 0.3370797 | N2 | 39 |
| 0.3370812 | N2 | 39 |
| 0.3370816 | N2 | 39 |
| 0.3370826 | N2 | 39 |
| 0.3370919 | N2 | 39 |
| 0.3370986 | N2 | 39 |
| 0.3371037 | N2 | 39 |
| 0.3371075 | N2 | 39 |
| 0.3371082 | N2 | 39 |
| 0.3371113 | N2 | 39 |
| 0.3371121 | N2 | 39 |
| 0.3371135 | N2 | 39 |
| 0.3371143 | N2 | 39 |
| 0.3371172 | N2 | 39 |
| 0.3371266 | N2 | 39 |
| 0.3371307 | N2 | 39 |
| 0.3371366 | N2 | 39 |
| 0.3371392 | N2 | 39 |
| 0.3371421 | N2 | 39 |
| 0.3371429 | N2 | 39 |
| 0.337496 | KR2+ | 6 |
| 0.3378256 | NE+ | 2 |
| 0.338128 | O3+ | 23 |
| 0.338133 | O3+ | 23 |
| 0.338554 | O3+ | 23 |
| 0.3386428 | N2 | 39 |
| 0.3392799 | NE+ | 2 |
| 0.3392861 | CL2+ | 27 |
| 0.339320 | NE+ | 2 |
| 0.3393444 | CL2+ | 27 |
| 0.3420 | I2 | 50 |
| 0.3423 | I2 | 50 |
| 0.3424 | I2 | 50 |
| 0.3428 | I2 | 50 |
| 0.345134 | B+ | 17 |
| 0.3454248 | XE2+ | 7 |
| 0.347867 | N3+ | 21 |
| 0.348195 | NE+ | 2 |

| WAVELENGTH [MICROMETER] | ACTIVE MEDIUM | PAGE |
|---|---|---|
| 0.348296 | N3+ | 21 |
| 0.348322 | XE3+ | 7 |
| 0.34875 | XEF | 31 |
| 0.3497332 | S2+ | 24 |
| 0.3507420 | KR2+ | 6 |
| 0.35091 | XEF | 31 |
| 0.35097 | XEF | 31 |
| 0.351112 | AR2+ | 4 |
| 0.35114 | XEF | 31 |
| 0.351418 | AR2+ | 4 |
| 0.3530016 | CL2+ | 27 |
| 0.35305 | XEF | 31 |
| 0.35354 | XEF | 31 |
| 0.3560632 | CL2+ | 27 |
| 0.356423 | KR2+ | 6 |
| 0.3575460 | N2 | 39 |
| 0.3575798 | N2 | 39 |
| 0.3575980 | N2 | 39 |
| 0.3576112 | N2 | 39 |
| 0.3576194 | N2 | 39 |
| 0.3576250 | N2 | 39 |
| 0.3576320 | N2 | 39 |
| 0.3576571 | N2 | 39 |
| 0.357661 | AR+ | 4 |
| 0.3576613 | N2 | 39 |
| 0.3576778 | N2 | 39 |
| 0.3576899 | N2 | 40 |
| 0.3576955 | N2 | 40 |
| 0.358744 | AL+ | 17 |
| 0.359661 | XE2+ | 7 |
| 0.360 | FE | 9 |
| 0.360210 | CL2+ | 27 |
| 0.361283 | CL2+ | 27 |
| 0.362268 | CL2+ | 27 |
| 0.363789 | AR2+ | 4 |
| 0.363954 | PB | 20 |
| 0.3645478 | XE3+ | 7 |
| 0.365 | HG | 16 |
| 0.365-0.5 | S2 | 47 |
| 0.366921 | XE2+ | 7 |
| 0.37052 | AR2+ | 4 |
| 0.3709354 | S2+ | 24 |
| 0.371309 | NE+ | 2 |
| 0.3720436 | CL2+ | 27 |
| 0.374571 | XE2+ | 7 |
| 0.3748770 | CL2+ | 27 |
| 0.374949 | O+ | 23 |
| 0.375426 | O2+ | 23 |
| 0.375467 | O2+ | 23 |
| 0.375979 | XE3+ | 7 |
| 0.375988 | O2+ | 23 |
| 0.3780990 | XE2+ | 7 |
| 0.379532 | AR2+ | 4 |
| 0.380322 | XE3+ | 7 |
| 0.3804 | N2 | 40 |
| 0.385 | FE | 9 |
| 0.385829 | AR2+ | 4 |
| 0.395 | FE | 9 |
| 0.397302 | XE3+ | 7 |
| 0.399501 | N+ | 21 |
| 0.402472 | F+ | 26 |
| 0.4045 | K | 10 |
| 0.405005 | XE2+ | 7 |
| 0.405779 | PB | 20 |
| 0.4058 | N2 | 40 |
| 0.406048 | XE2+ | 7 |
| 0.4062 | PB | 20 |
| 0.406737 | KR2+ | 6 |
| 0.40859 | AG+ | 12 |
| 0.409732 | N2+ | 21 |
| 0.410338 | N2+ | 21 |
| 0.413133 | KR2+ | 6 |
| 0.413250 | CL+ | 27 |
| 0.414671 | AR2+ | 4 |
| 0.415444 | KR2+ | 6 |
| 0.417179 | KR2+ | 6 |
| 0.4172 | GA | 18 |
| 0.418298 | AR2+ | 4 |
| 0.4210 | RB | 10 |
| 0.421401 | XE2+ | 7 |
| 0.422658 | KR2+ | 6 |
| 0.424024 | XE2+ | 7 |
| 0.427259 | XE2+ | 7 |
| 0.4278 | N2+ | 40 |
| 0.428588 | XE2+ | 7 |
| 0.430577 | XE3+ | 7 |
| 0.431781 | KR+ | 6 |
| 0.4340 | H | 9 |
| 0.434738 | O+ | 23 |
| 0.435128 | O+ | 23 |
| 0.437075 | AR+ | 4 |
| 0.438654 | KR+ | 6 |
| 0.441314 | XE2+ | 7 |
| 0.441488 | O+ | 23 |
| 0.441563 | CD+ | 16 |
| 0.441697 | O+ | 23 |
| 0.442208 | P2+ | 21 |
| 0.443415 | XE2+ | 7 |
| 0.444329 | KR2+ | 6 |
| 0.446760 | SE+ | 25 |
| 0.448181 | AR+ | 4 |
| 0.448855 | I+ | 28 |
| 0.450248 | CO | 67 |
| 0.450382 | CO | 67 |
| 0.450501 | CO | 67 |
| 0.450600 | CU+ | 11 |
| 0.450602 | CO | 67 |
| 0.450690 | CO | 67 |
| 0.450762 | CO | 67 |
| 0.450821 | CO | 67 |
| 0.451076 | CO | 67 |
| 0.451088 | N2+ | 21 |
| 0.4511 | IN | 18 |
| 0.451487 | N2+ | 21 |

| WAVELENGTH [MICROMETER] | ACTIVE MEDIUM | PAGE |
|---|---|---|
| 0.4529 | FE | 9 |
| 0.453379 | I+ | 28 |
| 0.454505 | AR+ | 4 |
| 0.455259 | SI2+ | 19 |
| 0.4555 | CS | 11 |
| 0.455592 | CU+ | 11 |
| 0.455862 | XE3+ | 7 |
| 0.456084 | BI2+ | 23 |
| 0.456784 | SI2+ | 19 |
| 0.457720 | KR+ | 6 |
| 0.457935 | AR+ | 4 |
| 0.458285 | KR+ | 6 |
| 0.460303 | XE+ | 7 |
| 0.460434 | SE+ | 25 |
| 0.460552 | O+ | 23 |
| 0.460956 | AR+ | 4 |
| 0.461528 | KR+ | 6 |
| 0.461877 | SE+ | 25 |
| 0.461915 | KR+ | 6 |
| 0.463055 | N+ | 21 |
| 0.463386 | KR+ | 6 |
| 0.464745 | C2+ | 19 |
| 0.464759 | XE3+ | 7 |
| 0.464844 | SE+ | 25 |
| 0.464914 | O+ | 23 |
| 0.465016 | C2+ | 19 |
| 0.465016 | KR+ | 6 |
| 0.465073 | XE3+ | 7 |
| 0.465789 | AR+ | 4 |
| 0.467356 | CU+ | 11 |
| 0.467368 | XE2+ | 7 |
| 0.467440 | I+ | 28 |
| 0.4675 | BE | 13 |
| 0.467553 | I+ | 28 |
| 0.468041 | KR+ | 6 |
| 0.468082 | IN+ | 18 |
| 0.468199 | CU+ | 11 |
| 0.468354 | XE2+ | 7 |
| 0.469444 | KR+ | 6 |
| 0.471823 | SE+ | 25 |
| 0.4722 | BI | 23 |
| 0.472357 | XE2+ | 7 |
| 0.472686 | AR+ | 4 |
| 0.474042 | CL+ | 27 |
| 0.474097 | SE+ | 25 |
| 0.474266 | BR+ | 27 |
| 0.474895 | XE2+ | 7 |
| 0.476243 | KR+ | 6 |
| 0.476365 | SE+ | 25 |
| 0.476486 | AR+ | 4 |
| 0.476552 | SE+ | 25 |
| 0.476573 | KR+ | 6 |
| 0.476871 | CL+ | 27 |
| 0.478134 | CL+ | 27 |
| 0.47884 | AG+ | 12 |
| 0.479701 | HG2+ | 16 |
| 0.481947 | CO | 67 |
| 0.482155 | CO | 67 |
| 0.482348 | CO | 67 |
| 0.482517 | KR+ | 6 |
| 0.482524 | CO | 67 |
| 0.482685 | CO | 67 |
| 0.482829 | CO | 67 |
| 0.482956 | CO | 67 |
| 0.483 | XEF | 31 |
| 0.483067 | CO | 67 |
| 0.483162 | CO | 67 |
| 0.483467 | CO | 67 |
| 0.483503 | CO | 67 |
| 0.483523 | CO | 67 |
| 0.484063 | SE+ | 25 |
| 0.48429 | TE+ | 25 |
| 0.484496 | SE+ | 25 |
| 0.484659 | KR+ | 6 |
| 0.485497 | CU+ | 11 |
| 0.4861 | H | 9 |
| 0.486249 | XE+ | 7 |
| 0.486946 | XE2+ | 7 |
| 0.487986 | AR+ | 4 |
| 0.48820 | CD+ | 16 |
| 0.488730 | XE+ | 7 |
| 0.488903 | AR+ | 4 |
| 0.489 | SE | 25 |
| 0.489685 | CL+ | 27 |
| 0.490483 | CL+ | 27 |
| 0.490973 | CU+ | 11 |
| 0.49116 | ZN+ | 15 |
| 0.491781 | CL+ | 27 |
| 0.492404 | ZN+ | 15 |
| 0.492560 | S+ | 24 |
| 0.493165 | CU+ | 11 |
| 0.493467 | I+ | 28 |
| 0.49541 | C+ | 19 |
| 0.495414 | XE3+ | 7 |
| 0.496507 | AR+ | 4 |
| 0.496508 | XE3+ | 7 |
| 0.497566 | SE+ | 25 |
| 0.498692 | I+ | 28 |
| 0.499275 | SE+ | 25 |
| 0.49928 | AR2+ | 4 |
| 0.500772 | XE3+ | 7 |
| 0.50116 | S+ | 24 |
| 0.501261 | CU+ | 11 |
| 0.501424 | S+ | 24 |
| 0.501645 | KR2+ | 6 |
| 0.501716 | AR+ | 4 |
| 0.5018 | HGBR | 32 |
| 0.5020 | HGBR | 32 |
| 0.50204 | TE+ | 25 |
| 0.502129 | CU+ | 11 |
| 0.502240 | KR+ | 6 |
| 0.5023 | HGBR | 32 |
| 0.50259 | CD+ | 16 |
| 0.5026 | HGBR | 32 |

| WAVELENGTH [MICROMETER] | ACTIVE MEDIUM | PAGE |
|---|---|---|
| 0.50273 | AG+ | 12 |
| 0.503262 | S+ | 24 |
| 0.5039 | HGBR | 32 |
| 0.5042 | HGBR | 32 |
| 0.504492 | XE+ | 7 |
| 0.5046 | HGBR | 32 |
| 0.505178 | CU+ | 11 |
| 0.505463 | BR+ | 27 |
| 0.506064 | CU+ | 11 |
| 0.506204 | AR+ | 4 |
| 0.506865 | SE+ | 25 |
| 0.507829 | CL+ | 27 |
| 0.509650 | SE+ | 25 |
| 0.510310 | CL+ | 27 |
| 0.510554 | CU | 11 |
| 0.512573 | KR+ | 6 |
| 0.513175 | GE+ | 20 |
| 0.514179 | AR+ | 4 |
| 0.514214 | SE+ | 25 |
| 0.514532 | AR+ | 4 |
| 0.51457 | C+ | 19 |
| 0.5152 | TL | 18 |
| 0.515703 | XE3+ | 7 |
| 0.515905 | XE3+ | 7 |
| 0.516032 | S+ | 24 |
| 0.517598 | SE+ | 25 |
| 0.517865 | GE+ | 20 |
| 0.517959 | CO | 67 |
| 0.518211 | CO | 67 |
| 0.518238 | BR+ | 27 |
| 0.518442 | CO | 67 |
| 0.518653 | CO | 67 |
| 0.518843 | CO | 67 |
| 0.519001 | CO | 67 |
| 0.519154 | CO | 67 |
| 0.519281 | CO | 67 |
| 0.519363 | CO | 67 |
| 0.519387 | CO | 67 |
| 0.519472 | CO | 67 |
| 0.519488 | CO | 67 |
| 0.519595 | CO | 67 |
| 0.519680 | CO | 67 |
| 0.519743 | CO | 67 |
| 0.519786 | CO | 67 |
| 0.519807 | CO | 67 |
| 0.519807 | CO | 67 |
| 0.520831 | KR+ | 6 |
| 0.5210 | HG2+ | 16 |
| 0.521408 | I+ | 28 |
| 0.521627 | I+ | 28 |
| 0.521776 | CL+ | 27 |
| 0.521962 | S+ | 24 |
| 0.522136 | CL+ | 27 |
| 0.522751 | SE+ | 25 |
| 0.523826 | BR+ | 27 |
| 0.523893 | XE2+ | 7 |
| 0.5250 | NA2 | 37 |
| 0.525307 | SE+ | 25 |
| 0.525363 | SE+ | 25 |
| 0.525637 | XE3+ | 7 |
| 0.52564 | TE+ | 25 |
| 0.525992 | XE+ | 7 |
| 0.526015 | XE3+ | 7 |
| 0.526042 | XE+ | 7 |
| 0.526195 | XE+ | 7 |
| 0.526333 | NA2 | 37 |
| 0.527111 | SE+ | 25 |
| 0.5272 | BE | 13 |
| 0.5279 | NA2 | 37 |
| 0.528690 | AR+ | 4 |
| 0.529816 | NA2 | 37 |
| 0.529952 | NA2 | 37 |
| 0.5300 | XEO | 31 |
| 0.530535 | SE+ | 25 |
| 0.530865 | KR+ | 6 |
| 0.531387 | XE+ | 7 |
| 0.532088 | S+ | 24 |
| 0.5326 | NA2 | 37 |
| 0.533203 | BR+ | 27 |
| 0.533749 | CD+ | 16 |
| 0.5338 | NA2 | 37 |
| 0.534106 | MN | 28 |
| 0.534131 | XE3+ | 7 |
| 0.534283 | NA2 | 37 |
| 0.534331 | XE3+ | 7 |
| 0.5345 | NA2 | 37 |
| 0.534583 | S+ | 24 |
| 0.534930 | NA2 | 37 |
| 0.535 | CA | 14 |
| 0.53503 | TL | 18 |
| 0.535297 | XE3+ | 7 |
| 0.536902 | NA2 | 37 |
| 0.53721 | PB+ | 20 |
| 0.5375 | NA2 | 37 |
| 0.5376 | NA2 | 37 |
| 0.537804 | CD+ | 16 |
| 0.537814 | NA2 | 37 |
| 0.5381 | NA2 | 37 |
| 0.538497 | NA2 | 37 |
| 0.538520 | AS+ | 22 |
| 0.538635 | NA2 | 37 |
| 0.539216 | CL+ | 27 |
| 0.5394 | NA2 | 37 |
| 0.539461 | XE3+ | 7 |
| 0.539525 | XE+ | 7 |
| 0.540 | FE | 9 |
| 0.54006 | NE | 2 |
| 0.540104 | XE2+ | 7 |
| 0.540244 | NA2 | 37 |
| 0.540736 | I+ | 28 |
| 0.541311 | NA2 | 37 |
| 0.5417 | NA2 | 37 |
| 0.5419 | I+ | 28 |
| 0.541915 | XE+ | 7 |

| WAVELENGTH [MICROMETER] | ACTIVE MEDIUM | PAGE |
|---|---|---|
| 0.542036 | MN | 26 |
| 0.5421 | NA2 | 37 |
| 0.542874 | S+ | 24 |
| 0.543287 | S+ | 24 |
| 0.544150 | NA2 | 37 |
| 0.544694 | NA2 | 37 |
| 0.54498 | TE+ | 25 |
| 0.5453 | NA2 | 37 |
| 0.545388 | S+ | 24 |
| 0.54540 | TE+ | 25 |
| 0.5459 | NA2 | 37 |
| 0.5461 | HG | 16 |
| 0.547064 | MN | 26 |
| 0.5472 | NA2 | 37 |
| 0.547374 | S+ | 24 |
| 0.5474 | NA2 | 37 |
| 0.54791 | TE+ | 25 |
| 0.5485 | NA2 | 37 |
| 0.5490 | NA2 | 37 |
| 0.5491 | NA2 | 37 |
| 0.549158 | NA2 | 37 |
| 0.549695 | AS+ | 22 |
| 0.549773 | AS+ | 22 |
| 0.549933 | XE3+ | 7 |
| 0.55020 | BR2 | 49 |
| 0.550220 | AR2+ | 4 |
| 0.5504 | NA2 | 37 |
| 0.55053 | BR2 | 49 |
| 0.550990 | S+ | 24 |
| 0.55163 | AU+ | 13 |
| 0.551677 | MN | 26 |
| 0.55221 | AU+ | 13 |
| 0.552242 | SE+ | 25 |
| 0.552442 | XE2+ | 7 |
| 0.553776 | MN | 26 |
| 0.5543 | I2 | 50 |
| 0.5550 | I2 | 50 |
| 0.5550 | XEO | 31 |
| 0.555809 | AS+ | 22 |
| 0.5562 | NA2 | 37 |
| 0.556511 | S+ | 24 |
| 0.556693 | SE+ | 25 |
| 0.5567 | I2 | 50 |
| 0.5568 | NA2 | 37 |
| 0.5571 | TE2 | 48 |
| 0.5575 | TE2 | 48 |
| 0.5575 | TE2 | 47 |
| 0.55762 | HGCL | 32 |
| 0.55762 | TE+ | 25 |
| 0.55764 | TE+ | 25 |
| 0.5577 | ARO | 31 |
| 0.5578 | TE2 | 48 |
| 0.55781 | KRO | 31 |
| 0.5579 | TE2 | 47 |
| 0.558 | FE | 9 |
| 0.55835 | HGCL | 32 |
| 0.558362 | CO | 67 |
| 0.558411 | CO | 67 |
| 0.558748 | CO | 67 |
| 0.5589 | SN+ | 20 |
| 0.559058 | CO | 67 |
| 0.559116 | SE+ | 25 |
| 0.55923 | XE3+ | 8 |
| 0.559237 | O2+ | 23 |
| 0.559312 | I+ | 28 |
| 0.559343 | CO | 67 |
| 0.5596 | NA2 | 37 |
| 0.559602 | CO | 67 |
| 0.560043 | CO | 67 |
| 0.560224 | CO | 67 |
| 0.560380 | CO | 67 |
| 0.560509 | CO | 67 |
| 0.560563 | CO | 67 |
| 0.560613 | CO | 67 |
| 0.560693 | CO | 67 |
| 0.560700 | CO | 67 |
| 0.560811 | CO | 68 |
| 0.560898 | CO | 68 |
| 0.560958 | CO | 68 |
| 0.560993 | CO | 68 |
| 0.562313 | SE+ | 25 |
| 0.562569 | I+ | 28 |
| 0.5626 | TE2 | 48 |
| 0.563 | FE | 9 |
| 0.5638 | TE2 | 48 |
| 0.564012 | S+ | 24 |
| 0.56405 | TE+ | 25 |
| 0.5642 | TE2 | 48 |
| 0.5643 | TE2 | 47 |
| 0.5646 | TE2 | 48 |
| 0.5647 | TE2 | 47 |
| 0.564716 | S+ | 24 |
| 0.5649 | TE2 | 48 |
| 0.5650 | TE2 | 47 |
| 0.56516 | AS+ | 22 |
| 0.565938 | XE+ | 8 |
| 0.56662 | TE+ | 25 |
| 0.566663 | N+ | 21 |
| 0.567601 | N+ | 21 |
| 0.56773 | HG+ | 16 |
| 0.567808 | I+ | 28 |
| 0.567956 | N+ | 21 |
| 0.5680 | I2 | 50 |
| 0.568188 | KR+ | 6 |
| 0.5696 | TE2 | 48 |
| 0.5697 | I2 | 50 |
| 0.569788 | SE+ | 25 |
| 0.5700 | CU | 11 |
| 0.5701 | TE2 | 48 |
| 0.5701 | TE2 | 48 |
| 0.57081 | TE+ | 25 |
| 0.5711 | TE2 | 48 |
| 0.5714 | TE2 | 48 |
| 0.5715 | TE2 | 47 |

| WAVELENGTH [MICROMETER] | ACTIVE MEDIUM | PAGE |
|---|---|---|
| 0.5715 | TE2 | 46 |
| 0.5719 | TE2 | 48 |
| 0.5719 | TE2 | 47 |
| 0.571921 | BI+ | 23 |
| 0.5720 | TE2 | 47 |
| 0.5721 | TE2 | 46 |
| 0.5724 | TE2 | 47 |
| 0.572691 | XE+ | 6 |
| 0.57416 | TE+ | 25 |
| 0.5745 | I2 | 50 |
| 0.574762 | SE+ | 25 |
| 0.575103 | XE+ | 8 |
| 0.575298 | KR+ | 6 |
| 0.57559 | TE+ | 25 |
| 0.57563 | TE+ | 25 |
| 0.576072 | I+ | 26 |
| 0.5764 | I2 | 50 |
| 0.5766 | TE2 | 46 |
| 0.5767 | TE2 | 47 |
| 0.5773 | TE2 | 46 |
| 0.5774 | TE2 | 48 |
| 0.5780 | TE2 | 46 |
| 0.578213 | CU | 11 |
| 0.5783 | TE2 | 46 |
| 0.5784 | TE2 | 47 |
| 0.5785 | TE2 | 48 |
| 0.5786 | TE2 | 46 |
| 0.5787 | TE2 | 48 |
| 0.5790 | TE2 | 47 |
| 0.5793 | TE2 | 46 |
| 0.5794 | TE2 | 46 |
| 0.5797 | TE2 | 46 |
| 0.5798 | TE2 | 48 |
| 0.579918 | SN+ | 20 |
| 0.58048 | BR2 | 49 |
| 0.58090 | BR2 | 49 |
| 0.5815 | I2 | 50 |
| 0.581935 | S+ | 24 |
| 0.5830 | I2 | 50 |
| 0.583790 | AS+ | 22 |
| 0.5841 | TE2 | 46 |
| 0.584268 | SE+ | 25 |
| 0.5849 | TE2 | 46 |
| 0.5851 | TE2 | 47 |
| 0.5851 | TE2 | 46 |
| 0.58511 | TE+ | 25 |
| 0.58525 | NE | 2 |
| 0.5857 | TE2 | 46 |
| 0.5859 | TE2 | 47 |
| 0.586 | CA | 14 |
| 0.5865 | TE2 | 46 |
| 0.5866 | NA | 10 |
| 0.586627 | SE+ | 25 |
| 0.5869 | TE2 | 48 |
| 0.5870 | TE2 | 46 |
| 0.5874 | TE2 | 46 |
| 0.5880 | I2 | 50 |
| 0.5890 | NA | 10 |
| 0.589330 | XE+ | 8 |
| 0.58944 | ZN+ | 15 |
| 0.5905 | I2 | 50 |
| 0.5924 | TE2 | 48 |
| 0.5927 | TE2 | 47 |
| 0.5929 | BI2 | 46 |
| 0.5934 | TE2 | 48 |
| 0.5936 | TE2 | 47 |
| 0.59361 | TE+ | 25 |
| 0.59393 | NE | 2 |
| 0.59448 | NE | 2 |
| 0.5949 | TL | 18 |
| 0.595565 | XE3+ | 8 |
| 0.5969 | I2 | 50 |
| 0.597111 | XE+ | 8 |
| 0.59726 | TE+ | 25 |
| 0.59747 | TE+ | 25 |
| 0.6.27 | NH3 | 108 |
| 0.6.69 | NH3 | 108 |
| 0.6002 | TE2 | 48 |
| 0.6004 | TE2 | 47 |
| 0.6005 | TE2 | 48 |
| 0.6005 | TE2 | 48 |
| 0.6008 | TE2 | 47 |
| 0.6009 | TE2 | 48 |
| 0.6009 | TE2 | 48 |
| 0.60145 | TE+ | 25 |
| 0.6021 | ZN+ | 15 |
| 0.602421 | P+ | 21 |
| 0.6025 | I2 | 50 |
| 0.603421 | P+ | 21 |
| 0.603717 | KR2+ | 6 |
| 0.60381 | KR+ | 6 |
| 0.604325 | P+ | 21 |
| 0.604369 | CO | 68 |
| 0.60461 | NE | 2 |
| 0.6048 | I2 | 50 |
| 0.604816 | CO | 68 |
| 0.605232 | CO | 68 |
| 0.605596 | SE+ | 25 |
| 0.605619 | CO | 68 |
| 0.606296 | CO | 68 |
| 0.606583 | SE+ | 25 |
| 0.606588 | CO | 68 |
| 0.606848 | CO | 68 |
| 0.606893 | I+ | 28 |
| 0.607075 | CO | 68 |
| 0.607270 | CO | 68 |
| 0.607436 | CO | 68 |
| 0.607566 | CO | 68 |
| 0.607584 | CO | 68 |
| 0.607663 | CO | 68 |
| 0.607731 | CO | 68 |
| 0.607850 | CO | 68 |
| 0.607934 | CO | 68 |
| 0.607967 | CO | 68 |

| WAVELENGTH [MICROMETER] | ACTIVE MEDIUM | PAGE |
|---|---|---|
| 0.608007 | CU | 68 |
| 0.6082 | TE2 | 48 |
| 0.60823 | TE+ | 25 |
| 0.6085 | TE2 | 48 |
| 0.6085 | TE2 | 48 |
| 0.6087 | TE2 | 47 |
| 0.608786 | P+ | 21 |
| 0.6089 | TE2 | 48 |
| 0.609361 | XE+ | 8 |
| 0.609473 | CL+ | 27 |
| 0.610196 | SE+ | 25 |
| 0.6102 | CA | 14 |
| 0.610253 | ZN+ | 15 |
| 0.6110 | I2 | 50 |
| 0.611756 | BR+ | 27 |
| 0.61180 | NE | 2 |
| 0.6122 | CA | 14 |
| 0.61272 | BR2 | 49 |
| 0.612749 | I+ | 28 |
| 0.61299 | SB+ | 22 |
| 0.61316 | BR2 | 49 |
| 0.61318 | BR2 | 49 |
| 0.61368 | BR2 | 49 |
| 0.614172 | BA+ | 15 |
| 0.61431 | NE | 2 |
| 0.61499 | HG+ | 16 |
| 0.6160 | BI2 | 46 |
| 0.6162 | TE2 | 46 |
| 0.6162 | CA | 14 |
| 0.6165 | TE2 | 48 |
| 0.6165 | TE2 | 48 |
| 0.616577 | P+ | 21 |
| 0.6168 | TE2 | 47 |
| 0.616878 | BR+ | 27 |
| 0.616880 | KR+ | 6 |
| 0.6170 | TE2 | 48 |
| 0.617027 | AS+ | 22 |
| 0.617482 | I2 | 50 |
| 0.61766 | XE2+ | 8 |
| 0.617676 | I2 | 50 |
| 0.617868 | I2 | 50 |
| 0.617947 | I2 | 50 |
| 0.618193 | I2 | 50 |
| 0.618267 | I2 | 50 |
| 0.618441 | I2 | 50 |
| 0.618535 | I2 | 50 |
| 0.6198 | I2 | 50 |
| 0.6204 | TE2 | 48 |
| 0.620486 | I+ | 28 |
| 0.62123 | AU+ | 13 |
| 0.62307 | TE+ | 25 |
| 0.623824 | XE2+ | 8 |
| 0.6239 | BI2 | 46 |
| 0.6239651 | F | 26 |
| 0.62454 | TE+ | 25 |
| 0.6258 | I2 | 50 |
| 0.627081 | XE+ | 8 |
| 0.6278 | TE2 | 48 |
| 0.627818 | AU | 13 |
| 0.62865 | XE3+ | 8 |
| 0.6287 | TE2 | 48 |
| 0.6288 | TE2 | 48 |
| 0.6295 | TE2 | 48 |
| 0.6300 | BI2 | 46 |
| 0.631022 | KR2+ | 6 |
| 0.631276 | KR+ | 6 |
| 0.63282 | NE | 2 |
| 0.6330 | I2 | 50 |
| 0.6339 | BI2 | 46 |
| 0.633997 | I+ | 28 |
| 0.63435 | XE2+ | 8 |
| 0.634724 | SI+ | 19 |
| 0.6348508 | F | 26 |
| 0.63497 | TE+ | 25 |
| 0.63518 | NE | 2 |
| 0.6352 | I2 | 50 |
| 0.63548 | CD+ | 16 |
| 0.63601 | CD+ | 16 |
| 0.63612 | BR2 | 49 |
| 0.63654 | BR2 | 49 |
| 0.63654 | BR2 | 49 |
| 0.63705 | BR2 | 49 |
| 0.6371 | TE2 | 48 |
| 0.637148 | SI+ | 19 |
| 0.6379 | TE2 | 48 |
| 0.638075 | SR | 14 |
| 0.6381 | TE2 | 48 |
| 0.6388 | TE2 | 48 |
| 0.64011 | NE | 2 |
| 0.64027 | AG+ | 12 |
| 0.6413651 | F | 26 |
| 0.6414 | BI2 | 46 |
| 0.641661 | KR+ | 6 |
| 0.6422 | BI2 | 46 |
| 0.644425 | SE+ | 25 |
| 0.644981 | CA | 14 |
| 0.645350 | SN+ | 20 |
| 0.6465 | TE2 | 48 |
| 0.647088 | KR+ | 6 |
| 0.6473 | TE2 | 48 |
| 0.6477 | TE2 | 48 |
| 0.64831 | AR+ | 4 |
| 0.6484 | TE2 | 48 |
| 0.6490 | I2 | 50 |
| 0.649048 | SE+ | 25 |
| 0.649690 | BA+ | 15 |
| 0.65015 | HG2+ | 16 |
| 0.65100 | KR+ | 6 |
| 0.6511 | I2 | 50 |
| 0.651174 | AS+ | 22 |
| 0.651618 | I+ | 28 |
| 0.652865 | XE+ | 8 |
| 0.653495 | SE+ | 25 |
| 0.6561 | TE2 | 48 |

| WAVELENGTH [MICROMETER] | ACTIVE MEDIUM | PAGE |
|---|---|---|
| 0.6569 | TE2 | 48 |
| 0.657012 | KR+ | 6 |
| 0.6574 | TE2 | 48 |
| 0.6576 | BI2 | 46 |
| 0.65780 | C+ | 19 |
| 0.6579 | SN | 20 |
| 0.6581 | TE2 | 48 |
| 0.658103 | CO | 68 |
| 0.6582 | BI2 | 46 |
| 0.65851 | TE+ | 25 |
| 0.658521 | I+ | 28 |
| 0.658623 | CO | 68 |
| 0.659105 | CO | 68 |
| 0.6592 | I2 | 50 |
| 0.659547 | CO | 68 |
| 0.659948 | CO | 68 |
| 0.66029 | KR+ | 6 |
| 0.6603 | BI2 | 46 |
| 0.660306 | CO | 68 |
| 0.660630 | CO | 68 |
| 0.660910 | CO | 68 |
| 0.660971 | CO | 68 |
| 0.661151 | CO | 68 |
| 0.661243 | CO | 68 |
| 0.661353 | CO | 68 |
| 0.661512 | CO | 68 |
| 0.661634 | CO | 68 |
| 0.661919 | CO | 68 |
| 0.662003 | CO | 68 |
| 0.662033 | CO | 68 |
| 0.662235 | I+ | 28 |
| 0.66402 | U+ | 23 |
| 0.6645 | EU+ | 29 |
| 0.6645 | I2 | 50 |
| 0.66486 | TE+ | 25 |
| 0.6650 | BI2 | 46 |
| 0.667193 | SI+ | 19 |
| 0.667193 | P | 21 |
| 0.66761 | TE+ | 25 |
| 0.66943 | XE+ | 8 |
| 0.669950 | XE3+ | 8 |
| 0.67014 | AU+ | 13 |
| 0.6717 | CA | 14 |
| 0.672136 | O | 23 |
| 0.6730 | AR+ | 4 |
| 0.67408 | BR2 | 49 |
| 0.67455 | BR2 | 49 |
| 0.67456 | BR2 | 49 |
| 0.67506 | BR2 | 49 |
| 0.676442 | KR+ | 6 |
| 0.67838 | C+ | 19 |
| 0.6809 | BI2 | 46 |
| 0.682523 | I+ | 28 |
| 0.684405 | SN+ | 20 |
| 0.68613 | AR+ | 4 |
| 0.687084 | KR+ | 6 |
| 0.68851 | TE+ | 26 |
| 0.69029 | AU+ | 13 |
| 0.690477 | I+ | 28 |
| 0.691996 | AL+ | 17 |
| 0.6936 | I2 | 50 |
| 0.69403 | AU+ | 13 |
| 0.6950 | TL | 18 |
| 0.696635 | F | 26 |
| 0.7006 | BI2 | 46 |
| 0.7013 | BI2 | 46 |
| 0.703299 | I+ | 28 |
| 0.703745 | F | 26 |
| 0.70391 | TE+ | 26 |
| 0.70394 | F | 26 |
| 0.704206 | AL+ | 17 |
| 0.705656 | AL+ | 17 |
| 0.706389 | SE+ | 25 |
| 0.7065 | HG+ | 16 |
| 0.706517 | HE | 1 |
| 0.706521 | HE | 1 |
| 0.70723 | XE+ | 8 |
| 0.710272 | AS+ | 22 |
| 0.7114 | I2 | 51 |
| 0.712033 | BA | 15 |
| 0.712788 | F | 26 |
| 0.71298 | F | 26 |
| 0.713897 | I+ | 28 |
| 0.714894 | XE2+ | 8 |
| 0.720237 | F | 26 |
| 0.72043 | F | 26 |
| 0.72291 | PB | 20 |
| 0.72369 | CD+ | 16 |
| 0.72558 | CU+ | 11 |
| 0.7292 | BI2 | 46 |
| 0.7301 | BI2 | 46 |
| 0.73048 | NE | 2 |
| 0.7310102 | F | 26 |
| 0.7335 | BI2 | 47 |
| 0.73466 | HG+ | 16 |
| 0.7364 | BI2 | 47 |
| 0.7366 | BI2 | 47 |
| 0.7376 | BI2 | 47 |
| 0.739199 | SE+ | 25 |
| 0.7398 | BI2 | 47 |
| 0.7398688 | F | 26 |
| 0.73999 | CU+ | 11 |
| 0.740434 | CU+ | 11 |
| 0.7408 | BI2 | 47 |
| 0.74181 | HG+ | 16 |
| 0.74257 | F | 26 |
| 0.7435764 | KR+ | 6 |
| 0.74382 | CU+ | 11 |
| 0.7439 | BI2 | 47 |
| 0.74582 | BR2 | 49 |
| 0.74638 | BR2 | 49 |
| 0.74641 | BR2 | 49 |
| 0.7468 | BI2 | 47 |
| 0.74704 | BR2 | 49 |

| WAVELENGTH [MICROMETER] | ACTIVE MEDIUM | PAGE |
|---|---|---|
| 0.7471 | BI2 | 47 |
| 0.747137 | AL+ | 17 |
| 0.7475 | BI2 | 47 |
| 0.747879 | ZN+ | 15 |
| 0.7482 | BI2 | 47 |
| 0.7482187 | N2 | 40 |
| 0.74827 | F | 26 |
| 0.748274 | N2 | 40 |
| 0.7485941 | N2 | 40 |
| 0.7486135 | N2 | 40 |
| 0.7486253 | N2 | 40 |
| 0.7486413 | N2 | 40 |
| 0.7487409 | N2 | 40 |
| 0.7488046 | N2 | 40 |
| 0.7488246 | N2 | 40 |
| 0.7489107 | N2 | 40 |
| 0.748914 | F | 26 |
| 0.7489626 | N2 | 40 |
| 0.7489809 | N2 | 40 |
| 0.7490096 | N2 | 40 |
| 0.7490317 | N2 | 40 |
| 0.7491510 | N2 | 40 |
| 0.7491705 | N2 | 40 |
| 0.7492379 | N2 | 40 |
| 0.7493082 | N2 | 40 |
| 0.7493716 | N2 | 40 |
| 0.7493910 | N2 | 40 |
| 0.7495086 | N2 | 40 |
| 0.7495465 | N2 | 40 |
| 0.7495660 | N2 | 40 |
| 0.7496024 | N2 | 40 |
| 0.7497256 | N2 | 40 |
| 0.7497524 | N2 | 40 |
| 0.7497726 | N2 | 40 |
| 0.7498898 | N2 | 40 |
| 0.7499013 | N2 | 40 |
| 0.7499327 | N2 | 40 |
| 0.7499593 | N2 | 40 |
| 0.7499825 | N2 | 40 |
| 0.7500071 | N2 | 40 |
| 0.7500646 | N2 | 40 |
| 0.7500734 | N2 | 40 |
| 0.7501056 | N2 | 40 |
| 0.7501295 | N2 | 40 |
| 0.7501404 | N2 | 40 |
| 0.7501553 | N2 | 40 |
| 0.7502139 | N2 | 40 |
| 0.7502729 | N2 | 40 |
| 0.7502768 | N2 | 40 |
| 0.7503 | AR | 4 |
| 0.7503035 | N2 | 40 |
| 0.7503371 | N2 | 40 |
| 0.7503416 | N2 | 40 |
| 0.7503642 | N2 | 40 |
| 0.7503669 | N2 | 40 |
| 0.7503697 | N2 | 40 |
| 0.7503838 | N2 | 40 |
| 0.7503960 | N2 | 40 |
| 0.7503994 | N2 | 41 |
| 0.7504106 | N2 | 41 |
| 0.7504160 | N2 | 41 |
| 0.7504184 | N2 | 41 |
| 0.7504274 | N2 | 41 |
| 0.7504598 | N2 | 41 |
| 0.7504768 | N2 | 41 |
| 0.7505113 | N2 | 41 |
| 0.7505710 | N2 | 41 |
| 0.7505903 | N2 | 41 |
| 0.7506063 | N2 | 41 |
| 0.7506356 | N2 | 41 |
| 0.7508145 | N2 | 41 |
| 0.7509890 | N2 | 41 |
| 0.7510133 | N2 | 41 |
| 0.7510923 | N2 | 41 |
| 0.7511592 | N2 | 41 |
| 0.7511799 | N2 | 41 |
| 0.7512003 | N2 | 41 |
| 0.7512569 | N2 | 41 |
| 0.7513357 | N2 | 41 |
| 0.7514079 | N2 | 41 |
| 0.75150 | F | 26 |
| 0.7515446 | N2 | 41 |
| 0.7515650 | N2 | 41 |
| 0.7517728 | N2 | 41 |
| 0.7518013 | N2 | 41 |
| 0.752464 | H2 | 34 |
| 0.752546 | KR | 6 |
| 0.7543 | BI2 | 47 |
| 0.7551 | BI2 | 47 |
| 0.7552235 | F | 26 |
| 0.75558 | AU+ | 13 |
| 0.7574329 | N2 | 41 |
| 0.756105 | N2 | 41 |
| 0.756423 | N2 | 41 |
| 0.7566439 | N2 | 41 |
| 0.7567693 | N2 | 41 |
| 0.756848 | ZN+ | 15 |
| 0.7569668 | N2 | 41 |
| 0.7591960 | N2 | 41 |
| 0.75929 | AU+ | 13 |
| 0.7593908 | N2 | 41 |
| 0.7594941 | N2 | 41 |
| 0.7597269 | N2 | 41 |
| 0.759870 | N2 | 41 |
| 0.75990 | BI2+ | 23 |
| 0.76005 | AU+ | 13 |
| 0.7603 | KR | 6 |
| 0.7603477 | N2 | 41 |
| 0.7606374 | N2 | 41 |
| 0.76067 | AU+ | 13 |
| 0.7607626 | N2 | 41 |
| 0.7608801 | N2 | 41 |
| 0.7609853 | N2 | 41 |
| 0.7610759 | N2 | 41 |

| WAVELENGTH [MICROMETER] | ACTIVE MEDIUM | PAGE |
|---|---|---|
| 0.7611082 | N2 | 41 |
| 0.7611514 | N2 | 41 |
| 0.7612105 | N2 | 41 |
| 0.7612528 | N2 | 41 |
| 0.761290 | ZN+ | 15 |
| 0.7613260 | N2 | 41 |
| 0.7613612 | N2 | 41 |
| 0.7615347 | N2 | 41 |
| 0.7616994 | N2 | 41 |
| 0.7617357 | N2 | 41 |
| 0.761850 | I+ | 28 |
| 0.76186 | XE+ | 8 |
| 0.7619 | RB | 10 |
| 0.7619288 | N2 | 41 |
| 0.7620844 | N2 | 41 |
| 0.7620943 | N2 | 41 |
| 0.7621161 | N2 | 41 |
| 0.7622235 | N2 | 41 |
| 0.7622565 | N2 | 42 |
| 0.7622959 | N2 | 42 |
| 0.7623256 | N2 | 42 |
| 0.7623264 | N2 | 42 |
| 0.7623311 | N2 | 42 |
| 0.7623582 | N2 | 42 |
| 0.7623686 | N2 | 42 |
| 0.7623918 | N2 | 42 |
| 0.7624220 | N2 | 42 |
| 0.7624690 | N2 | 42 |
| 0.7624924 | N2 | 42 |
| 0.7625115 | N2 | 42 |
| 0.7625445 | N2 | 42 |
| 0.7625709 | N2 | 42 |
| 0.7625770 | N2 | 42 |
| 0.7625812 | N2 | 42 |
| 0.7625906 | N2 | 42 |
| 0.7626007 | N2 | 42 |
| 0.7626044 | N2 | 42 |
| 0.7626114 | N2 | 42 |
| 0.7626180 | N2 | 42 |
| 0.7626207 | N2 | 42 |
| 0.7626360 | N2 | 42 |
| 0.7626560 | N2 | 42 |
| 0.7626700 | N2 | 42 |
| 0.7626749 | N2 | 42 |
| 0.7626826 | N2 | 42 |
| 0.7627806 | N2 | 42 |
| 0.7628854 | N2 | 42 |
| 0.7629102 | N2 | 42 |
| 0.7630305 | N2 | 42 |
| 0.7631880 | N2 | 42 |
| 0.7632446 | N2 | 42 |
| 0.7633348 | N2 | 42 |
| 0.7633985 | N2 | 42 |
| 0.7634546 | N2 | 42 |
| 0.7634779 | N2 | 42 |
| 0.76351 | AU+ | 13 |
| 0.7635474 | N2 | 42 |
| 0.7636126 | N2 | 42 |
| 0.7636904 | N2 | 42 |
| 0.7637586 | N2 | 42 |
| 0.7638274 | N2 | 42 |
| 0.7639571 | N2 | 42 |
| 0.7639715 | N2 | 42 |
| 0.7640383 | N2 | 42 |
| 0.7640794 | N2 | 42 |
| 0.7641929 | N2 | 42 |
| 0.7642478 | N2 | 42 |
| 0.7644612 | N2 | 42 |
| 0.766470 | CU+ | 11 |
| 0.7665 | K | 10 |
| 0.767482 | SE+ | 25 |
| 0.7699 | K | 10 |
| 0.771206 | N2 | 42 |
| 0.772404 | SE+ | 25 |
| 0.7724562 | N2 | 42 |
| 0.7725 | S | 24 |
| 0.7730032 | N2 | 42 |
| 0.77325 | ZN+ | 15 |
| 0.7735040 | N2 | 42 |
| 0.773578 | I+ | 28 |
| 0.773868 | CU+ | 11 |
| 0.7739632 | N2 | 42 |
| 0.7743859 | N2 | 42 |
| 0.7752354 | N2 | 42 |
| 0.775270 | N2 | 42 |
| 0.7753652 | N2 | 43 |
| 0.775470 | F | 26 |
| 0.775786 | ZN+ | 15 |
| 0.7758 | RB | 10 |
| 0.777 | SE | 25 |
| 0.777874 | CU+ | 11 |
| 0.779615 | SE+ | 25 |
| 0.7800 | RB | 10 |
| 0.780022 | F | 26 |
| 0.78017 | TE+ | 26 |
| 0.780519 | CU+ | 11 |
| 0.780766 | CU+ | 11 |
| 0.782566 | CU+ | 11 |
| 0.782763 | XE+ | 8 |
| 0.783881 | SE+ | 25 |
| 0.78443 | CD+ | 16 |
| 0.784503 | CU+ | 11 |
| 0.784563 | P | 21 |
| 0.784930 | NA2 | 37 |
| 0.786590 | NA2 | 37 |
| 0.789583 | CU+ | 12 |
| 0.789740 | NA2 | 37 |
| 0.789790 | NA2 | 37 |
| 0.790257 | CU+ | 12 |
| 0.791783 | NA2 | 37 |
| 0.79217 | TE+ | 26 |
| 0.792947 | NA2 | 37 |
| 0.79314 | KR+ | 6 |
| 0.793697 | NA2 | 37 |

| WAVELENGTH [MICROMETER] | ACTIVE MEDIUM | PAGE |
|---|---|---|
| 0.794442 | CU+ | 12 |
| 0.79447 | HG+ | 16 |
| 0.7945 | RB | 10 |
| 0.79624 | NI+ | 9 |
| 0.797474 | NA2 | 37 |
| 0.79754 | NI+ | 9 |
| 0.797657 | NA2 | 37 |
| 0.798800 | XE+ | 8 |
| 0.798817 | CU+ | 12 |
| 0.799091 | NA2 | 37 |
| 0.799322 | KR+ | 6 |
| 0.799660 | NA2 | 37 |
| 0.80011 | NA2 | 38 |
| 0.8002 | NA2 | 38 |
| 0.80054 | AG+ | 12 |
| 0.800840 | NA2 | 38 |
| 0.80154 | NA2 | 38 |
| 0.803650 | NA2 | 38 |
| 0.803931 | NA2 | 38 |
| 0.804447 | NA2 | 38 |
| 0.805366 | NA2 | 38 |
| 0.805611 | NA2 | 38 |
| 0.8066 | NA2 | 38 |
| 0.80669 | CD+ | 16 |
| 0.80689 | BI2+ | 23 |
| 0.806943 | NA2 | 38 |
| 0.80715 | NA2 | 38 |
| 0.80859 | NA2 | 38 |
| 0.808858 | CU+ | 12 |
| 0.8096 | CU+ | 12 |
| 0.810433 | KR | 6 |
| 0.8144 | I2 | 51 |
| 0.817007 | I+ | 28 |
| 0.819228 | CU+ | 12 |
| 0.823162 | XE+ | 8 |
| 0.825381 | I+ | 28 |
| 0.82547 | AG+ | 12 |
| 0.8263 | AG+ | 12 |
| 0.82729 | AU+ | 13 |
| 0.8277 | CU+ | 12 |
| 0.827752 | D2 | 36 |
| 0.828037 | KR+ | 6 |
| 0.828321 | CU+ | 12 |
| 0.830952 | SE+ | 25 |
| 0.83244 | AG+ | 12 |
| 0.833270 | XE+ | 8 |
| 0.834950 | H2 | 34 |
| 0.8358 | I2 | 51 |
| 0.83795 | AG+ | 12 |
| 0.84032 | AG+ | 12 |
| 0.840919 | XE+ | 8 |
| 0.844628 | U | 23 |
| 0.844638 | U | 23 |
| 0.844672 | U | 23 |
| 0.844680 | U | 23 |
| 0.84634 | NE | 2 |
| 0.851104 | CU+ | 12 |
| 0.8521 | CS | 11 |
| 0.85300 | CD+ | 16 |
| 0.854209 | CA+ | 14 |
| 0.85498 | HG+ | 16 |
| 0.85716 | XE2+ | 8 |
| 0.8578 | I2 | 51 |
| 0.858251 | XE+ | 8 |
| 0.85878 | KR+ | 6 |
| 0.8594 | N | 21 |
| 0.86046 | TE+ | 26 |
| 0.8622 | HG+ | 16 |
| 0.86284 | N | 21 |
| 0.86353 | NE | 2 |
| 0.865331 | N2 | 43 |
| 0.865492 | N2 | 43 |
| 0.866089 | N2 | 43 |
| 0.866214 | CA+ | 14 |
| 0.866256 | N2 | 43 |
| 0.866345 | N2 | 43 |
| 0.866572 | N2 | 43 |
| 0.86676 | N2 | 43 |
| 0.8669223 | N2 | 43 |
| 0.866959 | N2 | 43 |
| 0.8671332 | N2 | 43 |
| 0.867554 | N2 | 43 |
| 0.8677 | HG+ | 16 |
| 0.868281 | N2 | 43 |
| 0.8682937 | N2 | 43 |
| 0.868374 | N2 | 43 |
| 0.868762 | N2 | 43 |
| 0.86901 | KR+ | 6 |
| 0.869136 | N2 | 43 |
| 0.8692580 | N2 | 43 |
| 0.869490 | N2 | 43 |
| 0.8696366 | N2 | 43 |
| 0.8697945 | N2 | 43 |
| 0.8698263 | N2 | 43 |
| 0.8699397 | N2 | 43 |
| 0.8700670 | N2 | 43 |
| 0.8700684 | N2 | 43 |
| 0.8701481 | N2 | 43 |
| 0.8701718 | N2 | 43 |
| 0.8702541 | N2 | 43 |
| 0.8702681 | N2 | 43 |
| 0.8703093 | N2 | 43 |
| 0.870331 | N2 | 43 |
| 0.8703457 | N2 | 43 |
| 0.8704549 | N2 | 43 |
| 0.8707478 | N2 | 43 |
| 0.8710118 | N2 | 43 |
| 0.8710273 | N2 | 43 |
| 0.8712956 | N2 | 43 |
| 0.8713533 | N2 | 43 |
| 0.871450 | N2 | 43 |
| 0.8715519 | N2 | 43 |
| 0.871617 | XE+ | 8 |
| 0.871644 | N2 | 43 |

| WAVELENGTH [MICROMETER] | ACTIVE MEDIUM | PAGE | WAVELENGTH [MICROMETER] | ACTIVE MEDIUM | PAGE |
|---|---|---|---|---|---|
| 0.8716718 | N2 | 43 | 0.886256 | N2 | 44 |
| 0.8717377 | N2 | 43 | 0.886278 | N2 | 44 |
| 0.8717970 | N2 | 43 | 0.88653 | NE | 2 |
| 0.8718571 | N2 | 43 | 0.886697 | N2 | 44 |
| 0.8718654 | N2 | 43 | 0.88676 | AU+ | 13 |
| 0.8719537 | N2 | 43 | 0.886799 | N2 | 44 |
| 0.8719562 | N2 | 43 | 0.887121 | N2 | 44 |
| 0.8719562 | N2 | 43 | 0.887531 | N2 | 44 |
| 0.8719791 | N2 | 43 | 0.887613 | H2 | 34 |
| 0.8720251 | N2 | 43 | 0.887761 | I+ | 28 |
| 0.8720284 | N2 | 43 | 0.88778 | CD+ | 16 |
| 0.8720308 | N2 | 43 | 0.887918 | N2 | 44 |
| 0.8720419 | N2 | 43 | 0.8880521 | N2 | 44 |
| 0.8720848 | N2 | 43 | 0.888288 | N2 | 44 |
| 0.8721155 | N2 | 43 | 0.8884527 | N2 | 44 |
| 0.8721327 | N2 | 44 | 0.8886204 | N2 | 44 |
| 0.8721718 | N2 | 44 | 0.8886378 | N2 | 44 |
| 0.8721971 | N2 | 44 | 0.8887756 | N2 | 44 |
| 0.8722007 | N2 | 44 | 0.8889111 | N2 | 44 |
| 0.8722220 | N2 | 44 | 0.8889738 | N2 | 44 |
| 0.8722341 | N2 | 44 | 0.8890243 | N2 | 44 |
| 0.8722569 | N2 | 44 | 0.8891133 | N2 | 44 |
| 0.8722836 | N2 | 44 | 0.8891769 | N2 | 44 |
| 0.8723057 | N2 | 44 | 0.8892149 | N2 | 44 |
| 0.8726333 | N2 | 44 | 0.8892940 | N2 | 44 |
| 0.8728430 | N2 | 44 | 0.8896001 | N2 | 44 |
| 0.8730453 | N2 | 44 | 0.889882 | H2 | 34 |
| 0.8732394 | N2 | 44 | 0.8898930 | N2 | 44 |
| 0.87338 | TE+ | 26 | 0.8899078 | N2 | 44 |
| 0.8734247 | N2 | 44 | 0.8901733 | N2 | 44 |
| 0.8735995 | N2 | 44 | 0.8902420 | N2 | 44 |
| 0.8737644 | N2 | 44 | 0.8902711 | N2 | 45 |
| 0.8739162 | N2 | 44 | 0.890372 | N2 | 45 |
| 0.8740559 | N2 | 44 | 0.8904419 | N2 | 45 |
| 0.8742917 | N2 | 44 | 0.890566 | N2 | 45 |
| 0.87476 | AG+ | 12 | 0.8906097 | N2 | 45 |
| 0.8764 | CS | 11 | 0.8906649 | N2 | 45 |
| 0.87717 | NE | 2 | 0.8906994 | N2 | 45 |
| 0.877186 | AR+ | 4 | 0.8907920 | N2 | 45 |
| 0.8772 | AG+ | 12 | 0.8908808 | N2 | 45 |
| 0.8804 | I2 | 51 | 0.8908878 | N2 | 45 |
| 0.880428 | I+ | 28 | 0.8909451 | N2 | 45 |
| 0.88058 | KR | 6 | 0.8909527 | N2 | 45 |
| 0.8813 | I2 | 51 | 0.8909750 | N2 | 45 |
| 0.884129 | N2 | 44 | 0.8910132 | N2 | 45 |
| 0.8845349 | N2 | 44 | 0.8910480 | N2 | 45 |
| 0.8846598 | N2 | 44 | 0.8910612 | N2 | 45 |
| 0.884758 | N2 | 44 | 0.8911001 | N2 | 45 |
| 0.884920 | N2 | 44 | 0.8911063 | N2 | 45 |
| 0.885026 | N2 | 44 | 0.8911280 | N2 | 45 |
| 0.885261 | N2 | 44 | 0.8911502 | N2 | 45 |
| 0.885460 | N2 | 44 | 0.8911538 | N2 | 45 |
| 0.8856271 | N2 | 44 | 0.8911608 | N2 | 45 |
| 0.885649 | N2 | 44 | 0.8911898 | N2 | 45 |
| 0.8858470 | N2 | 44 | 0.8912139 | N2 | 45 |
| 0.8861195 | N2 | 44 | 0.8918033 | N2 | 45 |
| 0.886153 | N2 | 44 | 0.8920184 | N2 | 45 |

| WAVELENGTH [MICROMETER] | ACTIVE MEDIUM | PAGE |
|---|---|---|
| 0.8922249 | N2 | 45 |
| 0.8924223 | N2 | 45 |
| 0.8926099 | N2 | 45 |
| 0.8927865 | N2 | 45 |
| 0.8929 | KR | 6 |
| 0.8929509 | N2 | 45 |
| 0.8931019 | N2 | 45 |
| 0.8933580 | N2 | 45 |
| 0.8943 | CS | 11 |
| 0.89721 | TE+ | 26 |
| 0.89886 | NE | 2 |
| 0.89982 | TE+ | 26 |
| 0.9.3 | NH3 | 106 |
| 0.9.6 | NH3 | 106 |
| 0.9.7 | NH3 | 106 |
| 0.9.75 | CH3F | 120 |
| 0.9.9 | NH3 | 106 |
| 0.9037 | I2 | 51 |
| 0.904539 | XE | 8 |
| 0.90455 | N | 21 |
| 0.9047 | I2 | 51 |
| 0.905930 | XE+ | 8 |
| 0.9060 | I2 | 51 |
| 0.912297 | AR | 4 |
| 0.917201 | HD | 37 |
| 0.9218 | MG | 14 |
| 0.9244 | MG | 14 |
| 0.92493 | SE+ | 25 |
| 0.926539 | XE+ | 8 |
| 0.9274 | I2 | 51 |
| 0.92874 | NE+ | 2 |
| 0.9288 | I2 | 51 |
| 0.928854 | XE+ | 8 |
| 0.9295 | I2 | 51 |
| 0.9305 | I2 | 51 |
| 0.93779 | TE+ | 26 |
| 0.93862 | N | 21 |
| 0.93921 | N | 21 |
| 0.93968 | HG+ | 16 |
| 0.944156 | D2 | 36 |
| 0.9451 | CL | 27 |
| 0.9518 | I2 | 51 |
| 0.952367 | D2 | 36 |
| 0.953005 | D2 | 36 |
| 0.9545 | I2 | 51 |
| 0.9555 | I2 | 51 |
| 0.965389 | N2 | 45 |
| 0.965778 | AR | 4 |
| 0.965846 | N2 | 45 |
| 0.966599 | N2 | 45 |
| 0.967270 | N2 | 45 |
| 0.967758 | N2 | 45 |
| 0.967943 | N2 | 45 |
| 0.968061 | N2 | 45 |
| 0.969552 | N2 | 45 |
| 0.969859 | XE+ | 8 |
| 0.969879 | N2 | 45 |

| WAVELENGTH [MICROMETER] | ACTIVE MEDIUM | PAGE |
|---|---|---|
| 0.9766 | I2 | 51 |
| 0.979970 | XE | 8 |
| 0.98 | I | 28 |
| 0.9898 | EU+ | 29 |
| 0.995515 | SE+ | 25 |
| 0.9963 | I2 | 51 |
| 0.9973 | I2 | 51 |
| 1.0019 | I2 | 51 |
| 1.002 | EU+ | 29 |
| 1.0053 | I2 | 51 |
| 1.008 | P | 21 |
| 1.01 | CS | 11 |
| 1.01 | I | 28 |
| 1.016 | EU+ | 29 |
| 1.0225 | I2 | 51 |
| 1.0245 | I2 | 51 |
| 1.0255 | I2 | 51 |
| 1.0274 | I2 | 51 |
| 1.0295 | NE | 2 |
| 1.03 | I | 28 |
| 1.0322 | YB | 30 |
| 1.033014 | SR+ | 14 |
| 1.040881 | SE+ | 25 |
| 1.043874 | N2 | 45 |
| 1.044548 | N2 | 45 |
| 1.045 | AS | 22 |
| 1.045169 | N2 | 45 |
| 1.0455 | S | 24 |
| 1.045806 | N2 | 45 |
| 1.046404 | N2 | 45 |
| 1.046956 | N2 | 45 |
| 1.0470 | AR | 4 |
| 1.047482 | N2 | 45 |
| 1.047979 | N2 | 45 |
| 1.048249 | N2 | 45 |
| 1.048461 | N2 | 45 |
| 1.048922 | N2 | 46 |
| 1.049348 | N2 | 46 |
| 1.049766 | N2 | 46 |
| 1.050161 | N2 | 46 |
| 1.050519 | N2 | 46 |
| 1.050800 | N2 | 46 |
| 1.051005 | N2 | 46 |
| 1.051118 | N2 | 46 |
| 1.052548 | N2 | 46 |
| 1.052911 | N2 | 46 |
| 1.053382 | N2 | 46 |
| 1.0534 | I2 | 51 |
| 1.053760 | N2 | 46 |
| 1.0568 | N | 21 |
| 1.0583 | HG+ | 16 |
| 1.06 | I | 28 |
| 1.061 | SN | 20 |
| 1.061 | AS | 22 |
| 1.0611 | N | 21 |
| 1.062 | SN+ | 20 |
| 1.0621 | NE | 2 |

| WAVELENGTH [MICROMETER] | ACTIVE MEDIUM | PAGE |
|---|---|---|
| 1.0623 | N | 21 |
| 1.063385 | XE+ | 8 |
| 1.0636 | S | 24 |
| 1.0691 | C | 19 |
| 1.074 | SN+ | 20 |
| 1.0775 | I2 | 51 |
| 1.0788 | I2 | 51 |
| 1.0798 | NE | 2 |
| 1.0844 | NE | 2 |
| 1.086 | S2 | 47 |
| 1.0915 | S2 | 47 |
| 1.0915 | MG | 14 |
| 1.0917 | S2 | 47 |
| 1.091797 | SR+ | 14 |
| 1.0920 | S2 | 47 |
| 1.0923 | S2 | 47 |
| 1.092344 | AR+ | 4 |
| 1.0941 | S2 | 47 |
| 1.0946 | S2 | 47 |
| 1.0950 | XE+ | 8 |
| 1.0952 | MG | 14 |
| 1.099 | S2 | 47 |
| 1.09963 | CN | 63 |
| 1.09965 | CN | 63 |
| 1.09966 | CN | 63 |
| 1.09974 | CN | 63 |
| 1.09974 | CN | 63 |
| 1.09987 | CN | 63 |
| 1.100 | S2 | 47 |
| 1.10007 | CN | 63 |
| 1.10031 | CN | 63 |
| 1.10061 | CN | 63 |
| 1.10082 | CN | 63 |
| 1.10096 | CN | 63 |
| 1.10136 | CN | 63 |
| 1.10232 | CN | 63 |
| 1.10288 | CN | 63 |
| 1.10348 | CN | 63 |
| 1.10414 | CN | 63 |
| 1.10445 | CN | 64 |
| 1.10485 | CN | 63 |
| 1.10521 | CN | 64 |
| 1.10603 | CN | 64 |
| 1.1068 | I2 | 51 |
| 1.10689 | CN | 64 |
| 1.1069 | NO | 62 |
| 1.10726 | CN | 63 |
| 1.1073 | I2 | 51 |
| 1.10782 | CN | 64 |
| 1.10879 | CN | 64 |
| 1.10981 | CN | 64 |
| 1.11090 | CN | 64 |
| 1.11200 | CN | 64 |
| 1.11321 | CN | 64 |
| 1.11441 | NE | 2 |
| 1.116 | P | 21 |
| 1.116220 | H2 | 34 |
| 1.11768 | HG | 16 |
| 1.1177 | NE | 2 |
| 1.119 | P | 21 |
| 1.1207 | I2 | 51 |
| 1.1214 | I2 | 51 |
| 1.1215 | I2 | 51 |
| 1.122205 | H2 | 34 |
| 1.1224 | I2 | 51 |
| 1.124 | AS | 22 |
| 1.1255 | I2 | 51 |
| 1.1303 | BA | 15 |
| 1.1327 | I2 | 51 |
| 1.1336 | I2 | 51 |
| 1.1347 | I2 | 51 |
| 1.1349 | I2 | 51 |
| 1.1350 | I2 | 51 |
| 1.1382 | NA | 10 |
| 1.1390 | NE | 2 |
| 1.1404 | NA | 10 |
| 1.1409 | NE | 2 |
| 1.1448 | AR | 4 |
| 1.1454 | I2 | 51 |
| 1.14582 | KR | 6 |
| 1.1464 | I2 | 51 |
| 1.1519 | AS | 22 |
| 1.152 | AS | 22 |
| 1.1521 | AS | 22 |
| 1.15235 | NE | 2 |
| 1.1525901 | NE | 2 |
| 1.154 | P | 21 |
| 1.1587 | S2 | 47 |
| 1.1601 | NE | 2 |
| 1.1614 | NE | 2 |
| 1.17 | K | 10 |
| 1.17673 | NE | 2 |
| 1.178 | P | 21 |
| 1.1789 | NE | 2 |
| 1.1984 | SI | 19 |
| 1.1985 | NE | 2 |
| 1.2034 | SI | 19 |
| 1.2066 | NE | 2 |
| 1.2096 | BE | 13 |
| 1.21396 | AR | 4 |
| 1.217 | I2 | 51 |
| 1.2222 | HG | 16 |
| 1.2237 | NO | 62 |
| 1.2246 | HG | 17 |
| 1.230598 | N2 | 46 |
| 1.231430 | N2 | 46 |
| 1.232219 | N2 | 46 |
| 1.232962 | N2 | 46 |
| 1.233671 | N2 | 46 |
| 1.234332 | N2 | 46 |
| 1.234969 | N2 | 46 |
| 1.24028 | AR | 4 |
| 1.2434 | K | 10 |
| 1.2460 | NE | 2 |

| WAVELENGTH [MICROMETER] | ACTIVE MEDIUM | PAGE |
|---|---|---|
| 1.2523 | K | 10 |
| 1.2545 | HG | 17 |
| 1.2548 | YB | 30 |
| 1.25663 | I2 | 52 |
| 1.25697 | I2 | 52 |
| 1.258678 | SE+ | 25 |
| 1.26359 | I2 | 52 |
| 1.26392 | I2 | 52 |
| 1.2689 | NE | 2 |
| 1.27022 | AR | 4 |
| 1.2714 | YB+ | 30 |
| 1.274 | I2 | 51 |
| 1.2760 | HG | 17 |
| 1.27918 | I2 | 52 |
| 1.27980 | I2 | 52 |
| 1.28212 | I2 | 52 |
| 1.28245 | I2 | 52 |
| 1.28468 | I2 | 52 |
| 1.28513 | I2 | 52 |
| 1.2870 | I2 | 51 |
| 1.28757 | I2 | 52 |
| 1.28789 | I2 | 52 |
| 1.2887 | NE | 2 |
| 1.28966 | I2 | 52 |
| 1.28998 | MN | 28 |
| 1.29007 | I2 | 52 |
| 1.2912 | NE | 2 |
| 1.2925 | I2 | 51 |
| 1.294 | I2 | 51 |
| 1.294 | AS | 22 |
| 1.2981 | HG | 17 |
| 1.304 | TM | 30 |
| 1.304 | I2 | 51 |
| 1.30545 | I2 | 52 |
| 1.305662 | H2 | 34 |
| 1.30572 | I2 | 52 |
| 1.30656 | I2 | 52 |
| 1.30784 | I2 | 52 |
| 1.30900 | I2 | 52 |
| 1.30926 | I2 | 52 |
| 1.310058 | TM | 30 |
| 1.31131 | I2 | 52 |
| 1.31166 | I2 | 52 |
| 1.31216 | I2 | 52 |
| 1.31241 | I2 | 52 |
| 1.31374 | I2 | 52 |
| 1.31407 | I2 | 52 |
| 1.31498 | I2 | 52 |
| 1.315 | I | 28 |
| 1.31523 | I2 | 52 |
| 1.3153 | I2 | 51 |
| 1.316109 | H2 | 34 |
| 1.31775 | KR | 6 |
| 1.3192 | I2 | 51 |
| 1.320 | I2 | 51 |
| 1.3282 | I2 | 51 |
| 1.3291 | I2 | 51 |
| 1.32938 | MN | 28 |
| 1.32952 | I2 | 52 |
| 1.33065 | I2 | 52 |
| 1.3310 | I2 | 51 |
| 1.33190 | MN | 28 |
| 1.3324 | I2 | 51 |
| 1.3333 | I2 | 51 |
| 1.3349 | I2 | 51 |
| 1.33546 | I2 | 52 |
| 1.33567 | I2 | 52 |
| 1.33573 | I2 | 52 |
| 1.33619 | I2 | 52 |
| 1.33661 | I2 | 52 |
| 1.33681 | I2 | 52 |
| 1.338 | I2 | 52 |
| 1.338008 | TM | 30 |
| 1.34142 | I2 | 52 |
| 1.34145 | I2 | 52 |
| 1.34173 | I2 | 52 |
| 1.34173 | I2 | 52 |
| 1.34192 | I2 | 52 |
| 1.34193 | I2 | 52 |
| 1.34248 | I2 | 52 |
| 1.34256 | I2 | 52 |
| 1.34295 | N | 21 |
| 1.3453 | YB+ | 30 |
| 1.35818 | N | 21 |
| 1.360 | CS | 11 |
| 1.361 | EU+ | 29 |
| 1.36225 | KR | 6 |
| 1.36267 | MN | 28 |
| 1.3655 | HG | 17 |
| 1.36562 | XE | 8 |
| 1.3675 | HG | 17 |
| 1.37 | RB | 10 |
| 1.3759 | AG+ | 12 |
| 1.376 | CS | 11 |
| 1.3859 | CL | 27 |
| 1.38642 | MN | 28 |
| 1.3891 | CL | 27 |
| 1.3968 | NI | 9 |
| 1.39975 | MN | 28 |
| 1.40 | CD | 16 |
| 1.402 | S | 24 |
| 1.40948 | AR | 4 |
| 1.412 | AS | 22 |
| 1.41830 | CN | 64 |
| 1.41849 | CN | 64 |
| 1.41876 | CN | 64 |
| 1.41911 | CN | 64 |
| 1.41954 | CN | 64 |
| 1.42005 | CN | 64 |
| 1.42065 | CN | 64 |
| 1.42081 | CN | 64 |
| 1.42132 | CN | 64 |
| 1.42207 | CN | 64 |
| 1.42289 | CN | 64 |

| WAVELENGTH [MICROMETER] | ACTIVE MEDIUM | PAGE |
|---|---|---|
| 1.42380 | CN | 64 |
| 1.42478 | CN | 64 |
| 1.4255 | AS | 22 |
| 1.42583 | CN | 64 |
| 1.42696 | CN | 64 |
| 1.4276 | NE | 2 |
| 1.4280 | YB | 30 |
| 1.42808 | CN | 64 |
| 1.42945 | CN | 64 |
| 1.43 | CD | 16 |
| 1.4304 | NE | 2 |
| 1.4321 | NE | 2 |
| 1.4330 | NE | 2 |
| 1.433973 | TM | 30 |
| 1.4346 | NE | 2 |
| 1.4368 | NE | 2 |
| 1.44269 | KR | 6 |
| 1.448509 | TM | 30 |
| 1.45 | CD | 16 |
| 1.4540 | C | 19 |
| 1.4542 | I | 26 |
| 1.45423 | N | 21 |
| 1.4550 | NI | 9 |
| 1.463 | AS | 22 |
| 1.47 | CS | 11 |
| 1.47648 | KR | 6 |
| 1.477 | EU+ | 29 |
| 1.477548 | D2 | 36 |
| 1.4787 | YB | 30 |
| 1.48 | RB | 10 |
| 1.484450 | NE | 2 |
| 1.486926 | NE | 2 |
| 1.487247 | NE | 2 |
| 1.488759 | NE | 2 |
| 1.489954 | NE | 2 |
| 1.493623 | NE | 2 |
| 1.4966 | KR | 6 |
| 1.50 | H2 | 34 |
| 1.500 | TM | 30 |
| 1.5000 | BA | 15 |
| 1.5046 | AR | 4 |
| 1.5231 | NE | 2 |
| 1.5295 | HG | 17 |
| 1.53 | RB | 10 |
| 1.5330 | KR | 6 |
| 1.5422 | S | 24 |
| 1.553 | I | 26 |
| 1.5555 | HG+ | 17 |
| 1.564193 | H2 | 34 |
| 1.571 | P | 21 |
| 1.581950 | H2 | 34 |
| 1.5883 | SI | 19 |
| 1.5900 TO | F | 26 |
| 1.5982 | AG+ | 12 |
| 1.60519 | XE | 8 |
| 1.6180 | AR | 4 |
| 1.619395 | AR | 4 |
| 1.63 | H2 | 34 |
| 1.637914 | TM | 30 |
| 1.64 | CD | 16 |
| 1.6462 | AG+ | 12 |
| 1.648 | P | 21 |
| 1.6498 | YB+ | 30 |
| 1.652 | AR | 4 |
| 1.6543 | S | 24 |
| 1.66 | EU | 29 |
| 1.6656 | AG+ | 12 |
| 1.675404 | TM | 30 |
| 1.68533 | KR | 6 |
| 1.68965 | KR | 6 |
| 1.6920 | HG | 17 |
| 1.6936 | KR | 6 |
| 1.6941 | AR | 4 |
| 1.6942 | HG | 17 |
| 1.7073 | HG | 17 |
| 1.71099 | HG | 17 |
| 1.7162 | NE | 2 |
| 1.7203 | AG+ | 12 |
| 1.7319 | TM | 30 |
| 1.73254 | XE | 8 |
| 1.7329 | HG | 17 |
| 1.7345 | AG+ | 12 |
| 1.7363 | GA | 18 |
| 1.7438 | CU+ | 12 |
| 1.7454 | YB | 30 |
| 1.7478 | AG+ | 12 |
| 1.7596 | EU | 29 |
| 1.7674 | AG+ | 12 |
| 1.7708 | CU+ | 12 |
| 1.7843 | KR | 6 |
| 1.791437 | AR | 4 |
| 1.7977 | YB | 30 |
| 1.8004 | CU+ | 12 |
| 1.8049 | AS | 22 |
| 1.8057 | YB+ | 30 |
| 1.807 | AS | 22 |
| 1.8130 | HG | 17 |
| 1.8185 | KR | 6 |
| 1.8196 | CU | 12 |
| 1.82 | BA | 15 |
| 1.8210 | NE | 2 |
| 1.8228 | CU | 12 |
| 1.8253 | NE | 2 |
| 1.8276 | NE | 2 |
| 1.8304 | NE | 2 |
| 1.8308 | ZN+ | 15 |
| 1.836 | DF | 55 |
| 1.8380 | AG | 12 |
| 1.8403 | NE | 2 |
| 1.8408 | AG+ | 12 |
| 1.844 | DF | 55 |
| 1.8448 | CU+ | 12 |
| 1.8463 | AG+ | 12 |
| 1.854 | DF | 55 |

| WAVELENGTH [MICROMETER] | ACTIVE MEDIUM | PAGE |
|---|---|---|
| 1.8591 | NE | 2 |
| 1.8597 | NE | 2 |
| 1.8685 | HE | 1 |
| 1.8725 | AG+ | 12 |
| 1.8732 | IN | 18 |
| 1.8751 | H | 9 |
| 1.8795 | AG+ | 12 |
| 1.894 | P | 21 |
| 1.8979 | AG+ | 12 |
| 1.9017 | BA | 15 |
| 1.912 | SM | 29 |
| 1.9154 | CU+ | 12 |
| 1.9211 | KR | 6 |
| 1.9328 | CU+ | 12 |
| 1.9370 | AG | 12 |
| 1.9479 | CU+ | 12 |
| 1.9543 | HE | 1 |
| 1.9574 | NE | 2 |
| 1.9577 | NE | 2 |
| 1.958443 | TM | 30 |
| 1.9712 | CU+ | 12 |
| 1.9714 | AG+ | 12 |
| 1.973 | TM | 30 |
| 1.9750 | AS | 22 |
| 1.9755 | CL | 27 |
| 1.9809 | GE | 20 |
| 1.9823 | AG+ | 12 |
| 1.9830 | YB | 30 |
| 1.9853 | CA | 14 |
| 1.994160 | TM | 30 |
| 2.0006 | CU+ | 12 |
| 2.0036 | YB | 30 |
| 2.0195 | V | 23 |
| 2.0199 | CL | 27 |
| 2.0200 | GE | 20 |
| 2.02623 | XE | 8 |
| 2.0277 | AS | 22 |
| 2.0350 | NE | 2 |
| 2.0353 | NE | 2 |
| 2.04240 | KR | 6 |
| 2.0482 | SM | 29 |
| 2.05813 | HE | 1 |
| 2.060 | P | 21 |
| 2.0603 | HE | 1 |
| 2.0616 | AR | 4 |
| 2.0645 | C | 19 |
| 2.0796 | AG+ | 12 |
| 2.0986 | AR | 5 |
| 2.1041 | NE | 2 |
| 2.107 | TM | 30 |
| 2.1165 | KR | 6 |
| 2.1181 | YB | 30 |
| 2.1332 | AR | 5 |
| 2.1480 | YB+ | 30 |
| 2.1534 | AR | 5 |
| 2.1568 | BA | 15 |
| 2.1708 | NE | 2 |
| 2.19020 | KR | 6 |
| 2.2038 | AR | 5 |
| 2.2077 | AR | 5 |
| 2.2475 | KR | 6 |
| 2.254 | RB | 10 |
| 2.28 | H2O | 104 |
| 2.2854 | BR | 27 |
| 2.293 | RB | 10 |
| 2.31339 | AR | 5 |
| 2.3193 | XE | 8 |
| 2.3254 | BA | 15 |
| 2.3260 | NE | 2 |
| 2.3474 | CO | 68 |
| 2.3511 | BR | 27 |
| 2.37 | NE | 2 |
| 2.3769 | CO | 68 |
| 2.3779 | IN | 18 |
| 2.384515 | TM | 30 |
| 2.3951 | NE | 2 |
| 2.3966 | AR | 5 |
| 2.40415 | MG+ | 14 |
| 2.41 | HF | 53 |
| 2.41245 | MG+ | 14 |
| 2.4219 | NE | 2 |
| 2.4250 | NE | 2 |
| 2.4260 | KR | 6 |
| 2.43 | HF | 53 |
| 2.4344 | CO | 68 |
| 2.4363 | S | 24 |
| 2.4377 | YB+ | 30 |
| 2.4380 | CO | 68 |
| 2.4460 | AS | 22 |
| 2.4466 | CL | 27 |
| 2.4473 | V | 23 |
| 2.45 | HF | 53 |
| 2.4696 | CO | 68 |
| 2.4758 | BA | 15 |
| 2.48 | HF | 53 |
| 2.4825 | XE | 8 |
| 2.5008 | AR | 5 |
| 2.5019 | CO | 68 |
| 2.51528 | XE | 8 |
| 2.52342 | KR | 6 |
| 2.5350 | CO | 68 |
| 2.5393 | NE | 2 |
| 2.5487 | AR | 5 |
| 2.5504 | AR | 5 |
| 2.551 | HF | 53 |
| 2.5515 | BA | 15 |
| 2.5524 | NE | 2 |
| 2.5627 | AR | 5 |
| 2.5661 | AR | 5 |
| 2.5669 | CO | 68 |
| 2.579 | HF | 53 |
| 2.5811 | EU | 29 |
| 2.5924 | BA | 15 |
| 2.5966 | I | 28 |

| WAVELENGTH [MICROMETER] | ACTIVE MEDIUM | PAGE | WAVELENGTH [MICROMETER] | ACTIVE MEDIUM | PAGE |
|---|---|---|---|---|---|
| 2.6036 | CO | 68 | 2.8375 | BR | 27 |
| 2.6072 | NO | 62 | 2.8446 | CO | 68 |
| 2.6084 | HF | 53 | 2.8476 | CO | 68 |
| 2.6260 | KR | 6 | 2.8507 | CO | 68 |
| 2.6269 | XE | 8 | 2.8542 | HF | 54 |
| 2.6281 | KR | 6 | 2.8590 | XE | 8 |
| 2.6380 | NO | 62 | 2.86134 | KR | 6 |
| 2.6392 | CO | 68 | 2.862 | AR | 5 |
| 2.6396 | HF | 53 | 2.8656 | KR | 6 |
| 2.65146 | XE | 8 | 2.8657 | HF | 53 |
| 2.652 | O | 23 | 2.8705 | HF | 54 |
| 2.6542 | AR | 5 | 2.8776 | AR | 5 |
| 2.6601 | XE | 8 | 2.8836 | AR | 5 |
| 2.6665 | XE | 8 | 2.8890 | HF | 54 |
| 2.6668 | HF | 53 | 2.8892 | CO | 68 |
| 2.6726 | HF | 53 | 2.89 | O | 23 |
| 2.6756 | CO | 68 | 2.8923 | CO | 68 |
| 2.6836 | AR | 5 | 2.9057 | BA | 15 |
| 2.6886 | CO | 68 | 2.9103 | HF | 53 |
| 2.6914 | CO | 68 | 2.9111 | HF | 54 |
| 2.6963 | HF | 54 | 2.9221 | HF | 54 |
| 2.6998 | SM | 29 | 2.9227 | BA | 15 |
| 2.7075238 | HF | 53 | 2.9257 | HF | 54 |
| 2.7129 | CO | 68 | 2.9273 | AR | 5 |
| 2.714 | BR | 27 | 2.9288 | CO | 68 |
| 2.71529 | AR | 5 | 2.9319 | CO | 69 |
| 2.7174 | EU | 29 | 2.93432 | OH | 64 |
| 2.72 | K | 10 | 2.9351 | CO | 69 |
| 2.7262 | CO | 68 | 2.9448 | NE | 2 |
| 2.7275 | HF | 54 | 2.95 | CS | 11 |
| 2.7290 | CO | 68 | 2.9539 | HF | 54 |
| 2.7319 | CO | 68 | 2.9549 | HF | 54 |
| 2.7357 | AR | 5 | 2.9573 | HF | 53 |
| 2.7440 | HF | 53 | 2.9644 | HF | 54 |
| 2.7511 | CO | 68 | 2.9663 | SM | 29 |
| 2.7572 | I | 28 | 2.9668 | NE | 2 |
| 2.7574 | NE | 2 | 2.96999 | OH | 64 |
| 2.7604 | HF | 54 | 2.9725 | CO | 69 |
| 2.7647 | CO | 68 | 2.9757 | CO | 69 |
| 2.7676 | CO | 68 | 2.9788 | AR | 5 |
| 2.7705 | CO | 68 | 2.9789 | CO | 69 |
| 2.7799 | S | 24 | 2.9805 | NE | 2 |
| 2.7819 | NE | 2 | 2.9805 | AS | 22 |
| 2.7826 | HF | 53 | 2.9836 | KR | 6 |
| 2.79 | RB | 10 | 2.9870 | KR | 6 |
| 2.7902 | HF | 54 | 2.9896 | HF | 54 |
| 2.7903 | CO | 68 | 2.9989 | HF | 54 |
| 2.7952 | HF | 54 | 3.0052 | HF | 54 |
| 2.8042 | CO | 68 | 3.0064 | HF | 53 |
| 2.8071 | CO | 68 | 3.010 | CS | 11 |
| 2.8101 | CO | 68 | 3.0111 | SR | 14 |
| 2.8195 | AR | 5 | 3.0174 | CO | 69 |
| 2.8213 | HF | 54 | 3.0206 | CO | 69 |
| 2.8231 | HF | 53 | 3.026 | HF | 54 |
| 2.8238 | AR | 5 | 3.0260 | NE | 2 |
| 2.8306 | CO | 68 | 3.0268 | NE | 2 |
| 2.8319 | HF | 54 | 3.0360 | I | 28 |

| WAVELENGTH [MICROMETER] | ACTIVE MEDIUM | PAGE |
|---|---|---|
| 3.0454 | AR | 5 |
| 3.0461 | HF | 54 |
| 3.0482 | HF | 54 |
| 3.0528 | KR | 6 |
| 3.0582 | HF | 53 |
| 3.065 | HF | 54 |
| 3.0664 | KR | 6 |
| 3.0664 | CL | 27 |
| 3.0668 | CO | 69 |
| 3.07877 | OH | 64 |
| 3.0935 | HF | 54 |
| 3.095 | CS | 11 |
| 3.0958 | HF | 54 |
| 3.0982 | HF | 54 |
| 3.0988 | AR | 5 |
| 3.1069 | XE | 8 |
| 3.1125 | HF | 53 |
| 3.11677 | OH | 64 |
| 3.1325 | AR | 5 |
| 3.1338 | AR | 5 |
| 3.1350 | HF | 54 |
| 3.140 | K | 10 |
| 3.1411 | HF | 54 |
| 3.1454 | HF | 54 |
| 3.1480 | HF | 54 |
| 3.1492 | HF | 54 |
| 3.15 | K | 10 |
| 3.1508 | KR | 6 |
| 3.15697 | OH | 64 |
| 3.160 | K | 10 |
| 3.1640 | HF | 54 |
| 3.1695 | HF | 53 |
| 3.1720 | TE | 26 |
| 3.1738 | PB | 20 |
| 3.1912 | HF | 54 |
| 3.2029 | HF | 54 |
| 3.2040 | CS | 11 |
| 3.2151 | HF | 54 |
| 3.2206 | HF | 54 |
| 3.2292 | HF | 53 |
| 3.23615 | OH | 64 |
| 3.2363 | I | 28 |
| 3.2438 | HF | 54 |
| 3.258 | HF | 54 |
| 3.2603 | HF | 54 |
| 3.2739 | XE | 8 |
| 3.27653 | OH | 64 |
| 3.2882 | CD+ | 16 |
| 3.2919 | HF | 53 |
| 3.29462 | N2 | 46 |
| 3.2991 | HF | 54 |
| 3.30166 | N2 | 46 |
| 3.3044 | HF | 54 |
| 3.30755 | N2 | 46 |
| 3.3085 | XE | 8 |
| 3.31239 | N2 | 46 |
| 3.31616 | N2 | 46 |
| 3.3173 | NE | 2 |
| 3.31892 | N2 | 46 |
| 3.333 | HF | 54 |
| 3.3333 | NE | 2 |
| 3.3353 | NE | 2 |
| 3.3401 | KR | 6 |
| 3.3411 | KR | 6 |
| 3.3500 | NE | 2 |
| 3.3510 | NE | 2 |
| 3.3666 | XE | 8 |
| 3.377 | HF | 54 |
| 3.3804 | NE | 2 |
| 3.3840 | NE | 2 |
| 3.3892 | S | 24 |
| 3.3903 | NE | 2 |
| 3.391 | NE | 2 |
| 3.3913 | NE | 2 |
| 3.4014 | XE | 8 |
| 3.4046 | C | 19 |
| 3.428 | OCS | 101 |
| 3.4296 | I | 28 |
| 3.4335 | XE | 8 |
| 3.4471 | NE | 2 |
| 3.4489 | NE | 2 |
| 3.45212 | N2 | 46 |
| 3.45852 | N2 | 46 |
| 3.46377 | N2 | 46 |
| 3.4654 | SM | 29 |
| 3.46804 | N2 | 46 |
| 3.47127 | N2 | 46 |
| 3.475 | NE | 2 |
| 3.4789 | NE | 2 |
| 3.4873 | KR | 6 |
| 3.4885 | KR | 6 |
| 3.489 | CS | 11 |
| 3.493 | DF | 56 |
| 3.5070 | XE | 8 |
| 3.5155 | C | 19 |
| 3.521 | DF | 56 |
| 3.5361 | SM | 29 |
| 3.550 | DF | 56 |
| 3.5728 | HCL | 57 |
| 3.581 | DF | 56 |
| 3.5835 | NE | 2 |
| 3.6026 | HCL | 57 |
| 3.612 | DF | 56 |
| 3.613 | CS | 11 |
| 3.6210 | XE | 8 |
| 3.6312 | AR | 5 |
| 3.6337 | HCL | 57 |
| 3.6362 | HCL | 57 |
| 3.6363 | DF | 56 |
| 3.645 | DF | 56 |
| 3.6509 | XE | 8 |
| 3.6515 | NE | 2 |
| 3.6660 | HCL | 57 |
| 3.6665 | DF | 56 |

| WAVELENGTH [MICROMETER] | ACTIVE MEDIUM | PAGE | WAVELENGTH [MICROMETER] | ACTIVE MEDIUM | PAGE |
|---|---|---|---|---|---|
| 3.6685 | HCL | 57 | 3.9145 | DF | 56 |
| 3.67794 | MG | 14 | 3.9149 | HCL | 57 |
| 3.6788 | XE | 8 | 3.9155 | DF | 56 |
| 3.679 | DF | 56 | 3.9181 | HCL | 57 |
| 3.68154 | MG | 14 | 3.9205 | HCL | 57 |
| 3.6849 | XE | 8 | 3.9272 | DF | 56 |
| 3.6983 | DF | 56 | 3.93 | HG | 17 |
| 3.6996 | HCL | 57 | 3.9467 | DF | 56 |
| 3.70138 | AR | 5 | 3.9536 | HCL | 57 |
| 3.7021 | HCL | 57 | 3.956 | KR | 6 |
| 3.7071 | HCL | 57 | 3.9560 | HCL | 57 |
| 3.708 | AR | 5 | 3.9565 | DF | 56 |
| 3.7098 | HCL | 57 | 3.9572 | DF | 56 |
| 3.71439 | AR | 5 | 3.9573 | KR | 6 |
| 3.716 | DF | 56 | 3.9578 | BA | 15 |
| 3.7310 | DF | 56 | 3.9654 | DF | 56 |
| 3.7341 | HCL | 57 | 3.9806 | NE | 2 |
| 3.7370 | HCL | 57 | 3.9843 | DF | 56 |
| 3.7383 | HCL | 57 | 3.9909 | HCL | 57 |
| 3.7406 | HCL | 57 | 3.9955 | XE | 8 |
| 3.752 | DF | 56 | 3.9995 | DF | 56 |
| 3.7563 | DF | 56 | 4.0032 | DF | 56 |
| 3.7651 | DF | 56 | 4.0054 | HCL | 57 |
| 3.7707 | HCL | 57 | 4.0054 | DF | 56 |
| 3.7710 | HCL | 57 | 4.0069 | BA | 15 |
| 3.7735 | HCL | 57 | 4.0170 | HBR | 60 |
| 3.7736 | NE | 2 | 4.0176 | HBR | 60 |
| 3.774 | KR | 6 | 4.0196 | XE | 8 |
| 3.7878 | DF | 56 | 4.0212 | DF | 56 |
| 3.7901 | DF | 56 | 4.0295 | HCL | 57 |
| 3.7942 | N | 21 | 4.0399 | HCL | 57 |
| 3.8007 | DF | 56 | 4.0435 | DF | 56 |
| 3.8050 | HCL | 57 | 4.0464 | DF | 56 |
| 3.8074 | HCL | 57 | 4.0470 | HBR | 60 |
| 3.8081 | HCL | 57 | 4.0475 | HBR | 60 |
| 3.8125 | TL | 18 | 4.0502 | DF | 56 |
| 3.8154 | N | 21 | 4.0595 | DF | 56 |
| 3.8206 | DF | 56 | 4.068 | KR | 6 |
| 3.8298 | DF | 56 | 4.0759 | HCL | 57 |
| 3.8375 | DF | 56 | 4.0783 | HBR | 60 |
| 3.8401 | HCL | 57 | 4.0788 | HBR | 60 |
| 3.8425 | HCL | 57 | 4.0893 | DF | 56 |
| 3.8503 | DF | 56 | 4.0895 | DF | 56 |
| 3.8509 | HCL | 57 | 4.1107 | HBR | 60 |
| 3.8536 | HCL | 57 | 4.1112 | HBR | 60 |
| 3.8547 | DF | 56 | 4.1135 | HCL | 58 |
| 3.86573 | MG | 14 | 4.1337 | DF | 56 |
| 3.8686 | XE | 8 | 4.1368 | SM | 29 |
| 3.8707 | DF | 56 | 4.1369 | DF | 56 |
| 3.8757 | DF | 56 | 4.142 | KR | 6 |
| 3.8768 | HCL | 57 | 4.1442 | HBR | 60 |
| 3.8817 | DF | 56 | 4.1448 | HBR | 60 |
| 3.8840 | HCL | 57 | 4.1516 | XE | 8 |
| 3.8864 | HCL | 57 | 4.1653 | HBR | 60 |
| 3.8903 | DF | 56 | 4.1658 | HBR | 60 |
| 3.8940 | XE | 8 | 4.1796 | HBR | 60 |
| 3.9133 | DF | 56 | 4.1798 | DF | 56 |

| WAVELENGTH [MICROMETER] | ACTIVE MEDIUM | PAGE |
|---|---|---|
| 4.1862 | DF | 56 |
| 4.1970 | HBR | 60 |
| 4.1975 | HBR | 60 |
| 4.20018 | MG | 14 |
| 4.2033 | AR | 5 |
| 4.2171 | NE | 2 |
| 4.22 | CS | 11 |
| 4.2295 | HBR | 60 |
| 4.2633 | HBR | 60 |
| 4.2639 | HBR | 60 |
| 4.2988 | HBR | 60 |
| 4.2994 | HBR | 60 |
| 4.314 | CO2 | 86 |
| 4.3202 | EU | 29 |
| 4.3203 | CO2 | 86 |
| 4.3249 | CO2 | 86 |
| 4.3250 | HBR | 60 |
| 4.3255 | HBR | 60 |
| 4.3276 | CO2 | 86 |
| 4.33 | BA | 15 |
| 4.331 | I | 28 |
| 4.3354 | HBR | 60 |
| 4.3359 | HBR | 60 |
| 4.340 | CO2 | 86 |
| 4.346 | CO2 | 86 |
| 4.354 | CO2 | 86 |
| 4.3549 | CO2 | 86 |
| 4.3579 | HBR | 60 |
| 4.3580 | CO2 | 86 |
| 4.3585 | HBR | 60 |
| 4.3612 | CO2 | 86 |
| 4.36269 | MG | 14 |
| 4.3644 | CO2 | 86 |
| 4.3677 | CO2 | 86 |
| 4.371 | CO2 | 86 |
| 4.3711 | CO2 | 86 |
| 4.3736 | KR | 6 |
| 4.3745 | CO2 | 86 |
| 4.3755 | KR | 6 |
| 4.377 | CO2 | 87 |
| 4.3779 | CO2 | 86 |
| 4.3814 | CO2 | 86 |
| 4.382 | CO2 | 87 |
| 4.3849 | CO2 | 86 |
| 4.385 | CO2 | 87 |
| 4.392 | CO2 | 87 |
| 4.3925 | HBR | 60 |
| 4.3931 | HBR | 60 |
| 4.398 | CO2 | 87 |
| 4.4281 | HBR | 60 |
| 4.4307 | HBR | 60 |
| 4.4652 | HBR | 60 |
| 4.4658 | HBR | 60 |
| 4.5041 | HBR | 60 |
| 4.5047 | HBR | 60 |
| 4.5330 | HBR | 60 |
| 4.5335 | HBR | 60 |
| 4.5381 | XE | 8 |
| 4.5607 | O | 23 |
| 4.5632 | O | 24 |
| 4.5665 | XE | 8 |
| 4.5691 | HBR | 60 |
| 4.5694 | XE | 8 |
| 4.5696 | HBR | 61 |
| 4.60535 | HE | 1 |
| 4.60567 | HE[3] | 1 |
| 4.6070 | HBR | 60 |
| 4.6076 | HBR | 61 |
| 4.6097 | XE | 8 |
| 4.6146 | SN | 20 |
| 4.6463 | HBR | 60 |
| 4.6467 | HBR | 61 |
| 4.6706 | BA | 15 |
| 4.6935 | EU | 29 |
| 4.7138 | AR | 5 |
| 4.7156 | BA | 15 |
| 4.7171 | BA | 15 |
| 4.735872 | CO | 69 |
| 4.745130 | CO | 69 |
| 4.754501 | CO | 69 |
| 4.763984 | CO | 69 |
| 4.767821 | CO | 69 |
| 4.77 | H2O | 104 |
| 4.773562 | CO | 69 |
| 4.776892 | CO | 69 |
| 4.783295 | CO | 69 |
| 4.786076 | CO | 69 |
| 4.793123 | CO | 69 |
| 4.795373 | CO | 69 |
| 4.8009 | YB | 30 |
| 4.803067 | CO | 69 |
| 4.804785 | CO | 69 |
| 4.813129 | CO | 69 |
| 4.814312 | CO | 69 |
| 4.823310 | CO | 69 |
| 4.823954 | CO | 69 |
| 4.833609 | CO | 69 |
| 4.833714 | CO | 69 |
| 4.843591 | CO | 69 |
| 4.844029 | CO | 69 |
| 4.846781 | CO | 69 |
| 4.853586 | CO | 69 |
| 4.854569 | CO | 69 |
| 4.856233 | CO | 69 |
| 4.858 | I | 28 |
| 4.8619 | I | 28 |
| 4.863700 | CO | 69 |
| 4.865231 | CO | 69 |
| 4.8656 | SM | 29 |
| 4.865803 | CO | 69 |
| 4.873935 | CO | 69 |
| 4.875490 | CO | 69 |
| 4.8760 | KR | 6 |
| 4.876016 | CO | 69 |

| WAVELENGTH [MICROMETER] | ACTIVE MEDIUM | PAGE |
|---|---|---|
| 4.880759 | CO | 70 |
| 4.8819 | KR | 6 |
| 4.884291 | CO | 69 |
| 4.885296 | CO | 69 |
| 4.890016 | CO | 70 |
| 4.894769 | CO | 69 |
| 4.895221 | CO | 69 |
| 4.899391 | CO | 70 |
| 4.905267 | CO | 69 |
| 4.905369 | CO | 69 |
| 4.908883 | CO | 70 |
| 4.9146 | AR | 5 |
| 4.915434 | CO | 69 |
| 4.916094 | CO | 69 |
| 4.918494 | CO | 70 |
| 4.9199 | AR | 5 |
| 4.925723 | CO | 69 |
| 4.926943 | CO | 69 |
| 4.928238 | CO | 70 |
| 4.936136 | CO | 69 |
| 4.938078 | CO | 70 |
| 4.943828 | CO | 70 |
| 4.946672 | CO | 69 |
| 4.948052 | CO | 70 |
| 4.9496 | AR | 5 |
| 4.953240 | CO | 70 |
| 4.957333 | CO | 69 |
| 4.958148 | CO | 70 |
| 4.962772 | CO | 70 |
| 4.968120 | CO | 69 |
| 4.968369 | CO | 70 |
| 4.972425 | CO | 70 |
| 4.978711 | CO | 70 |
| 4.979035 | CO | 69 |
| 4.982220 | CO | 70 |
| 4.989181 | CO | 70 |
| 4.990016 | CO | 70 |
| 4.992099 | CO | 70 |
| 4.9983 | KR | 6 |
| 4.999775 | CO | 70 |
| 4.9999 | KR | 6 |
| 5.001277 | CO | 70 |
| 5.002121 | CO | 70 |
| 5.008369 | CO | 71 |
| 5.010497 | CO | 70 |
| 5.012268 | CO | 70 |
| 5.012578 | CO | 70 |
| 5.017940 | CO | 71 |
| 5.021347 | CO | 70 |
| 5.022539 | CO | 70 |
| 5.0230 | XE | 8 |
| 5.02338 | AR | 5 |
| 5.023976 | CO | 70 |
| 5.02441 | XE | 8 |
| 5.027635 | CO | 71 |
| 5.0309 | BA | 15 |
| 5.032321 | CO | 70 |
| 5.032938 | CO | 70 |
| 5.035544 | CO | 70 |
| 5.037454 | CO | 71 |
| 5.043435 | CO | 70 |
| 5.043462 | CO | 70 |
| 5.0445 | DCL | 59 |
| 5.047242 | CO | 70 |
| 5.047397 | CO | 71 |
| 5.0514 | DCL | 59 |
| 5.054117 | CO | 70 |
| 5.054676 | CO | 70 |
| 5.057467 | CO | 71 |
| 5.059073 | CO | 70 |
| 5.0647 | EU | 29 |
| 5.064899 | CO | 70 |
| 5.066048 | CO | 70 |
| 5.067663 | CO | 71 |
| 5.071040 | CO | 70 |
| 5.0743 | DCL | 59 |
| 5.074432 | CO | 71 |
| 5.075812 | CO | 70 |
| 5.077554 | CO | 70 |
| 5.077988 | CO | 71 |
| 5.0811 | DCL | 59 |
| 5.083144 | CO | 70 |
| 5.084166 | CO | 71 |
| 5.0848 | ZN+ | 15 |
| 5.086856 | CO | 70 |
| 5.088440 | CO | 71 |
| 5.092002 | CO | 72 |
| 5.094028 | CO | 71 |
| 5.095386 | CO | 70 |
| 5.096828 | CO | 73 |
| 5.098033 | CO | 70 |
| 5.099023 | CO | 71 |
| 5.100430 | CO | 72 |
| 5.104017 | CO | 71 |
| 5.104413 | CO | 73 |
| 5.1049 | DCL | 59 |
| 5.1059 | TL | 18 |
| 5.107766 | CO | 70 |
| 5.108980 | CO | 72 |
| 5.109343 | CO | 70 |
| 5.109734 | CO | 71 |
| 5.1118 | DCL | 59 |
| 5.112118 | CO | 73 |
| 5.114134 | CO | 71 |
| 5.119944 | CO | 73 |
| 5.120267 | CO | 70 |
| 5.1203 | AR | 5 |
| 5.1205 | AR | 5 |
| 5.120577 | CO | 71 |
| 5.120787 | CO | 70 |
| 5.122753 | CO | 74 |
| 5.124408 | CO | 71 |
| 5.127892 | CO | 73 |
| 5.129726 | CO | 74 |

| WAVELENGTH [MICROMETER] | ACTIVE MEDIUM | PAGE |
|---|---|---|
| 5.1298 | KR | 6 |
| 5.131252 | CO | 70 |
| 5.131555 | CO | 71 |
| 5.132949 | CO | 70 |
| 5.134757 | CO | 71 |
| 5.135961 | CO | 73 |
| 5.1363 | DCL | 59 |
| 5.136819 | CO | 74 |
| 5.142062 | CO | 72 |
| 5.142663 | CO | 71 |
| 5.1431 | DCL | 59 |
| 5.144032 | CO | 74 |
| 5.144084 | CO | 70 |
| 5.145264 | CO | 71 |
| 5.145754 | CO | 70 |
| 5.1511 | DCL | 59 |
| 5.151366 | CO | 74 |
| 5.151996 | CO | 72 |
| 5.153909 | CO | 71 |
| 5.155902 | CO | 71 |
| 5.155938 | CO | 70 |
| 5.158822 | CO | 74 |
| 5.162000 | CO | 72 |
| 5.165289 | CO | 71 |
| 5.166400 | CO | 74 |
| 5.166672 | CO | 71 |
| 5.167931 | CO | 71 |
| 5.1688 | DCL | 59 |
| 5.1696 | NE | 2 |
| 5.172164 | CO | 72 |
| 5.174100 | CO | 74 |
| 5.176806 | CO | 71 |
| 5.177575 | CO | 71 |
| 5.179608 | CO | 74 |
| 5.180064 | CO | 71 |
| 5.1811 | DCL | 59 |
| 5.181924 | CO | 74 |
| 5.182459 | CO | 72 |
| 5.186476 | CO | 74 |
| 5.1879 | DCL | 59 |
| 5.188460 | CO | 71 |
| 5.188617 | CO | 71 |
| 5.189871 | CO | 74 |
| 5.192338 | CO | 71 |
| 5.192888 | CO | 72 |
| 5.193387 | CO | 74 |
| 5.199792 | CO | 71 |
| 5.200254 | CO | 71 |
| 5.200459 | CO | 74 |
| 5.201370 | CO | 72 |
| 5.203447 | CO | 72 |
| 5.204755 | CO | 71 |
| 5.207655 | CO | 74 |
| 5.211102 | CO | 71 |
| 5.211315 | CO | 72 |
| 5.2118 | DCL | 59 |
| 5.21187 | CO | 71 |
| 5.214142 | CO | 72 |
| 5.214974 | CO | 74 |
| 5.217312 | CO | 71 |
| 5.2186 | DCL | 59 |
| 5.221392 | CO | 72 |
| 5.222417 | CO | 74 |
| 5.222555 | CO | 72 |
| 5.224262 | CO | 71 |
| 5.224972 | CO | 72 |
| 5.229984 | CO | 74 |
| 5.230020 | CO | 71 |
| 5.231603 | CO | 72 |
| 5.234145 | CO | 72 |
| 5.235937 | CO | 72 |
| 5.236479 | CO | 71 |
| 5.237677 | CO | 74 |
| 5.241946 | CO | 72 |
| 5.242872 | CO | 71 |
| 5.2435 | DCL | 59 |
| 5.245496 | CO | 74 |
| 5.245874 | CO | 72 |
| 5.247038 | CO | 72 |
| 5.248840 | CO | 71 |
| 5.2503 | DCL | 59 |
| 5.252424 | CO | 73 |
| 5.255870 | CO | 71 |
| 5.257745 | CO | 72 |
| 5.258279 | CO | 72 |
| 5.261343 | CO | 71 |
| 5.263039 | CO | 73 |
| 5.269018 | CO | 71 |
| 5.269659 | CO | 72 |
| 5.269759 | CO | 72 |
| 5.272122 | CO | 73 |
| 5.273789 | CO | 73 |
| 5.273997 | CO | 71 |
| 5.2760 | DCL | 59 |
| 5.2811 | EU | 29 |
| 5.281183 | CO | 72 |
| 5.281916 | CO | 72 |
| 5.282243 | CO | 73 |
| 5.282316 | CO | 71 |
| 5.2829 | DCL | 59 |
| 5.284678 | CO | 73 |
| 5.2865 | AS | 22 |
| 5.286796 | CO | 71 |
| 5.292498 | CO | 73 |
| 5.292846 | CO | 72 |
| 5.294218 | CO | 72 |
| 5.295704 | CO | 73 |
| 5.2985 | KR | 6 |
| 5.299744 | CO | 71 |
| 5.3004 | KR | 6 |
| 5.302890 | CO | 73 |
| 5.303284 | CO | 81 |
| 5.304651 | CO | 72 |
| 5.306666 | CO | 72 |

| WAVELENGTH [MICROMETER] | ACTIVE MEDIUM | PAGE | WAVELENGTH [MICROMETER] | ACTIVE MEDIUM | PAGE |
|---|---|---|---|---|---|
| 5.306870 | CO | 73 | 5.377654 | CO | 76 |
| 5.3097 | DCL | 59 | 5.377903 | CO | 75 |
| 5.312842 | CO | 71 | 5.378545 | CO | 72 |
| 5.313418 | CO | 73 | 5.379526 | CO | 73 |
| 5.314516 | CO | 81 | 5.3799 | DCL | 59 |
| 5.316600 | CO | 72 | 5.381078 | CO | 76 |
| 5.318176 | CO | 73 | 5.384241 | CO | 81 |
| 5.319261 | CO | 72 | 5.384494 | CO | 72 |
| 5.324085 | CO | 73 | 5.385585 | CO | 76 |
| 5.3243 | NE | 2 | 5.386640 | CO | 74 |
| 5.3244 | DCL | 59 | 5.386899 | CO | 75 |
| 5.3249 | NE | 2 | 5.387807 | CO | 76 |
| 5.325877 | CO | 81 | 5.387850 | CO | 81 |
| 5.326091 | CO | 71 | 5.3889 | DCL | 59 |
| 5.3284 | BI | 23 | 5.389039 | CO | 73 |
| 5.328437 | CO | 81 | 5.3897 | AR | 5 |
| 5.328694 | CO | 72 | 5.391045 | CO | 73 |
| 5.329624 | CO | 73 | 5.391382 | CO | 72 |
| 5.332005 | CO | 72 | 5.393649 | CO | 76 |
| 5.334893 | CO | 73 | 5.394666 | CO | 76 |
| 5.337376 | CO | 81 | 5.3956 | DCL | 59 |
| 5.339330 | CO | 81 | 5.395810 | CO | 81 |
| 5.339493 | CO | 71 | 5.396031 | CO | 75 |
| 5.340935 | CO | 72 | 5.397499 | CO | 74 |
| 5.341216 | CO | 73 | 5.398001 | CO | 72 |
| 5.343257 | CO | 75 | 5.398801 | CO | 81 |
| 5.3443 | DCL | 59 | 5.401364 | CO | 73 |
| 5.344601 | CO | 74 | 5.401656 | CO | 76 |
| 5.344899 | CO | 72 | 5.401846 | CO | 76 |
| 5.345838 | CO | 73 | 5.402707 | CO | 73 |
| 5.350355 | CO | 81 | 5.4033 | NE | 2 |
| 5.351719 | CO | 75 | 5.404372 | CO | 72 |
| 5.352955 | CO | 73 | 5.405 | NE | 2 |
| 5.353049 | CO | 71 | 5.405298 | CO | 75 |
| 5.353322 | CO | 72 | 5.407516 | CO | 81 |
| 5.354648 | CO | 76 | 5.408504 | CO | 74 |
| 5.354901 | CO | 74 | 5.408780 | CO | 76 |
| 5.3551 | XE | 8 | 5.409888 | CO | 81 |
| 5.3562 | DCL | 59 | 5.410177 | CO | 76 |
| 5.356924 | CO | 73 | 5.411665 | CO | 72 |
| 5.357945 | CO | 72 | 5.413837 | CO | 73 |
| 5.360314 | CO | 75 | 5.414516 | CO | 73 |
| 5.362186 | CO | 76 | 5.414703 | CO | 75 |
| 5.3629 | DCL | 59 | 5.416030 | CO | 76 |
| 5.364837 | CO | 73 | 5.417516 | CO | 72 |
| 5.365340 | CO | 74 | 5.418642 | CO | 76 |
| 5.365859 | CO | 72 | 5.419652 | CO | 74 |
| 5.367272 | CO | 81 | 5.421112 | CO | 81 |
| 5.368153 | CO | 73 | 5.423415 | CO | 76 |
| 5.369042 | CO | 75 | 5.424245 | CO | 75 |
| 5.369854 | CO | 76 | 5.426463 | CO | 73 |
| 5.371143 | CO | 72 | 5.426477 | CO | 73 |
| 5.372811 | CO | 81 | 5.427242 | CO | 76 |
| 5.374479 | CO | 76 | 5.4292 | EU | 29 |
| 5.375920 | CO | 74 | 5.429350 | CO | 74 |
| 5.376865 | CO | 73 | 5.4295 | DCL | 59 |
| 5.377032 | CO | 81 | 5.430815 | CO | 72 |

| WAVELENGTH [MICROMETER] | ACTIVL MEDIUM | PAGE |
|---|---|---|
| 5.430932 | CO | 76 |
| 5.430943 | CO | 74 |
| 5.4322471 | CO | 81 |
| 5.433926 | CO | 75 |
| 5.435978 | CO | 76 |
| 5.437800 | CO | 82 |
| 5.438584 | CO | 76 |
| 5.438584 | CO | 73 |
| 5.439219 | CO | 73 |
| 5.439978 | CO | 74 |
| 5.442381 | CO | 74 |
| 5.4439721 | CO | 81 |
| 5.444270 | CO | 72 |
| 5.444850 | CO | 76 |
| 5.446369 | CO | 76 |
| 5.448674 | CO | 82 |
| 5.450750 | CO | 74 |
| 5.450840 | CO | 73 |
| 5.452172 | CO | 73 |
| 5.453966 | CO | 74 |
| 5.454289 | CO | 76 |
| 5.4556001 | CO | 81 |
| 5.4577 | DCL | 59 |
| 5.457884 | CO | 72 |
| 5.459684 | CO | 82 |
| 5.461666 | CO | 74 |
| 5.462344 | CO | 76 |
| 5.463249 | CO | 73 |
| 5.465259 | CO | 73 |
| 5.465699 | CO | 74 |
| 5.4666 | AR | 5 |
| 5.4673911 | CO | 81 |
| 5.4680 | AR | 5 |
| 5.470535 | CO | 76 |
| 5.470831 | CO | 82 |
| 5.471632 | CO | 72 |
| 5.472735 | CO | 74 |
| 5.4735 | XE | 6 |
| 5.475810 | CO | 73 |
| 5.477582 | CO | 74 |
| 5.478502 | CO | 73 |
| 5.4793111 | CO | 81 |
| 5.4796 | BA | 15 |
| 5.482117 | CO | 82 |
| 5.483932 | CO | 74 |
| 5.485591 | CO | 72 |
| 5.488524 | CO | 73 |
| 5.489622 | CO | 74 |
| 5.4913771 | CO | 81 |
| 5.491904 | CO | 73 |
| 5.4935 | DCL | 59 |
| 5.493543 | CO | 82 |
| 5.495289 | CO | 75 |
| 5.4972 | I | 28 |
| 5.499688 | CO | 72 |
| 5.500027 | CO | 82 |
| 5.501394 | CO | 73 |

| WAVELENGTH [MICROMETER] | ACTIVE MEDIUM | PAGE |
|---|---|---|
| 5.501808 | CO | 74 |
| 5.5035831 | CO | 81 |
| 5.505110 | CO | 82 |
| 5.505464 | CO | 73 |
| 5.505651 | CO | 75 |
| 5.506793 | CO | 75 |
| 5.5084 | DCL | 59 |
| 5.510954 | CO | 82 |
| 5.514147 | CO | 74 |
| 5.514421 | CO | 73 |
| 5.516474 | CO | 75 |
| 5.516821 | CO | 82 |
| 5.518447 | CO | 75 |
| 5.519186 | CO | 73 |
| 5.522022 | CO | 82 |
| 5.526640 | CO | 74 |
| 5.527444 | CO | 75 |
| 5.527606 | CO | 73 |
| 5.528674 | CO | 82 |
| 5.530252 | CO | 75 |
| 5.5304 | DCL | 59 |
| 5.533070 | CO | 73 |
| 5.533232 | CO | 82 |
| 5.538558 | CO | 75 |
| 5.539268 | CO | 74 |
| 5.540673 | CO | 82 |
| 5.540950 | CO | 74 |
| 5.542208 | CO | 75 |
| 5.5423 | DCL | 59 |
| 5.544580 | CO | 82 |
| 5.5457 | CA | 14 |
| 5.547118 | CO | 73 |
| 5.549827 | CO | 75 |
| 5.552093 | CO | 74 |
| 5.552816 | CO | 82 |
| 5.554315 | CO | 75 |
| 5.554455 | CO | 74 |
| 5.556074 | CO | 82 |
| 5.561252 | CO | 75 |
| 5.561330 | CO | 73 |
| 5.5636 | BA | 15 |
| 5.565057 | CO | 74 |
| 5.566589 | CO | 75 |
| 5.567715 | CO | 82 |
| 5.568122 | CO | 74 |
| 5.5685 | KR | 6 |
| 5.572816 | CO | 75 |
| 5.5739 | XE | 8 |
| 5.574745 | CO | 82 |
| 5.575710 | CO | 73 |
| 5.5776 | DCL | 59 |
| 5.578179 | CO | 74 |
| 5.579010 | CO | 75 |
| 5.579476 | CO | 82 |
| 5.583870 | CO | 76 |
| 5.583870 | CO | 77 |
| 5.584536 | CO | 75 |

| WAVELENGTH [MICROMETER] | ACTIVE MEDIUM | PAGE |
|---|---|---|
| 5.5848 | KR | 6 |
| 5.585871 | CO | 82 |
| 5.591415 | CO | 82 |
| 5.591463 | CO | 74 |
| 5.591588 | CO | 75 |
| 5.594893 | CO | 77 |
| 5.594893 | CO | 76 |
| 5.5956 | C | 19 |
| 5.596412 | CO | 75 |
| 5.597139 | CO | 82 |
| 5.6019 | XE | 6 |
| 5.603490 | CO | 82 |
| 5.604325 | CO | 75 |
| 5.604909 | CO | 74 |
| 5.606068 | CO | 76 |
| 5.606068 | CO | 77 |
| 5.608443 | CO | 75 |
| 5.608550 | CO | 82 |
| 5.6137 | DCL | 59 |
| 5.615709 | CO | 82 |
| 5.617221 | CO | 75 |
| 5.617391 | CO | 76 |
| 5.618519 | CO | 74 |
| 5.620109 | CO | 82 |
| 5.620630 | CO | 75 |
| 5.628080 | CO | 82 |
| 5.628872 | CO | 76 |
| 5.6290 | KR | 6 |
| 5.630278 | CO | 75 |
| 5.631813 | CO | 82 |
| 5.632293 | CO | 74 |
| 5.632975 | CO | 75 |
| 5.640505 | CO | 76 |
| 5.640600 | CO | 82 |
| 5.643498 | CO | 75 |
| 5.643666 | CO | 82 |
| 5.645485 | CO | 75 |
| 5.646235 | CO | 74 |
| 5.651276 | CO | 82 |
| 5.652298 | CO | 76 |
| 5.653272 | CO | 82 |
| 5.655664 | CO | 82 |
| 5.656881 | CO | 75 |
| 5.658148 | CO | 75 |
| 5.662607 | CO | 82 |
| 5.664243 | CO | 76 |
| 5.6652 | NE | 2 |
| 5.666098 | CO | 82 |
| 5.667813 | CO | 82 |
| 5.670430 | CO | 75 |
| 5.670974 | CO | 75 |
| 5.674081 | CO | 82 |
| 5.676348 | CO | 76 |
| 5.680114 | CO | 82 |
| 5.683961 | CO | 75 |
| 5.684145 | CO | 75 |
| 5.685704 | CO | 82 |
| 5.686687 | CO | 76 |
| 5.688612 | CO | 76 |
| 5.692566 | CO | 82 |
| 5.697113 | CO | 75 |
| 5.697477 | CO | 82 |
| 5.698029 | CO | 75 |
| 5.698230 | CO | 76 |
| 5.701037 | CO | 76 |
| 5.705170 | CO | 82 |
| 5.7053 | NE | 2 |
| 5.709398 | CO | 82 |
| 5.709930 | CO | 76 |
| 5.710431 | CO | 75 |
| 5.712083 | CO | 75 |
| 5.713623 | CO | 76 |
| 5.717932 | CO | 82 |
| 5.721472 | CO | 82 |
| 5.721786 | CO | 76 |
| 5.723916 | CO | 75 |
| 5.726379 | CO | 76 |
| 5.729687 | CO | 83 |
| 5.730846 | CO | 82 |
| 5.733698 | CO | 82 |
| 5.733804 | CO | 76 |
| 5.737568 | CO | 75 |
| 5.739293 | CO | 76 |
| 5.741227 | CO | 83 |
| 5.745984 | CO | 76 |
| 5.746076 | CO | 82 |
| 5.746341 | CO | 77 |
| 5.751390 | CO | 75 |
| 5.752373 | CO | 76 |
| 5.752912 | CO | 83 |
| 5.7534 | GA | 18 |
| 5.757944 | CO | 77 |
| 5.758322 | CO | 76 |
| 5.758611 | CO | 82 |
| 5.764755 | CO | 83 |
| 5.765384 | CO | 75 |
| 5.765620 | CO | 76 |
| 5.769382 | CO | 77 |
| 5.7706 | EU | 29 |
| 5.770827 | CO | 77 |
| 5.771303 | CO | 82 |
| 5.7758 | NE | 2 |
| 5.776747 | CO | 83 |
| 5.779036 | CO | 76 |
| 5.779551 | CO | 75 |
| 5.781150 | CO | 77 |
| 5.783496 | CO | 77 |
| 5.784148 | CO | 82 |
| 5.788893 | CO | 83 |
| 5.792622 | CO | 76 |
| 5.793071 | CO | 77 |
| 5.793891 | CO | 75 |
| 5.796340 | CO | 77 |
| 5.797155 | CO | 82 |

| WAVELENGTH [MICROMETER] | ACTIVE MEDIUM | PAGE |
|---|---|---|
| 5.798443 | CO | 83 |
| 5.801198 | CO | 83 |
| 5.8022 | AR | 5 |
| 5.8049 | DBR | 61 |
| 5.806380 | CO | 76 |
| 5.808408 | CO | 75 |
| 5.809343 | CO | 77 |
| 5.810042 | CO | 83 |
| 5.810319 | CO | 83 |
| 5.813656 | CO | 83 |
| 5.817413 | CO | 77 |
| 5.820310 | CO | 76 |
| 5.821796 | CO | 83 |
| 5.822516 | CO | 77 |
| 5.823103 | CO | 75 |
| 5.826267 | CO | 83 |
| 5.829830 | CO | 77 |
| 5.833703 | CO | 83 |
| 5.834415 | CO | 76 |
| 5.835859 | CO | 77 |
| 5.839055 | CO | 83 |
| 5.842407 | CO | 77 |
| 5.842415 | CO | 77 |
| 5.845769 | CO | 83 |
| 5.8461 | AR | 5 |
| 5.8462 | NO | 62 |
| 5.848696 | CO | 76 |
| 5.849374 | CO | 77 |
| 5.851981 | CO | 83 |
| 5.854229 | CO | 77 |
| 5.8549 | NO | 62 |
| 5.855168 | CO | 77 |
| 5.857987 | CO | 83 |
| 5.8584 | NO | 62 |
| 5.8620 | DBR | 61 |
| 5.8626 | DBR | 61 |
| 5.863063 | CO | 77 |
| 5.863154 | CO | 76 |
| 5.865079 | CO | 83 |
| 5.866221 | CO | 77 |
| 5.868096 | CO | 77 |
| 5.870367 | CO | 83 |
| 5.8706 | NO | 62 |
| 5.876926 | CO | 77 |
| 5.878384 | CO | 77 |
| 5.8789 | NO | 62 |
| 5.88 | HG | 17 |
| 5.881189 | CO | 77 |
| 5.882903 | CO | 83 |
| 5.8844 | NE | 2 |
| 5.8899 | BA | 15 |
| 5.890704 | CO | 77 |
| 5.890965 | CO | 77 |
| 5.892408 | CO | 83 |
| 5.8928 | DBR | 61 |
| 5.8944 | DBR | 61 |
| 5.894455 | CO | 77 |
| 5.895604 | CO | 83 |
| 5.903202 | CO | 77 |
| 5.9036 | NO | 62 |
| 5.904387 | CO | 83 |
| 5.905182 | CO | 77 |
| 5.907895 | CO | 77 |
| 5.9083 | NO | 62 |
| 5.908465 | CO | 83 |
| 5.915865 | CO | 77 |
| 5.916519 | CO | 83 |
| 5.919579 | CO | 77 |
| 5.921487 | CO | 83 |
| 5.921510 | CO | 77 |
| 5.9246 | DBR | 61 |
| 5.9261 | DBR | 61 |
| 5.928701 | CO | 77 |
| 5.928814 | CO | 83 |
| 5.934157 | CO | 77 |
| 5.934676 | CO | 83 |
| 5.935301 | CO | 77 |
| 5.941269 | CO | 83 |
| 5.941314 | CO | 78 |
| 5.941711 | CO | 77 |
| 5.9423 | NO | 62 |
| 5.9479 | EU | 29 |
| 5.948030 | CO | 83 |
| 5.948917 | CO | 77 |
| 5.949271 | CO | 77 |
| 5.953562 | CO | 78 |
| 5.953887 | CO | 83 |
| 5.9546 | NO | 62 |
| 5.954841 | CO | 77 |
| 5.9550 | NO | 62 |
| 5.9563 | NE | 2 |
| 5.9573 | DBR | 61 |
| 5.9590 | DBR | 61 |
| 5.961582 | CO | 83 |
| 5.9632 | NO | 62 |
| 5.963420 | CO | 77 |
| 5.964829 | CO | 83 |
| 5.965936 | CO | 78 |
| 5.966669 | CO | 83 |
| 5.9673 | NO | 62 |
| 5.968257 | CO | 77 |
| 5.975240 | CO | 83 |
| 5.9756 | NO | 62 |
| 5.976868 | CO | 83 |
| 5.977751 | CO | 77 |
| 5.978508 | CO | 78 |
| 5.979617 | CO | 83 |
| 5.9799 | NO | 62 |
| 5.981 | O | 24 |
| 5.9817 | HG | 17 |
| 5.981799 | CO | 77 |
| 5.9882 | NO | 62 |
| 5.989075 | CO | 83 |
| 5.991256 | CO | 78 |

| WAVELENGTH [MICROMETER] | ACTIVE MEDIUM | PAGE |
|---|---|---|
| 5.992264 | CO | 77 |
| 5.992728 | CO | 83 |
| 5.9931 | NO | 62 |
| 5.995510 | CO | 77 |
| 6.0010 | NO | 62 |
| 6.001439 | CO | 83 |
| 6.004170 | CO | 78 |
| 6.0054 | NO | 62 |
| 6.006059 | CO | 83 |
| 6.006963 | CO | 77 |
| 6.009406 | CO | 77 |
| 6.013970 | CO | 83 |
| 6.017267 | CO | 78 |
| 6.0192 | NO | 62 |
| 6.019459 | CO | 83 |
| 6.0209 | DBR | 62 |
| 6.0225 | DBR | 62 |
| 6.023483 | CO | 77 |
| 6.026667 | CO | 83 |
| 6.0267 | NO | 62 |
| 6.030541 | CO | 78 |
| 6.030715 | CO | 78 |
| 6.0324 | NO | 62 |
| 6.033077 | CO | 83 |
| 6.037744 | CO | 77 |
| 6.0386 | NO | 62 |
| 6.039534 | CO | 83 |
| 6.0402 | NO | 62 |
| 6.0419 | NO | 62 |
| 6.043187 | CO | 78 |
| 6.043400 | CO | 78 |
| 6.0515 | AR | 5 |
| 6.052189 | CO | 77 |
| 6.052565 | CO | 83 |
| 6.0529 | DBR | 62 |
| 6.0543 | NO | 62 |
| 6.0544 | DBR | 62 |
| 6.055827 | CO | 78 |
| 6.0576 | EU | 29 |
| 6.05760 | CO | 78 |
| 6.0628 | NO | 63 |
| 6.063494 | CO | 84 |
| 6.065767 | CO | 83 |
| 6.0673 | NO | 63 |
| 6.068653 | CO | 78 |
| 6.07142 | CO | 78 |
| 6.075932 | CO | 84 |
| 6.079138 | CO | 83 |
| 6.0801 | NO | 63 |
| 6.081656 | CO | 78 |
| 6.08542 | CO | 78 |
| 6.0858 | DBR | 62 |
| 6.0873 | DBR | 62 |
| 6.0884 | NO | 63 |
| 6.088540 | CO | 84 |
| 6.092683 | CO | 83 |
| 6.0934 | NO | 63 |
| 6.094841 | CO | 78 |
| 6.097625 | CO | 78 |
| 6.09960 | CO | 78 |
| 6.101318 | CO | 84 |
| 6.1015 | NO | 63 |
| 6.106404 | CO | 83 |
| 6.108206 | CO | 78 |
| 6.109989 | CO | 78 |
| 6.11397 | CO | 78 |
| 6.114263 | CO | 84 |
| 6.1200 | DBR | 62 |
| 6.120296 | CO | 83 |
| 6.1204 | NO | 63 |
| 6.1216 | DBR | 62 |
| 6.121753 | CO | 78 |
| 6.122525 | CO | 78 |
| 6.127383 | CO | 84 |
| 6.12852 | CO | 78 |
| 6.135248 | CO | 78 |
| 6.135488 | CO | 78 |
| 6.139855 | CO | 84 |
| 6.140673 | CO | 84 |
| 6.1417 | NO | 63 |
| 6.14327 | CO | 78 |
| 6.1460 | GA | 18 |
| 6.146152 | CO | 78 |
| 6.149406 | CO | 78 |
| 6.152366 | CO | 84 |
| 6.1538 | NO | 63 |
| 6.154138 | CO | 84 |
| 6.1546 | NO | 63 |
| 6.1546 | DBR | 62 |
| 6.1562 | DBR | 62 |
| 6.1576 | NO | 63 |
| 6.161240 | CO | 78 |
| 6.163511 | CO | 78 |
| 6.165049 | CO | 84 |
| 6.1663 | NO | 63 |
| 6.167780 | CO | 84 |
| 6.174509 | CO | 78 |
| 6.177806 | CO | 78 |
| 6.177904 | CO | 84 |
| 6.1792 | NO | 63 |
| 6.181597 | CO | 84 |
| 6.1838 | NO | 63 |
| 6.187966 | CO | 78 |
| 6.1903 | DBR | 62 |
| 6.190935 | CO | 84 |
| 6.191453 | CO | 78 |
| 6.1918 | DBR | 62 |
| 6.1921 | NO | 63 |
| 6.192291 | CO | 78 |
| 6.195595 | CO | 84 |
| 6.1972 | NO | 63 |
| 6.1973 | NO | 63 |
| 6.201608 | CO | 78 |
| 6.204067 | CO | 78 |

| WAVELENGTH [MICROMETER] | ACTIVE MEDIUM | PAGE |
|---|---|---|
| 6.204140 | CO | 84 |
| 6.2055 | NO | 63 |
| 6.206969 | CO | 78 |
| 6.209772 | CO | 84 |
| 6.2110 | NO | 63 |
| 6.215438 | CO | 78 |
| 6.216856 | CO | 78 |
| 6.217521 | CO | 84 |
| 6.2191 | NO | 63 |
| 6.221841 | CO | 78 |
| 6.2237 | DBR | 62 |
| 6.2249 | NO | 63 |
| 6.2272 | DBR | 62 |
| 6.2289 | DBR | 62 |
| 6.229461 | CO | 78 |
| 6.229843 | CO | 78 |
| 6.230817 | CO | 84 |
| 6.231077 | CO | 84 |
| 6.2328 | NO | 63 |
| 6.2381 | NO | 63 |
| 6.243012 | CO | 78 |
| 6.243577 | CO | 84 |
| 6.243673 | CO | 78 |
| 6.244817 | CO | 84 |
| 6.2511 | NO | 63 |
| 6.256366 | CO | 78 |
| 6.256511 | CO | 84 |
| 6.2566 | DBR | 62 |
| 6.258077 | CO | 78 |
| 6.2581 | DBR | 62 |
| 6.258735 | CO | 84 |
| 6.2602 | NO | 63 |
| 6.2645 | NO | 63 |
| 6.269624 | CO | 84 |
| 6.269915 | CO | 78 |
| 6.272677 | CO | 78 |
| 6.272833 | CO | 84 |
| 6.2778 | NO | 63 |
| 6.282915 | CO | 84 |
| 6.283649 | CO | 79 |
| 6.2865 | NO | 63 |
| 6.287118 | CO | 84 |
| 6.287472 | CO | 78 |
| 6.2913 | NO | 63 |
| 6.2916 | DBR | 62 |
| 6.2932 | DBR | 62 |
| 6.296385 | CO | 84 |
| 6.297578 | CO | 79 |
| 6.2998 | NO | 63 |
| 6.300749 | CO | 79 |
| 6.301586 | CO | 84 |
| 6.302466 | CO | 78 |
| 6.3051 | NO | 63 |
| 6.310036 | CO | 84 |
| 6.3103 | XE | 8 |
| 6.311705 | CO | 79 |
| 6.3136 | NO | 63 |
| 6.3137 | XE | 8 |
| 6.313813 | CO | 79 |
| 6.3191 | NO | 63 |
| 6.323871 | CO | 84 |
| 6.326023 | CO | 79 |
| 6.327064 | CO | 79 |
| 6.3274 | NO | 63 |
| 6.3279 | DBR | 62 |
| 6.3294 | DBR | 62 |
| 6.3336 | NO | 63 |
| 6.337216 | CO | 84 |
| 6.340507 | CO | 79 |
| 6.340538 | CO | 79 |
| 6.349964 | CO | 84 |
| 6.350412 | CO | 84 |
| 6.352094 | CO | 84 |
| 6.354144 | CO | 79 |
| 6.355252 | CO | 79 |
| 6.363789 | CO | 84 |
| 6.366467 | CO | 84 |
| 6.3672 | SE | 25 |
| 6.367975 | CO | 79 |
| 6.370176 | CO | 79 |
| 6.3764 | NO | 63 |
| 6.377348 | CO | 84 |
| 6.381067 | CO | 84 |
| 6.382004 | CO | 79 |
| 6.385283 | CO | 79 |
| 6.3894 | NO | 63 |
| 6.391091 | CO | 84 |
| 6.3980 | NO | 63 |
| 6.4 | K | 10 |
| 6.400168 | CO | 79 |
| 6.4031 | NO | 63 |
| 6.405026 | CO | 84 |
| 6.410658 | CO | 79 |
| 6.413510 | CO | 79 |
| 6.419144 | CO | 84 |
| 6.425284 | CO | 79 |
| 6.4262 | NO | 63 |
| 6.427033 | CO | 79 |
| 6.4321 | NO | 63 |
| 6.433454 | CO | 84 |
| 6.440113 | CO | 79 |
| 6.440760 | CO | 79 |
| 6.446846 | CO | 84 |
| 6.447956 | CO | 84 |
| 6.4546 | BA | 15 |
| 6.454686 | CO | 79 |
| 6.455148 | CO | 79 |
| 6.4567 | SR | 14 |
| 6.460499 | CO | 84 |
| 6.462649 | CO | 84 |
| 6.468812 | CO | 79 |
| 6.470368 | CO | 79 |
| 6.474336 | CO | 84 |
| 6.485838 | CO | 79 |

| WAVELENGTH [MICROMETER] | ACTIVE MEDIUM | PAGE |
|---|---|---|
| 6.488362 | CO | 84 |
| 6.49 | HG | 17 |
| 6.497679 | CO | 79 |
| 6.502437 | CO | 79 |
| 6.502585 | CO | 84 |
| 6.512417 | CO | 79 |
| 6.516055 | CO | 79 |
| 6.517002 | CO | 85 |
| 6.527363 | CO | 79 |
| 6.529870 | CO | 79 |
| 6.531610 | CO | 85 |
| 6.542518 | CO | 79 |
| 6.543890 | CO | 79 |
| 6.546418 | CO | 85 |
| 6.557884 | CO | 79 |
| 6.558117 | CO | 79 |
| 6.559859 | CO | 85 |
| 6.561421 | CO | 85 |
| 6.572552 | CO | 79 |
| 6.573463 | CO | 79 |
| 6.573982 | CO | 85 |
| 6.587194 | CO | 79 |
| 6.588305 | CO | 85 |
| 6.602046 | CO | 79 |
| 6.602826 | CO | 85 |
| 6.607691 | CO | 79 |
| 6.617109 | CO | 79 |
| 6.617547 | CO | 85 |
| 6.621596 | CO | 79 |
| 6.632387 | CO | 79 |
| 6.635713 | CO | 79 |
| 6.647880 | CO | 79 |
| 6.650037 | CO | 79 |
| 6.663590 | CO | 79 |
| 6.664574 | CO | 79 |
| 6.679326 | CO | 79 |
| 6.679519 | CO | 79 |
| 6.694287 | CO | 79 |
| 6.702067 | CO | 80 |
| 6.709469 | CO | 80 |
| 6.716066 | CO | 80 |
| 6.7198 | I | 28 |
| 6.730277 | CO | 80 |
| 6.7443 | AR | 5 |
| 6.744698 | CO | 80 |
| 6.759342 | CO | 80 |
| 6.7595 | TE | 28 |
| 6.774209 | CO | 80 |
| 6.7769 | NE | 2 |
| 6.789276 | CO | 80 |
| 6.804577 | CO | 80 |
| 6.8161 | O | 24 |
| 6.820100 | CO | 80 |
| 6.835854 | CO | 80 |
| 6.842243 | CO | 80 |
| 6.8470 | FE | 9 |
| 6.854639 | CO | 80 |
| 6.858 | O | 24 |
| 6.871970 | CO | 80 |
| 6.8731 | O | 24 |
| 6.8865 | NE | 2 |
| 6.887152 | CO | 80 |
| 6.902 | I | 28 |
| 6.902569 | CO | 80 |
| 6.918218 | CO | 80 |
| 6.934101 | CO | 80 |
| 6.9410 | AR | 5 |
| 6.942824 | CO | 80 |
| 6.9429 | AR | 5 |
| 6.957669 | CO | 80 |
| 6.972747 | CO | 80 |
| 6.9857 | NE | 3 |
| 6.988057 | CO | 80 |
| 7.003596 | CO | 80 |
| 7.019371 | CO | 80 |
| 7.035362 | CO | 80 |
| 7.0565 | KR | 6 |
| 7.076735 | CO | 80 |
| 7.092158 | CO | 80 |
| 7.093 | H2O | 104 |
| 7.107821 | CO | 80 |
| 7.123720 | CO | 80 |
| 7.139864 | CO | 80 |
| 7.156250 | CO | 80 |
| 7.1740 | PB | 20 |
| 7.1821 | CS | 11 |
| 7.199637 | CO | 80 |
| 7.204 | H2O | 104 |
| 7.2147 | AR | 5 |
| 7.215418 | CO | 80 |
| 7.231448 | CO | 80 |
| 7.247721 | CO | 80 |
| 7.264248 | CO | 80 |
| 7.285 | H2O | 104 |
| 7.297 | H2O | 104 |
| 7.310709 | CO | 80 |
| 7.3147 | XE | 8 |
| 7.3208 | NE | 3 |
| 7.326587 | CO | 80 |
| 7.342742 | CO | 80 |
| 7.359153 | CO | 80 |
| 7.3605 | KR | 7 |
| 7.375822 | CO | 80 |
| 7.390 | H2O | 104 |
| 7.392748 | CO | 80 |
| 7.4201 | NE | 3 |
| 7.4217 | NE | 3 |
| 7.425 | H2O | 104 |
| 7.4294 | XE | 8 |
| 7.441563 | CO | 80 |
| 7.4568 | H2O | 104 |
| 7.457822 | CO | 80 |
| 7.4679 | NE | 3 |
| 7.474367 | CO | 80 |

| WAVELENGTH [MICROMETER] | ACTIVE MEDIUM | PAGE | WAVELENGTH [MICROMETER] | ACTIVE MEDIUM | PAGE |
|---|---|---|---|---|---|
| 7.4779 | NE | 3 | 8.0380 | C2H2 | 114 |
| 7.491175 | CO | 80 | 8.0409 | C2H2 | 114 |
| 7.4973 | NE | 3 | 8.0442 | C2H2 | 114 |
| 7.508252 | CO | 80 | 8.049391 | CO | 81 |
| 7.525593 | CO | 80 | 8.0599 | NE | 3 |
| 7.5292 | NE | 3 | 8.067991 | CO | 81 |
| 7.543 | H2O | 104 | 8.086885 | CO | 81 |
| 7.543207 | CO | 80 | 8.106093 | CO | 81 |
| 7.5674 | NE | 3 | 8.1712 | NE | 3 |
| 7.5850 | NE | 3 | 8.18384 | N2 | 46 |
| 7.590 | H2O | 104 | 8.206919 | CO | 81 |
| 7.593623 | CO | 81 | 8.21106 | N2 | 46 |
| 7.5945 | H2O | 104 | 8.226011 | CO | 81 |
| 7.610553 | CO | 81 | 8.2388 | OCS | 101 |
| 7.6142 | NE | 3 | 8.2416 | OCS | 101 |
| 7.627777 | CO | 81 | 8.2439 | OCS | 101 |
| 7.6440 | NE | 3 | 8.245417 | CO | 81 |
| 7.645277 | CO | 81 | 8.2518 | OCS | 101 |
| 7.6489 | NE | 3 | 8.2543 | OCS | 101 |
| 7.663057 | CO | 81 | 8.2571 | OCS | 102 |
| 7.681105 | CO | 81 | 8.2595 | OCS | 102 |
| 7.6904 | NE | 3 | 8.2623 | OCS | 102 |
| 7.6994 | NE | 3 | 8.2645 | OCS | 102 |
| 7.7069 | H2O | 104 | 8.265146 | CO | 81 |
| 7.709 | H2O | 104 | 8.2673 | OCS | 102 |
| 7.7097 | H2O | 104 | 8.3347 | NE | 3 |
| 7.734191 | CO | 81 | 8.3472 | NE | 3 |
| 7.7389 | NE | 3 | 8.3625 | OCS | 102 |
| 7.740 | H2O | 104 | 8.3654 | OCS | 102 |
| 7.751571 | CO | 81 | 8.3685 | OCS | 102 |
| 7.7634 | NE | 3 | 8.3715 | OCS | 102 |
| 7.769230 | CO | 81 | 8.3746 | OCS | 102 |
| 7.7794 | NE | 3 | 8.3779 | OCS | 102 |
| 7.7856 | TE | 26 | 8.3809 | OCS | 102 |
| 7.787173 | CO | 81 | 8.3839 | OCS | 102 |
| 7.7982 | AR | 5 | 8.3870 | OCS | 102 |
| 7.8002 | AR | 5 | 8.3900 | OCS | 102 |
| 7.8042 | AR | 5 | 8.3930 | OCS | 102 |
| 7.805408 | CO | 81 | 8.3962 | OCS | 102 |
| 7.823924 | CO | 81 | 8.3999 | OCS | 102 |
| 7.8347 | NE | 3 | 8.4024 | OCS | 102 |
| 7.8693 | NE | 3 | 8.4055 | OCS | 102 |
| 7.879916 | CO | 81 | 8.4085 | OCS | 102 |
| 7.897740 | CO | 81 | 8.4117 | OCS | 102 |
| 7.9 | K | 10 | 8.4146 | OCS | 102 |
| 7.915858 | CO | 81 | 8.4178 | OCS | 102 |
| 7.934266 | CO | 81 | 8.4213 | OCS | 102 |
| 7.9399 | PB | 20 | 8.4243 | OCS | 102 |
| 7.9406 | NE | 3 | 8.4902 | FE | 9 |
| 7.952969 | CO | 81 | 8.53 | HE | 1 |
| 7.9824 | NE | 3 | 8.6388 | NE | 3 |
| 8.0066 | NE | 3 | 8.6528 | NE | 3 |
| 8.031109 | CO | 81 | 8.9877009 | CO2 | 93 |
| 8.0334 | C2H2 | 114 | 8.9949699 | CO2 | 93 |
| 8.0340 | C2H2 | 114 | 9.0024021 | CO2 | 93 |
| 8.0347 | C2H2 | 114 | 9.0040 | XE | 8 |
| 8.0356 | C2H2 | 114 | 9.0099989 | CO2 | 93 |

| WAVELENGTH [MICROMETER] | ACTIVE MEDIUM | PAGE |
|---|---|---|
| 9.0177619 | CO2 | 93 |
| 9.0256923 | CO2 | 93 |
| 9.0337917 | CO2 | 93 |
| 9.0420614 | CO2 | 93 |
| 9.0505029 | CO2 | 93 |
| 9.0591176 | CO2 | 93 |
| 9.0679068 | CO2 | 93 |
| 9.0702655 | CO2 | 87 |
| 9.0757663 | CO2 | 87 |
| 9.0768719 | CO2 | 93 |
| 9.0814571 | CO2 | 87 |
| 9.0860143 | CO2 | 93 |
| 9.0871 | NE | 3 |
| 9.0873410 | CO2 | 87 |
| 9.0934211 | CO2 | 87 |
| 9.0953353 | CO2 | 93 |
| 9.0997003 | CO2 | 87 |
| 9.1048363 | CO2 | 94 |
| 9.1061815 | CO2 | 87 |
| 9.1128676 | CO2 | 87 |
| 9.1145185 | CO2 | 94 |
| 9.1197615 | CO2 | 87 |
| 9.1243834 | CO2 | 94 |
| 9.1268660 | CO2 | 87 |
| 9.1341839 | CO2 | 87 |
| 9.1344321 | CO2 | 94 |
| 9.1417179 | CO2 | 87 |
| 9.1446661 | CO2 | 94 |
| 9.1494708 | CO2 | 87 |
| 9.1550865 | CO2 | 94 |
| 9.1574453 | CO2 | 87 |
| 9.1656440 | CO2 | 87 |
| 9.1656946 | CO2 | 94 |
| 9.1740695 | CO2 | 87 |
| 9.1764918 | CO2 | 94 |
| 9.1827244 | CO2 | 87 |
| 9.1874793 | CO2 | 94 |
| 9.1916114 | CO2 | 87 |
| 9.1986582 | CO2 | 94 |
| 9.2007329 | CO2 | 87 |
| 9.209171 | CO2 | 89 |
| 9.2100299 | CO2 | 94 |
| 9.2100915 | CO2 | 87 |
| 9.217773 | CO2 | 89 |
| 9.2196895 | CO2 | 87 |
| 9.2215955 | CO2 | 94 |
| 9.226615 | CO2 | 89 |
| 9.2295296 | CO2 | 87 |
| 9.235699 | CO2 | 89 |
| 9.2393102 | CO2 | 94 |
| 9.2396141 | CO2 | 87 |
| 9.245029 | CO2 | 89 |
| 9.2499453 | CO2 | 87 |
| 9.2513659 | CO2 | 94 |
| 9.254607 | CO2 | 89 |
| 9.2605258 | CO2 | 87 |
| 9.2636197 | CO2 | 94 |
| 9.264436 | CO2 | 89 |
| 9.2713577 | CO2 | 87 |
| 9.274517 | CO2 | 89 |
| 9.2760729 | CO2 | 94 |
| 9.2824434 | CO2 | 87 |
| 9.284854 | CO2 | 89 |
| 9.2887264 | CO2 | 94 |
| 9.2937852 | CO2 | 87 |
| 9.295448 | CO2 | 89 |
| 9.3015816 | CO2 | 94 |
| 9.3053853 | CO2 | 87 |
| 9.306302 | CO2 | 89 |
| 9.3146394 | CO2 | 94 |
| 9.316821 | CO2 | 89 |
| 9.3172460 | CO2 | 87 |
| 9.326 | I | 28 |
| 9.3279010 | CO2 | 94 |
| 9.328800 | CO2 | 89 |
| 9.3293695 | CO2 | 87 |
| 9.340448 | CO2 | 89 |
| 9.3413676 | CO2 | 94 |
| 9.3417579 | CO2 | 87 |
| 9.3462 | F | 26 |
| 9.352366 | CO2 | 89 |
| 9.3544134 | CO2 | 87 |
| 9.3550402 | CO2 | 94 |
| 9.364555 | CO2 | 89 |
| 9.3673380 | CO2 | 87 |
| 9.3689199 | CO2 | 94 |
| 9.377018 | CO2 | 89 |
| 9.3805340 | CO2 | 87 |
| 9.3830078 | CO2 | 94 |
| 9.389757 | CO2 | 89 |
| 9.3938 | H2O | 104 |
| 9.3940033 | CO2 | 87 |
| 9.3973048 | CO2 | 94 |
| 9.402774 | CO2 | 89 |
| 9.4118121 | CO2 | 94 |
| 9.4147242 | CO2 | 87 |
| 9.4265307 | CO2 | 94 |
| 9.4288857 | CO2 | 87 |
| 9.4414615 | CO2 | 94 |
| 9.4433275 | CO2 | 87 |
| 9.450554 | CO2 | 90 |
| 9.4566056 | CO2 | 94 |
| 9.4580515 | CO2 | 87 |
| 9.464848 | CO2 | 90 |
| 9.4719639 | CO2 | 94 |
| 9.4730598 | CO2 | 87 |
| 9.4747 | H2O | 104 |
| 9.479432 | CO2 | 90 |
| 9.4875375 | CO2 | 94 |
| 9.4883540 | CO2 | 87 |
| 9.4918511 | CO2 | 97 |
| 9.494307 | CO2 | 90 |
| 9.4999578 | CO2 | 97 |
| 9.5033271 | CO2 | 94 |

| WAVELENGTH [MICROMETER] | ACTIVE MEDIUM | PAGE |
|---|---|---|
| 9.5039361 | CO2 | 87 |
| 9.5082610 | CO2 | 97 |
| 9.509476 | CO2 | 90 |
| 9.5167627 | CO2 | 97 |
| 9.5193338 | CO2 | 94 |
| 9.5198079 | CO2 | 87 |
| 9.524939 | CO2 | 90 |
| 9.5254650 | CO2 | 97 |
| 9.5343699 | CO2 | 97 |
| 9.5355584 | CO2 | 94 |
| 9.5359711 | CO2 | 87 |
| 9.540700 | CO2 | 90 |
| 9.5434794 | CO2 | 97 |
| 9.5520019 | CO2 | 94 |
| 9.5524275 | CO2 | 87 |
| 9.5527954 | CO2 | 97 |
| 9.556760 | CO2 | 90 |
| 9.5623199 | CO2 | 97 |
| 9.5674 | H2O | 104 |
| 9.5686651 | CO2 | 94 |
| 9.5691788 | CO2 | 87 |
| 9.5720548 | CO2 | 97 |
| 9.573121 | CO2 | 90 |
| 9.5820021 | CO2 | 97 |
| 9.5855488 | CO2 | 94 |
| 9.5862267 | CO2 | 87 |
| 9.589785 | CO2 | 90 |
| 9.5921635 | CO2 | 97 |
| 9.5923994 | CO2 | 95 |
| 9.6016899 | CO2 | 95 |
| 9.6025410 | CO2 | 97 |
| 9.6026539 | CO2 | 94 |
| 9.6035727 | CO2 | 87 |
| 9.606753 | CO2 | 90 |
| 9.6112571 | CO2 | 95 |
| 9.6131363 | CO2 | 97 |
| 9.6199812 | CO2 | 94 |
| 9.6211040 | CO2 | 95 |
| 9.6212185 | CO2 | 87 |
| 9.6239513 | CO2 | 97 |
| 9.624027 | CO2 | 90 |
| 9.6312334 | CO2 | 95 |
| 9.6349876 | CO2 | 97 |
| 9.6375315 | CO2 | 94 |
| 9.6391656 | CO2 | 87 |
| 9.641609 | CO2 | 90 |
| 9.6416483 | CO2 | 95 |
| 9.6462476 | CO2 | 97 |
| 9.6523516 | CO2 | 95 |
| 9.655900 | CO2 | 90 |
| 9.6574156 | CO2 | 88 |
| 9.6577324 | CO2 | 97 |
| 9.6633460 | CO2 | 95 |
| 9.6694439 | CO2 | 97 |
| 9.6746345 | CO2 | 95 |
| 9.6750700 | CO2 | 88 |
| 9.677702 | CO2 | 90 |
| 9.6813838 | CO2 | 97 |
| 9.6862198 | CO2 | 95 |
| 9.6935538 | CO2 | 97 |
| 9.6948301 | CO2 | 88 |
| 9.696217 | CO2 | 90 |
| 9.6981048 | CO2 | 95 |
| 9.7002 | XE | 8 |
| 9.7059556 | CO2 | 97 |
| 9.7102921 | CO2 | 95 |
| 9.7139973 | CO2 | 88 |
| 9.715046 | CO2 | 90 |
| 9.7165908 | CO2 | 97 |
| 9.7227844 | CO2 | 95 |
| 9.7334730 | CO2 | 88 |
| 9.734191 | CO2 | 90 |
| 9.7355846 | CO2 | 95 |
| 9.7486953 | CO2 | 95 |
| 9.7512104 | CO2 | 97 |
| 9.7532586 | CO2 | 88 |
| 9.753653 | CO2 | 90 |
| 9.7621191 | CO2 | 95 |
| 9.7646751 | CO2 | 98 |
| 9.7733552 | CO2 | 88 |
| 9.773433 | CO2 | 90 |
| 9.7758587 | CO2 | 95 |
| 9.7763803 | CO2 | 98 |
| 9.7899166 | CO2 | 95 |
| 9.7923275 | CO2 | 98 |
| 9.793533 | CO2 | 90 |
| 9.7937640 | CO2 | 88 |
| 9.8042956 | CO2 | 95 |
| 9.8065182 | CO2 | 98 |
| 9.813954 | CO2 | 90 |
| 9.8144862 | CO2 | 88 |
| 9.8209538 | CO2 | 98 |
| 9.8355229 | CO2 | 88 |
| 9.8356358 | CO2 | 98 |
| 9.8505657 | CO2 | 98 |
| 9.8568751 | CO2 | 88 |
| 9.8571865 | CO2 | 95 |
| 9.8657447 | CO2 | 98 |
| 9.8730413 | CO2 | 95 |
| 9.8785439 | CO2 | 88 |
| 9.8811743 | CO2 | 98 |
| 9.8892309 | CO2 | 95 |
| 9.8968558 | CO2 | 98 |
| 9.9005300 | CO2 | 88 |
| 9.9057577 | CO2 | 95 |
| 9.9127905 | CO2 | 98 |
| 9.9226242 | CO2 | 95 |
| 9.9228344 | CO2 | 88 |
| 9.9289796 | CO2 | 98 |
| 9.9398328 | CO2 | 95 |
| 9.9454245 | CO2 | 98 |
| 9.9454579 | CO2 | 88 |
| 9.9573857 | CO2 | 95 |
| 9.9621263 | CO2 | 98 |

| WAVELENGTH [MICROMETER] | ACTIVE MEDIUM | PAGE |
|---|---|---|
| 9.9684012 | CO2 | 88 |
| 9.9752853 | CO2 | 95 |
| 9.9790862 | CO2 | 96 |
| 9.9916650 | CO2 | 88 |
| 9.9935340 | CO2 | 95 |
| 9.9963054 | CO2 | 96 |
| 9.9985568 | CO2 | 88 |
| 10.0049238 | CO2 | 88 |
| 10.0115934 | CO2 | 88 |
| 10.0121339 | CO2 | 95 |
| 10.0137851 | CO2 | 96 |
| 10.0152498 | CO2 | 88 |
| 10.0185643 | CO2 | 88 |
| 10.0258352 | CO2 | 88 |
| 10.0310874 | CO2 | 95 |
| 10.0315262 | CO2 | 96 |
| 10.0334048 | CO2 | 88 |
| 10.0391561 | CO2 | 88 |
| 10.0412720 | CO2 | 88 |
| 10.0494358 | CO2 | 88 |
| 10.0495299 | CO2 | 96 |
| 10.0503967 | CO2 | 95 |
| 10.0578953 | CO2 | 88 |
| 10.060 | NE | 3 |
| 10.0633844 | CO2 | 88 |
| 10.0666497 | CO2 | 88 |
| 10.0677972 | CO2 | 96 |
| 10.0700639 | CO2 | 95 |
| 10.0756984 | CO2 | 88 |
| 10.0850408 | CO2 | 88 |
| 10.0863291 | CO2 | 96 |
| 10.0879349 | CO2 | 88 |
| 10.0900911 | CO2 | 95 |
| 10.0946764 | CO2 | 88 |
| 10.1043466 | CO2 | 94 |
| 10.1046049 | CO2 | 88 |
| 10.1051265 | CO2 | 96 |
| 10.1104806 | CO2 | 95 |
| 10.1129549 | CO2 | 94 |
| 10.1148262 | CO2 | 88 |
| 10.1218628 | CO2 | 94 |
| 10.1241903 | CO2 | 96 |
| 10.1253400 | CO2 | 88 |
| 10.1310708 | CO2 | 94 |
| 10.1312343 | CO2 | 96 |
| 10.1361464 | CO2 | 88 |
| 10.1405794 | CO2 | 94 |
| 10.1435215 | CO2 | 96 |
| 10.146624 | CO2 | 90 |
| 10.1472454 | CO2 | 88 |
| 10.1503892 | CO2 | 94 |
| 10.1523543 | CO2 | 96 |
| 10.157295 | CO2 | 90 |
| 10.1586374 | CO2 | 88 |
| 10.1605007 | CO2 | 94 |
| 10.168257 | CO2 | 90 |
| 10.1703225 | CO2 | 88 |
| 10.1709149 | CO2 | 94 |
| 10.1738427 | CO2 | 96 |
| 10.179508 | CO2 | 90 |
| 10.1816326 | CO2 | 94 |
| 10.1823014 | CO2 | 88 |
| 10.191050 | CO2 | 90 |
| 10.1926548 | CO2 | 94 |
| 10.1945745 | CO2 | 88 |
| 10.1957013 | CO2 | 96 |
| 10.1978 | HF | 54 |
| 10.2 | NH3 | 108 |
| 10.202883 | CO2 | 90 |
| 10.2039826 | CO2 | 94 |
| 10.2071425 | CO2 | 88 |
| 10.215008 | CO2 | 90 |
| 10.2156172 | CO2 | 94 |
| 10.2200062 | CO2 | 88 |
| 10.227424 | CO2 | 90 |
| 10.2275598 | CO2 | 94 |
| 10.2331666 | CO2 | 88 |
| 10.2398120 | CO2 | 94 |
| 10.240133 | CO2 | 90 |
| 10.2466246 | CO2 | 88 |
| 10.2523751 | CO2 | 94 |
| 10.253135 | CO2 | 90 |
| 10.2603814 | CO2 | 88 |
| 10.2652509 | CO2 | 94 |
| 10.266431 | CO2 | 90 |
| 10.2744384 | CO2 | 88 |
| 10.2784410 | CO2 | 95 |
| 10.280023 | CO2 | 90 |
| 10.2887967 | CO2 | 88 |
| 10.288987 | CO2 | 91 |
| 10.2919473 | CO2 | 95 |
| 10.293911 | CO2 | 90 |
| 10.302426 | CO2 | 91 |
| 10.3034581 | CO2 | 88 |
| 10.3057717 | CO2 | 95 |
| 10.308097 | CO2 | 90 |
| 10.316157 | CO2 | 91 |
| 10.3184241 | CO2 | 88 |
| 10.322582 | CO2 | 90 |
| 10.330184 | CO2 | 91 |
| 10.3336965 | CO2 | 88 |
| 10.337367 | CO2 | 90 |
| 10.344505 | CO2 | 91 |
| 10.3456 | N2O | 99 |
| 10.3492772 | CO2 | 88 |
| 10.352455 | CO2 | 90 |
| 10.3532 | N2O | 99 |
| 10.359124 | CO2 | 91 |
| 10.3609 | N2O | 99 |
| 10.3651683 | CO2 | 88 |
| 10.367847 | CO2 | 90 |
| 10.3687 | N2O | 99 |
| 10.374040 | CO2 | 91 |
| 10.3765 | N2O | 99 |

| WAVELENGTH [MICROMETER] | ACTIVE MEDIUM | PAGE |
|---|---|---|
| 10.3813718 | CO2 | 88 |
| 10.383545 | CO2 | 90 |
| 10.3843 | N2O | 99 |
| 10.3875898 | CO2 | 95 |
| 10.389256 | CO2 | 91 |
| 10.3922 | N2O | 99 |
| 10.3978901 | CO2 | 89 |
| 10.399550 | CO2 | 90 |
| 10.400 | O | 24 |
| 10.4001 | N2O | 99 |
| 10.40331 | N2O | 100 |
| 10.4035372 | CO2 | 95 |
| 10.404773 | CO2 | 91 |
| 10.4081 | N2O | 99 |
| 10.41107 | N2O | 100 |
| 10.415866 | CO2 | 90 |
| 10.4161 | N2O | 99 |
| 10.41889 | N2O | 100 |
| 10.4198210 | CO2 | 95 |
| 10.420594 | CO2 | 91 |
| 10.4232632 | CO2 | 89 |
| 10.4242 | N2O | 99 |
| 10.42676 | N2O | 100 |
| 10.4323 | N2O | 99 |
| 10.43468 | N2O | 100 |
| 10.4364442 | CO2 | 95 |
| 10.4405 | N2O | 99 |
| 10.4405795 | CO2 | 89 |
| 10.44265 | N2O | 100 |
| 10.4487 | N2O | 99 |
| 10.449 | TL | 18 |
| 10.45067 | N2O | 100 |
| 10.4534098 | CO2 | 95 |
| 10.4570 | N2O | 99 |
| 10.4578 | HF | 54 |
| 10.458029 | CO2 | 90 |
| 10.4582196 | CO2 | 89 |
| 10.45874 | N2O | 100 |
| 10.4653 | N2O | 99 |
| 10.46686 | N2O | 100 |
| 10.4707211 | CO2 | 95 |
| 10.4737 | N2O | 99 |
| 10.47503 | N2O | 100 |
| 10.475449 | CO2 | 90 |
| 10.4761866 | CO2 | 89 |
| 10.4821 | N2O | 99 |
| 10.48325 | N2O | 100 |
| 10.4883813 | CO2 | 95 |
| 10.4906 | N2O | 99 |
| 10.49157 | N2O | 100 |
| 10.4928230 | CO2 | 96 |
| 10.493192 | CO2 | 90 |
| 10.4944835 | CO2 | 89 |
| 10.4991 | N2O | 99 |
| 10.49985 | N2O | 100 |
| 10.5 | NH3 | 108 |
| 10.5027068 | CO2 | 96 |
| 10.5063941 | CO2 | 95 |
| 10.5077 | N2O | 99 |
| 10.50816 | CO2 | 92 |
| 10.50823 | N2O | 100 |
| 10.51001 | CO2 | 92 |
| 10.51027 | CO2 | 93 |
| 10.511259 | CO2 | 90 |
| 10.5128754 | CO2 | 96 |
| 10.5131136 | CO2 | 89 |
| 10.5163 | N2O | 99 |
| 10.51666 | N2O | 100 |
| 10.52029 | CO2 | 92 |
| 10.52277 | CO2 | 92 |
| 10.5233288 | CO2 | 96 |
| 10.5247629 | CO2 | 95 |
| 10.5250 | N2O | 99 |
| 10.52513 | N2O | 100 |
| 10.529654 | CO2 | 90 |
| 10.53 | C2H4 | 115 |
| 10.53097 | CO2 | 91 |
| 10.5320802 | CO2 | 89 |
| 10.53273 | CO2 | 92 |
| 10.53367 | N2O | 100 |
| 10.5337 | N2O | 99 |
| 10.5340673 | CO2 | 96 |
| 10.53907 | CO2 | 93 |
| 10.54173 | CO2 | 93 |
| 10.54225 | N2O | 100 |
| 10.5425 | N2O | 99 |
| 10.54271 | CO2 | 93 |
| 10.5434916 | CO2 | 95 |
| 10.5450914 | CO2 | 96 |
| 10.54550 | CO2 | 92 |
| 10.548380 | CO2 | 90 |
| 10.54916 | CO2 | 91 |
| 10.54919 | CO2 | 92 |
| 10.55089 | N2O | 100 |
| 10.5513 | N2O | 99 |
| 10.5513868 | CO2 | 89 |
| 10.55376 | CO2 | 91 |
| 10.55455 | CO2 | 93 |
| 10.5564013 | CO2 | 96 |
| 10.55859 | CO2 | 92 |
| 10.55958 | N2O | 100 |
| 10.5602 | N2O | 99 |
| 10.5625841 | CO2 | 95 |
| 10.56284 | CO2 | 92 |
| 10.5639 | CO2 | 93 |
| 10.567440 | CO2 | 90 |
| 10.56762 | CO2 | 91 |
| 10.5679978 | CO2 | 97 |
| 10.56832 | N2O | 100 |
| 10.5692 | N2O | 99 |
| 10.5710372 | CO2 | 89 |
| 10.57170 | CO2 | 91 |
| 10.57201 | CO2 | 92 |
| 10.57678 | CO2 | 92 |

| WAVELENGTH [MICROMETER] | ACTIVE MEDIUM | PAGE |
|---|---|---|
| 10.57712 | N2O | 101 |
| 10.5781 | N2O | 99 |
| 10.5798813 | CO2 | 97 |
| 10.58112 | CO2 | 93 |
| 10.5819 | HF | 55 |
| 10.5820443 | CO2 | 95 |
| 10.58575 | CO2 | 92 |
| 10.58596 | N2O | 101 |
| 10.58646 | CO2 | 91 |
| 10.586838 | CO2 | 90 |
| 10.5872 | N2O | 99 |
| 10.5884190 | CO2 | 96 |
| 10.5900 | CO2 | 91 |
| 10.591025 | CO2 | 91 |
| 10.5910352 | CO2 | 89 |
| 10.5920528 | CO2 | 97 |
| 10.59486 | N2O | 101 |
| 10.5963 | N2O | 99 |
| 10.59982 | CO2 | 92 |
| 10.6 | NH3 | 108 |
| 10.6006264 | CO2 | 96 |
| 10.6018767 | CO2 | 95 |
| 10.60362 | CO2 | 93 |
| 10.60382 | N2O | 101 |
| 10.6045131 | CO2 | 97 |
| 10.6054 | N2O | 99 |
| 10.60556 | CO2 | 92 |
| 10.60562 | CO2 | 91 |
| 10.606578 | CO2 | 90 |
| 10.60858 | CO2 | 91 |
| 10.60885 | CO2 | 93 |
| 10.6113848 | CO2 | 89 |
| 10.61282 | N2O | 101 |
| 10.6131029 | CO2 | 96 |
| 10.61421 | CO2 | 92 |
| 10.6146 | N2O | 99 |
| 10.6172631 | CO2 | 97 |
| 10.62189 | N2O | 101 |
| 10.6220854 | CO2 | 95 |
| 10.6239 | N2O | 99 |
| 10.6258486 | CO2 | 96 |
| 10.626664 | CO2 | 90 |
| 10.6303040 | CO2 | 97 |
| 10.63100 | N2O | 101 |
| 10.6320902 | CO2 | 89 |
| 10.6332 | N2O | 99 |
| 10.6388640 | CO2 | 96 |
| 10.6426 | N2O | 99 |
| 10.6426750 | CO2 | 95 |
| 10.6436368 | CO2 | 97 |
| 10.647099 | CO2 | 90 |
| 10.6521496 | CO2 | 96 |
| 10.6531558 | CO2 | 89 |
| 10.6572630 | CO2 | 97 |
| 10.6614 | N2O | 99 |
| 10.6636502 | CO2 | 95 |
| 10.665124 | CO2 | 91 |
| 10.6657057 | CO2 | 96 |
| 10.667888 | CO2 | 90 |
| 10.6710 | N2O | 99 |
| 10.6711838 | CO2 | 97 |
| 10.6745861 | CO2 | 89 |
| 10.6795331 | CO2 | 96 |
| 10.6806 | N2O | 100 |
| 10.6850157 | CO2 | 95 |
| 10.6854009 | CO2 | 97 |
| 10.685646 | CO2 | 91 |
| 10.689036 | CO2 | 90 |
| 10.6903 | N2O | 100 |
| 10.6936325 | CO2 | 96 |
| 10.6963859 | CO2 | 89 |
| 10.6999 | N2O | 100 |
| 10.6999158 | CO2 | 97 |
| 10.7 | NH3 | 108 |
| 10.706519 | CO2 | 91 |
| 10.7067768 | CO2 | 95 |
| 10.7080047 | CO2 | 96 |
| 10.7097 | N2O | 100 |
| 10.710547 | CO2 | 91 |
| 10.7147303 | CO2 | 97 |
| 10.7185600 | CO2 | 89 |
| 10.7195 | N2O | 100 |
| 10.7226505 | CO2 | 96 |
| 10.72517 | N2O | 101 |
| 10.727749 | CO2 | 91 |
| 10.7289384 | CO2 | 95 |
| 10.7294 | N2O | 100 |
| 10.732425 | CO2 | 91 |
| 10.73489 | N2O | 101 |
| 10.7375710 | CO2 | 96 |
| 10.7393 | N2O | 100 |
| 10.7411135 | CO2 | 89 |
| 10.7439 | HF | 54 |
| 10.74468 | N2O | 101 |
| 10.7493 | N2O | 100 |
| 10.749339 | CO2 | 91 |
| 10.7527673 | CO2 | 96 |
| 10.75450 | N2O | 101 |
| 10.754676 | CO2 | 91 |
| 10.7593 | N2O | 100 |
| 10.7640517 | CO2 | 89 |
| 10.76439 | N2O | 101 |
| 10.7682404 | CO2 | 96 |
| 10.7694 | N2O | 100 |
| 10.771295 | CO2 | 91 |
| 10.77434 | N2O | 101 |
| 10.777305 | CO2 | 91 |
| 10.7796 | N2O | 100 |
| 10.7839918 | CO2 | 96 |
| 10.78435 | N2O | 101 |
| 10.7873802 | CO2 | 89 |
| 10.789077 | CO2 | 91 |
| 10.7898 | N2O | 100 |
| 10.793621 | CO2 | 91 |

| WAVELENGTH [MICROMETER] | ACTIVE MEDIUM | PAGE | WAVELENGTH [MICROMETER] | ACTIVE MEDIUM | PAGE |
|---|---|---|---|---|---|
| 10.79441 | N2O | 101 | 10.94145 | N2O | 101 |
| 10.8000 | N2O | 100 | 10.94235 | CO2 | 92 |
| 10.8000227 | CO2 | 90 | 10.942351 | CO2 | 91 |
| 10.800317 | CO2 | 91 | 10.9501 | N2O | 100 |
| 10.80453 | N2O | 101 | 10.95148 | CO2 | 92 |
| 10.8104 | N2O | 100 | 10.951486 | CO2 | 91 |
| 10.8111046 | CO2 | 89 | 10.95241 | N2O | 101 |
| 10.8117 | HF | 55 | 10.9613 | N2O | 100 |
| 10.81471 | N2O | 101 | 10.9621103 | CO2 | 89 |
| 10.816324 | CO2 | 91 | 10.96342 | N2O | 101 |
| 10.8163347 | CO2 | 90 | 10.96361 | CO2 | 92 |
| 10.8208 | N2O | 100 | 10.9726 | N2O | 100 |
| 10.823718 | CO2 | 91 | 10.97261 | CO2 | 92 |
| 10.82495 | N2O | 101 | 10.972615 | CO2 | 91 |
| 10.8312 | N2O | 100 | 10.9735 | CO2 | 92 |
| 10.8329293 | CO2 | 90 | 10.97450 | N2O | 101 |
| 10.8352307 | CO2 | 89 | 10.978 | NE | 3 |
| 10.83524 | N2O | 101 | 10.98 | C2H4 | 115 |
| 10.839408 | CO2 | 91 | 10.9839 | N2O | 100 |
| 10.8418 | N2O | 100 | 10.98526 | CO2 | 92 |
| 10.84560 | N2O | 101 | 10.985266 | CO2 | 91 |
| 10.847513 | CO2 | 91 | 10.98564 | N2O | 101 |
| 10.8498081 | CO2 | 90 | 10.9887835 | CO2 | 89 |
| 10.8523 | N2O | 100 | 10.99409 | CO2 | 92 |
| 10.85601 | N2O | 101 | 10.9950 | CO2 | 92 |
| 10.8597648 | CO2 | 89 | 10.9953 | N2O | 100 |
| 10.862879 | CO2 | 91 | 10.99664 | N2O | 101 |
| 10.8629 | N2O | 100 | 11.0 | NH3 | 108 |
| 10.86648 | N2O | 101 | 11.0044924 | CO2 | 97 |
| 10.8669730 | CO2 | 90 | 11.0050349 | CO2 | 96 |
| 10.871709 | CO2 | 91 | 11.0067 | N2O | 100 |
| 10.8736 | N2O | 100 | 11.00730 | CO2 | 92 |
| 10.87701 | N2O | 101 | 11.007301 | CO2 | 91 |
| 10.8844 | N2O | 100 | 11.0159060 | CO2 | 89 |
| 10.8844259 | CO2 | 90 | 11.01593 | CO2 | 92 |
| 10.8847131 | CO2 | 89 | 11.015934 | CO2 | 91 |
| 10.88760 | N2O | 101 | 11.0165 | CO2 | 92 |
| 10.890184 | CO2 | 91 | 11.0182 | N2O | 100 |
| 10.8952 | N2O | 100 | 11.0247180 | CO2 | 96 |
| 10.896312 | CO2 | 91 | 11.0249599 | CO2 | 97 |
| 10.89825 | N2O | 101 | 11.02974 | CO2 | 92 |
| 10.900964 | CO2 | 91 | 11.029744 | CO2 | 91 |
| 10.9021687 | CO2 | 90 | 11.0298 | N2O | 100 |
| 10.9061 | N2O | 100 | 11.0300 | CO2 | 92 |
| 10.90896 | N2O | 101 | 11.03813 | CO2 | 92 |
| 10.9100823 | CO2 | 89 | 11.0385 | CO2 | 92 |
| 10.9170 | N2O | 100 | 11.0415 | N2O | 100 |
| 10.91973 | N2O | 101 | 11.0434858 | CO2 | 89 |
| 10.921327 | CO2 | 91 | 11.0447094 | CO2 | 96 |
| 10.92146 | CO2 | 92 | 11.0457819 | CO2 | 97 |
| 10.921469 | CO2 | 91 | 11.05258 | CO2 | 92 |
| 10.9280 | N2O | 100 | 11.0535 | CO2 | 92 |
| 10.93056 | N2O | 101 | 11.0573 | HF | 54 |
| 10.93070 | CO2 | 92 | 11.06069 | CO2 | 92 |
| 10.930707 | CO2 | 91 | 11.0610 | CO2 | 92 |
| 10.9358790 | CO2 | 89 | 11.0650120 | CO2 | 96 |
| 10.9390 | N2O | 100 | 11.0669628 | CO2 | 97 |

| WAVELENGTH [MICROMETER] | ACTIVE MEDIUM | PAGE |
|---|---|---|
| 11.0715308 | CO2 | 89 |
| 11.07582 | CO2 | 92 |
| 11.0760 | CO2 | 92 |
| 11.08363 | CO2 | 92 |
| 11.083630 | CO2 | 91 |
| 11.0850 | CO2 | 92 |
| 11.0856290 | CO2 | 96 |
| 11.0885070 | CO2 | 97 |
| 11.09947 | CO2 | 92 |
| 11.1000 | CO2 | 92 |
| 11.1000493 | CO2 | 89 |
| 11.1065635 | CO2 | 96 |
| 11.10693 | CO2 | 92 |
| 11.1070 | CO2 | 92 |
| 11.1104194 | CO2 | 97 |
| 11.1235 | CO2 | 92 |
| 11.12354 | CO2 | 92 |
| 11.1278189 | CO2 | 96 |
| 11.1290499 | CO2 | 89 |
| 11.13062 | CO2 | 92 |
| 11.1315 | CO2 | 92 |
| 11.1327046 | CO2 | 97 |
| 11.14803 | CO2 | 92 |
| 11.1485 | CO2 | 93 |
| 11.1493987 | CO2 | 96 |
| 11.15468 | CO2 | 92 |
| 11.1553678 | CO2 | 97 |
| 11.1555 | CO2 | 93 |
| 11.1585415 | CO2 | 89 |
| 11.1713063 | CO2 | 96 |
| 11.17295 | CO2 | 92 |
| 11.1736 | CO2 | 93 |
| 11.1784142 | CO2 | 97 |
| 11.1790 | CO2 | 93 |
| 11.17914 | CO2 | 92 |
| 11.1885334 | CO2 | 89 |
| 11.1935457 | CO2 | 96 |
| 11.1980 | CO2 | 93 |
| 11.19830 | CO2 | 92 |
| 11.2035 | CO2 | 93 |
| 11.20398 | CO2 | 92 |
| 11.2161205 | CO2 | 96 |
| 11.2190349 | CO2 | 89 |
| 11.2235 | CO2 | 93 |
| 11.22408 | CO2 | 92 |
| 11.2295 | CO2 | 93 |
| 11.2390350 | CO2 | 96 |
| 11.2495 | CO2 | 93 |
| 11.2500559 | CO2 | 89 |
| 11.2545 | CO2 | 93 |
| 11.2622931 | CO2 | 96 |
| 11.2770 | CO2 | 93 |
| 11.2805 | CO2 | 93 |
| 11.2858992 | CO2 | 96 |
| 11.289 | XE | 8 |
| 11.296 | XE | 8 |
| 11.3098579 | CO2 | 96 |
| 11.329 | CO2 | 98 |
| 11.3341736 | CO2 | 96 |
| 11.346 | CO2 | 98 |
| 11.3588512 | CO2 | 96 |
| 11.364 | CO2 | 98 |
| 11.382 | CO2 | 98 |
| 11.3838956 | CO2 | 96 |
| 11.400 | CO2 | 98 |
| 11.4033 | HF | 54 |
| 11.4093120 | CO2 | 96 |
| 11.446 | NH3 | 108 |
| 11.459 | NH3 | 108 |
| 11.482 | CS2 | 102 |
| 11.489 | CS2 | 102 |
| 11.503 | CS2 | 102 |
| 11.510 | CS2 | 102 |
| 11.517 | CS2 | 102 |
| 11.524 | CS2 | 102 |
| 11.526 | NH3 | 108 |
| 11.531 | CS2 | 102 |
| 11.538 | CS2 | 103 |
| 11.5408 | HF | 55 |
| 11.545 | CS2 | 103 |
| 11.5547 | NH3 | 108 |
| 11.596 | CS2 | 102 |
| 11.677 | CO2 | 98 |
| 11.700 | CO2 | 98 |
| 11.721 | NH3 | 108 |
| 11.723 | CO2 | 98 |
| 11.7298463 | CO2 | 97 |
| 11.746 | CO2 | 98 |
| 11.770 | CO2 | 98 |
| 11.7854 | HF | 54 |
| 11.794 | CO2 | 98 |
| 11.80 | NH3 | 108 |
| 11.811 | NH3 | 108 |
| 11.8184661 | CO2 | 97 |
| 11.819 | CO2 | 98 |
| 11.83 | H2O | 104 |
| 11.8355948 | CO2 | 97 |
| 11.843 | CO2 | 98 |
| 11.8530426 | CO2 | 97 |
| 11.857 | NE | 3 |
| 11.868 | CO2 | 98 |
| 11.8708125 | CO2 | 97 |
| 11.8889076 | CO2 | 97 |
| 11.898 | NE | 3 |
| 11.9073308 | CO2 | 97 |
| 11.9202034 | CO2 | 96 |
| 11.9260857 | CO2 | 97 |
| 11.9385324 | CO2 | 96 |
| 11.9451757 | CO2 | 97 |
| 11.96 | H2O | 104 |
| 11.960 | CS2 | 103 |
| 11.9646042 | CO2 | 97 |
| 11.965 | CS2 | 103 |
| 11.9665825 | CO2 | 96 |

| WAVELENGTH [MICROMETER] | ACTIVE MEDIUM | PAGE |
|---|---|---|
| 11.9843752 | CO2 | 97 |
| 11.9856573 | CO2 | 96 |
| 11.986 | CS2 | 103 |
| 11.994 | NH3 | 109 |
| 12.010 | NH3 | 109 |
| 12.033 | SB | 22 |
| 12.078 | NH3 | 109 |
| 12.078 | NH3 | 109 |
| 12.0791 | NH3 | 109 |
| 12.1143 | NH3 | 109 |
| 12.138 | AR | 5 |
| 12.1558 | NH3 | 109 |
| 12.1846 | NH3 | 109 |
| 12.188 | AR | 5 |
| 12.2082 | HF | 54 |
| 12.217 | CS2 | 103 |
| 12.241 | CS2 | 103 |
| 12.245 | NH3 | 109 |
| 12.245 | NH3 | 109 |
| 12.249 | CS2 | 103 |
| 12.251 | NH3 | 109 |
| 12.2619 | HF | 55 |
| 12.263 | XE | 8 |
| 12.266 | NH3 | 109 |
| 12.273 | OH | 65 |
| 12.279 | OH | 65 |
| 12.280 | NH3 | 109 |
| 12.286 | NH3 | 109 |
| 12.316 | NH3 | 109 |
| 12.348 | NH3 | 109 |
| 12.5 | K | 10 |
| 12.520 | NH3 | 109 |
| 12.526 | NH3 | 109 |
| 12.541 | NH3 | 109 |
| 12.566 | NH3 | 109 |
| 12.591 | NH3 | 109 |
| 12.631 | NH3 | 109 |
| 12.660 | OH | 65 |
| 12.663 | OH | 65 |
| 12.6781 | HF | 54 |
| 12.689 | NH3 | 109 |
| 12.7006 | HF | 55 |
| 12.8115 | NH3 | 109 |
| 12.812 | NH3 | 109 |
| 12.831 | NE | 3 |
| 12.851 | NH3 | 109 |
| 12.876 | NH3 | 109 |
| 12.913 | XE | 8 |
| 12.921 | NH3 | 109 |
| 13.031 | NH3 | 109 |
| 13.073 | OH | 65 |
| 13.079 | OH | 65 |
| 13.088 | OH | 65 |
| 13.114 | NH3 | 109 |
| 13.124 | NH3 | 109 |
| 13.144 | CO2 | 93 |
| 13.145 | NH3 | 109 |
| 13.154 | CO2 | 93 |
| 13.159 | CO2 | 93 |
| 13.176 | NH3 | 109 |
| 13.185 | CO | 16 |
| 13.1877 | HF | 55 |
| 13.2009 | HF | 54 |
| 13.218 | NH3 | 109 |
| 13.2211 | HF | 55 |
| 13.269 | NH3 | 109 |
| 13.331 | NH3 | 109 |
| 13.411 | NH3 | 109 |
| 13.525 | OH | 65 |
| 13.538 | OH | 65 |
| 13.54 | CF3I | 114 |
| 13.541 | CO2 | 93 |
| 13.547 | OH | 65 |
| 13.557 | OH | 65 |
| 13.57 | CF3I | 114 |
| 13.576 | NH3 | 109 |
| 13.63 | CF3I | 114 |
| 13.632 | OH | 65 |
| 13.642 | OH | 65 |
| 13.7261 | NH3 | 109 |
| 13.7277 | HF | 55 |
| 13.736 | NE | 3 |
| 13.756 | NE | 3 |
| 13.7641 | HF | 55 |
| 13.821 | NH3 | 109 |
| 13.87 | CO2 | 93 |
| 13.8720 | HCL | 58 |
| 14.043 | OH | 65 |
| 14.059 | OH | 65 |
| 14.067 | OH | 65 |
| 14.081 | OH | 65 |
| 14.0994 | HCL | 58 |
| 14.1 | CO2 | 93 |
| 14.118 | OH | 65 |
| 14.129 | OH | 65 |
| 14.16 | CO2 | 93 |
| 14.19 | CO2 | 93 |
| 14.21 | CO2 | 93 |
| 14.2681 | HF | 55 |
| 14.3 | N[15]H | 110 |
| 14.3434 | HCL | 58 |
| 14.4406 | HF | 55 |
| 14.578 | CO | 16 |
| 14.620 | OH | 65 |
| 14.640 | OH | 65 |
| 14.646 | OH | 65 |
| 14.655 | OH | 65 |
| 14.662 | OH | 65 |
| 14.669 | OH | 65 |
| 14.78 | NH3 | 108 |
| 14.8 | N[15]H | 110 |
| 15.0163 | HF | 55 |
| 15.032 | AR | 5 |
| 15.037 | AR | 5 |

| WAVELENGTH [MICROMETER] | ACTIVE MEDIUM | PAGE |
|---|---|---|
| 15.04 | NH3 | 108 |
| 15.08 | NH3 | 108 |
| 15.08 | N[15]H | 108 |
| 15.1744 | HF | 55 |
| 15.2 | N[15]H | 110 |
| 15.256 | OH | 65 |
| 15.274 | OH | 65 |
| 15.289 | OH | 65 |
| 15.294 | OH | 65 |
| 15.313 | OH | 65 |
| 15.33 | CF4 | 112 |
| 15.41 | CF4 | 112 |
| 15.41 | NH3 | 108 |
| 15.47 | NH3 | 108 |
| 15.49 | CF4 | 112 |
| 15.49 | CF4 | 112 |
| 15.50 | CF4 | 113 |
| 15.55 | CF4 | 113 |
| 15.55 | CF4 | 113 |
| 15.56 | CF4 | 113 |
| 15.56 | CF4 | 113 |
| 15.58 | CF4 | 113 |
| 15.60 | CF4 | 113 |
| 15.61 | CF4 | 113 |
| 15.62 | CF4 | 113 |
| 15.62 | CF4 | 113 |
| 15.7 | N[15]H | 110 |
| 15.70 | CF4 | 113 |
| 15.74 | CF4 | 113 |
| 15.76 | CF4 | 113 |
| 15.77 | CF4 | 113 |
| 15.84 | CF4 | 113 |
| 15.84 | CF4 | 113 |
| 15.85 | CF4 | 113 |
| 15.85 | CF4 | 113 |
| 15.85 | CF4 | 113 |
| 15.85 | CF4 | 113 |
| 15.8782 | NH3 | 109 |
| 15.9005 | SF6 | 108 |
| 15.91 | N[15]H | 108 |
| 15.91 | CF4 | 113 |
| 15.94 | CF4 | 113 |
| 15.9452 | NH3 | 109 |
| 15.95 | K | 10 |
| 16.0 | N[15]H | 111 |
| 16.00 | CF4 | 113 |
| 16.0215 | HF | 55 |
| 16.03 | CF4 | 113 |
| 16.07 | CF4 | 113 |
| 16.10 | CF4 | 113 |
| 16.12 | CF4 | 113 |
| 16.18 | CF4 | 113 |
| 16.20 | CF4 | 113 |
| 16.2125 | HCL | 58 |
| 16.24 | CF4 | 113 |
| 16.24 | CF4 | 113 |
| 16.26 | CF4 | 113 |
| 16.27 | CF4 | 113 |
| 16.31 | CF4 | 113 |
| 16.35 | CF4 | 113 |
| 16.4 | NOCL | 112 |
| 16.40 | CF4 | 113 |
| 16.52 | NOCL | 112 |
| 16.57 | NOCL | 112 |
| 16.586 | CO2 | 93 |
| 16.596 | CO2 | 98 |
| 16.597 | CO2 | 93 |
| 16.6085 | HCL | 58 |
| 16.634 | NE | 3 |
| 16.664 | NE | 3 |
| 16.664 | HCL | 58 |
| 16.69 | NOCL | 112 |
| 16.7 | NOCL | 112 |
| 16.7 | NOCL | 112 |
| 16.75 | NOCL | 112 |
| 16.765 | HCL | 58 |
| 16.780 | CO2 | 98 |
| 16.85 | CF4 | 113 |
| 16.86 | NOCL | 112 |
| 16.889 | NE | 3 |
| 16.9 | NOCL | 112 |
| 16.927 | CO2 | 98 |
| 16.932 | H2O | 104 |
| 16.943 | NE | 3 |
| 16.970 | CO2 | 98 |
| 16.99 | NOCL | 112 |
| 17.023 | CO2 | 93 |
| 17.029 | CO2 | 93 |
| 17.0340 | HCL | 58 |
| 17.036 | CO2 | 93 |
| 17.048 | CO2 | 93 |
| 17.125 | HCL | 58 |
| 17.153 | NE | 3 |
| 17.184 | NE | 3 |
| 17.280 | CO2 | 98 |
| 17.370 | CO2 | 93 |
| 17.376 | CO2 | 93 |
| 17.390 | CO2 | 93 |
| 17.45 | C2D2 | 114 |
| 17.463 | CO2 | 98 |
| 17.4923 | HCL | 58 |
| 17.575 | HCL | 58 |
| 17.596 | CO2 | 98 |
| 17.61 | C2D2 | 114 |
| 17.639 | CO2 | 98 |
| 17.684 | CO2 | 98 |
| 17.730 | CO2 | 98 |
| 17.77 | C2D2 | 114 |
| 17.775 | CO2 | 98 |
| 17.8 | N[15]H | 111 |
| 17.800 | NE | 3 |
| 17.821 | CO2 | 98 |
| 17.837 | NE | 3 |
| 17.884 | NE | 3 |

| WAVELENGTH [MICROMETER] | ACTIVE MEDIUM | PAGE |
|---|---|---|
| 17.915 | CO2 | 98 |
| 17.962 | CO2 | 98 |
| 17.9874 | HCL | 58 |
| 17.997 | HCL | 58 |
| 18.010 | CO2 | 98 |
| 18.035 | HCL | 58 |
| 18.053 | CO2 | 98 |
| 18.09 | HF | 55 |
| 18.121 | OD | 66 |
| 18.138 | OD | 66 |
| 18.21 | NH3 | 108 |
| 18.3 | BCL3 | 107 |
| 18.392 | NE | 3 |
| 18.455 | OH | 65 |
| 18.492 | OH | 65 |
| 18.500 | XE | 6 |
| 18.502 | OH | 65 |
| 18.522 | HCL | 58 |
| 18.532 | OH | 65 |
| 18.555 | HCL | 58 |
| 18.590 | OD | 66 |
| 18.593 | HCL | 58 |
| 18.603 | OD | 66 |
| 18.624 | OD | 66 |
| 18.67 | C2D2 | 114 |
| 18.788 | OH | 65 |
| 18.79 | C2D2 | 114 |
| 18.79 | C2D2 | 114 |
| 18.8 | BCL3 | 107 |
| 18.8010 | HF | 55 |
| 18.828 | OH | 65 |
| 18.849 | OH | 65 |
| 18.85 | C2D2 | 114 |
| 18.878 | OH | 65 |
| 18.9250 | NH3 | 109 |
| 18.96 | C2D2 | 114 |
| 18.983 | OCS | 102 |
| 19.03 | C2D2 | 114 |
| 19.057 | OCS | 102 |
| 19.1 | BCL3 | 107 |
| 19.102 | OD | 66 |
| 19.1129 | HF | 55 |
| 19.121 | OD | 66 |
| 19.122 | HCL | 58 |
| 19.13 | C2D2 | 114 |
| 19.141 | OD | 66 |
| 19.145 | HCL | 58 |
| 19.161 | OD | 66 |
| 19.183 | HCL | 58 |
| 19.21 | C2D2 | 114 |
| 19.27 | C2D2 | 114 |
| 19.273 | OH | 66 |
| 19.321 | OH | 66 |
| 19.35 | HF | 55 |
| 19.399 | HBR | 61 |
| 19.4 | BCL3 | 107 |
| 19.5497 | NH3 | 109 |
| 19.55 | HF | 55 |
| 19.555 | OH | 65 |
| 19.594 | OH | 65 |
| 19.619 | OH | 65 |
| 19.650 | OH | 65 |
| 19.662 | OD | 66 |
| 19.67 | C2D2 | 114 |
| 19.681 | OD | 66 |
| 19.696 | OD | 66 |
| 19.7002 | HCL | 58 |
| 19.704 | OD | 66 |
| 19.783 | HCL | 58 |
| 19.821 | HCL | 58 |
| 19.915 | HF | 55 |
| 19.988 | HBR | 61 |
| 20.01 | C2D2 | 114 |
| 20.05 | OH | 65 |
| 20.1337 | HF | 55 |
| 20.2 | BCL3 | 107 |
| 20.271 | OD | 66 |
| 20.288 | OD | 66 |
| 20.296 | OD | 66 |
| 20.313 | OD | 66 |
| 20.3455 | HCL | 58 |
| 20.3513 | HF | 55 |
| 20.360 | HBR | 61 |
| 20.4106 | HCL | 58 |
| 20.44 | C2D2 | 114 |
| 20.474 | NE | 3 |
| 20.6 | BCL3 | 107 |
| 20.835 | HF | 55 |
| 20.87 | OH | 65 |
| 20.896 | HBR | 61 |
| 20.93 | OH | 65 |
| 20.9393 | HF | 55 |
| 20.949 | HBR | 61 |
| 20.9991 | HCL | 58 |
| 21.0470 | HCL | 58 |
| 21.1556 | HCL | 58 |
| 21.471 | NH3 | 108 |
| 21.48 | OH | 65 |
| 21.501 | HBR | 61 |
| 21.546 | HBR | 61 |
| 21.57 | OH | 65 |
| 21.6986 | HF | 55 |
| 21.746 | NE | 3 |
| 21.7685 | HF | 55 |
| 21.8127 | HCL | 58 |
| 21.9706 | HCL | 58 |
| 22.136 | HBR | 61 |
| 22.226 | HBR | 61 |
| 22.33 | OH | 65 |
| 22.4 | BCL3 | 107 |
| 22.45 | OH | 65 |
| 22.542 | NH3 | 108 |
| 22.563 | NH3 | 108 |
| 22.6514 | HCL | 58 |

| WAVELENGTH [MICROMETER] | ACTIVE MEDIUM | PAGE |
|---|---|---|
| 22.71 | NH3 | 108 |
| 22.830 | NE | 3 |
| 22.855 | HBR | 61 |
| 22.8637 | HCL | 58 |
| 23.0 | BCL3 | 107 |
| 23.13 | H2O | 104 |
| 23.14 | OH | 65 |
| 23.26 | OH | 65 |
| 23.365 | H2O | 104 |
| 23.436 | HBR | 61 |
| 23.5705 | HCL | 58 |
| 23.675 | NH3 | 108 |
| 23.8485 | HCL | 58 |
| 23.86 | NH3 | 108 |
| 24.07 | OH | 65 |
| 24.18 | OH | 66 |
| 24.3178 | HCL | 58 |
| 24.5833 | HCL | 58 |
| 24.6177 | HCL | 58 |
| 24.918 | NH3 | 108 |
| 24.9367 | HCL | 58 |
| 24.966 | H2O | 104 |
| 25.11 | OH | 65 |
| 25.12 | NH3 | 108 |
| 25.162 | H2O[18] | 104 |
| 25.28 | OH | 65 |
| 25.416 | NE | 3 |
| 25.4744 | NH3 | 109 |
| 25.7040 | HCL | 58 |
| 25.8839 | NH3 | 109 |
| 26.1046 | NH3 | 109 |
| 26.12 | OH | 66 |
| 26.1462 | HCL | 58 |
| 26.282 | NH3 | 108 |
| 26.30 | OH | 66 |
| 26.36 | D2O | 105 |
| 26.4416 | NH3 | 109 |
| 26.595 | H2O[18] | 104 |
| 26.660 | H2O | 104 |
| 26.7068 | NH3 | 109 |
| 26.937 | AR | 5 |
| 26.956 | AR | 5 |
| 27.47 | OH | 65 |
| 27.508 | HCL | 58 |
| 27.71 | OH | 65 |
| 27.8437 | NH3 | 109 |
| 27.970755 | H2O | 104 |
| 28.045 | NE | 3 |
| 28.054 | H2O | 104 |
| 28.270 | H2O | 104 |
| 28.295 | H2O[18] | 104 |
| 28.356 | H2O | 104 |
| 28.451 | H2O | 104 |
| 29.786 | HBR | 61 |
| 30.445 | HBR | 61 |
| 30.69 | NH3 | 108 |
| 30.948 | HBR | 61 |
| 31.368 | HBR | 61 |
| 31.47 | NH3 | 108 |
| 31.544 | NE | 3 |
| 31.849 | HBR | 61 |
| 31.919 | NE | 3 |
| 31.951 | NH3 | 108 |
| 32.007 | NE | 3 |
| 32.13 | NH3 | 108 |
| 32.469 | HBR | 61 |
| 32.507 | NE | 3 |
| 32.799 | HBR | 61 |
| 32.924 | H2O | 104 |
| 33.30 | H2S | 107 |
| 33.308 | H2O[18] | 104 |
| 33.329 | H2O | 104 |
| 33.409 | HBR | 61 |
| 33.815 | NE | 3 |
| 33.828 | NE | 3 |
| 33.896 | D2O | 105 |
| 34.2248 | NH3 | 109 |
| 34.543 | NE | 3 |
| 34.60 | H2O | 104 |
| 34.670 | NE | 3 |
| 34.8 | CD3OH | 126 |
| 35 | CD3OD | 128 |
| 35.017 | H2O | 104 |
| 35.081 | D2O | 105 |
| 35.1573 | NH3 | 109 |
| 35.383 | H2O[18] | 104 |
| 35.5011 | NH3 | 109 |
| 35.592 | NE | 3 |
| 35.833 | H2O | 104 |
| 36.02 | NH3 | 109 |
| 36.096 | D2O | 105 |
| 36.1686 | NH3 | 109 |
| 36.324 | D2O | 105 |
| 36.5 | HF | 55 |
| 36.526 | D2O | 105 |
| 36.606 | H2O | 104 |
| 37.221 | NE | 3 |
| 37.5 | CH3OH | 124 |
| 37.6 | CD3OH | 126 |
| 37.600 | H2S | 107 |
| 37.788 | D2O | 105 |
| 37.848 | H2O | 104 |
| 37.864 | D2O | 105 |
| 38.086 | H2O | 104 |
| 39.53 | D2O | 105 |
| 39.695 | H2O | 104 |
| 40.1 | CD3OH | 126 |
| 40.2 | CH3OH | 124 |
| 40.45 | H2O | 104 |
| 40.526 | HBR | 61 |
| 40.638 | H2O | 104 |
| 40.994 | D2O | 105 |
| 41 | CD3OD | 128 |
| 41.5 | CD3OH | 126 |

| WAVELENGTH [MICROMETER] | ACTIVE MEDIUM | PAGE |
|---|---|---|
| 41.730 | NE | 3 |
| 41.79 | D2O | 105 |
| 41.8 | CD3OH | 126 |
| 42.18 | CH3OH | 124 |
| 42.4 | HF | 55 |
| 42.51 | H2O | 104 |
| 43.4 | CH3OH | 124 |
| 43.47 | CH3OH | 124 |
| 43.9 | CD3OH | 126 |
| 45.517 | H2O | 104 |
| 45.91 | H2O | 104 |
| 46.7 | CH3OD | 128 |
| 47.244 | H2O | 104 |
| 47.39 | H2O | 104 |
| 47.468 | H2O | 104 |
| 47.687 | H2O | 104 |
| 48.19 | H2O | 104 |
| 48.366 | H2O[18] | 104 |
| 48.676 | H2O | 104 |
| 48.70 | H2S | 107 |
| 48.765 | H2O[18] | 104 |
| 48.80 | D2O | 105 |
| 49.0356 | NH3 | 109 |
| 49.06 | H2O | 104 |
| 49.430 | H2O[18] | 104 |
| 49.68 | RB | 10 |
| 49.8 | CD3OH | 126 |
| 50.5 | D2O | 105 |
| 50.69 | NE | 3 |
| 50.71 | D2O | 105 |
| 50.8 | HF | 55 |
| 50.93 | RB | 10 |
| 52.307 | H2S | 107 |
| 52.40 | NE | 3 |
| 52.9 | CD3OH | 126 |
| 53.47 | NE | 3 |
| 53.910 | H2O | 104 |
| 54.00 | NE | 3 |
| 54.10 | NE | 3 |
| 54.45 | NH3 | 109 |
| 54.73 | D2O | 105 |
| 55.000 | H2O | 105 |
| 55.088 | H2O | 105 |
| 55.39 | CH3OH | 124 |
| 55.51 | NE | 3 |
| 55.612 | H2S | 107 |
| 56.129 | H2O[18] | 105 |
| 56.830 | D2O | 105 |
| 56.8631 | NH3 | 110 |
| 57 | CH3OD | 128 |
| 57.34 | NE | 3 |
| 57.659 | H2O | 105 |
| 57.799 | H2O | 105 |
| 58.01 | NH3 | 110 |
| 58.1 | CH3OH | 124 |
| 60.224 | H2S | 107 |
| 60.25 | CH3OH | 124 |
| 60.8 | CD3OH | 126 |
| 61.182 | D2O | 105 |
| 61.413 | H2S | 107 |
| 62.5 | C2H6O2 | 132 |
| 62.6 | H2S | 107 |
| 63.25 | NH3 | 110 |
| 63.4 | HF | 55 |
| 64.5 | NH3 | 110 |
| 64.7274 | NH3 | 110 |
| 65.1 | CH3OH | 124 |
| 65.6 | CH3OH | 124 |
| 66 | D2O | 105 |
| 66.800 | H2O | 105 |
| 66.903 | H2O | 105 |
| 67.169 | H2O | 105 |
| 67.19 | HH3 | 110 |
| 67.24 | NH3 | 110 |
| 68.31 | NE | 3 |
| 69.1 | C2H6O2 | 132 |
| 69.5 | CH3OD | 128 |
| 69.70 | CH3OH | 124 |
| 70.1 | C2H6O2 | 132 |
| 70.3 | CH3OD | 128 |
| 70.511716 | CH3OH | 124 |
| 71.0 | CD3OH | 126 |
| 71.944 | D2O | 105 |
| 72.08 | NE | 3 |
| 72.427 | D2O | 105 |
| 72.6 | NH3 | 110 |
| 72.747780 | D2O | 105 |
| 72.76 | NH3 | 110 |
| 72.856 | H2O | 105 |
| 73.30 | CH3OH | 124 |
| 73.337 | D2O | 105 |
| 73.401 | H2O | 105 |
| 73.54 | H2S | 107 |
| 74.15 | NH3 | 110 |
| 74.341 | D2O | 105 |
| 74.526 | D2O | 105 |
| 75.2 | C2H6O2 | 132 |
| 75.578 | XE | 8 |
| 76.1 | CD3OH | 126 |
| 76.305 | D2O | 105 |
| 77.4 | C2H6O2 | 132 |
| 77.58 | PH3 | 111 |
| 77.92 | CH3OH | 124 |
| 78 | CD3OD | 128 |
| 78.16 | D2O | 105 |
| 78.28 | NH3 | 110 |
| 78.443329 | H2O | 105 |
| 79.091010 | H2O | 105 |
| 80.3 | CH3OH | 124 |
| 80.6 | CH3OH | 124 |
| 81.05 | H2S | 107 |
| 81.2 | CD3OH | 126 |
| 81.53 | NH3 | 110 |
| 81.554 | HCN | 103 |

| WAVELENGTH [MICROMETER] | ACTIVE MEDIUM | PAGE |
|---|---|---|
| 83 | D2O | 105 |
| 83.45 | H2S | 107 |
| 83.60 | NH3 | 110 |
| 83.730 | D2O | 105 |
| 83.77 | PH3 | 111 |
| 83.85 | NH3 | 110 |
| 84.111 | D2O | 105 |
| 84.278897 | D2O | 105 |
| 84.4 | HF | 55 |
| 84.64 | NH3 | 110 |
| 85.01 | NE | 3 |
| 85.31729 | CH3OH* | 125 |
| 85.564 | H2O | 105 |
| 85.59 | CH3OH | 124 |
| 86.11179 | CH3OH* | 125 |
| 86.301 | H2O | 105 |
| 86.4 | CD3OH | 126 |
| 86.471 | H2O | 105 |
| 86.93 | NE | 3 |
| 87.1 | NH3 | 110 |
| 87.323 | H2O | 105 |
| 87.469 | H2O | 105 |
| 87.580 | H2S | 107 |
| 88.05 | NH3 | 110 |
| 88.20 | NH3 | 110 |
| 88.47 | NE | 3 |
| 88.90 | NH3 | 110 |
| 89.68 | N[15]H | 111 |
| 89.76 | PH3 | 111 |
| 89.772 | H2O | 105 |
| 89.80 | PH3 | 111 |
| 89.82 | NE | 3 |
| 90.26 | PH3 | 111 |
| 90.5 | D2S | 107 |
| 90.50 | NH3 | 110 |
| 90.8 | C2H6O2 | 132 |
| 90.93 | NH3 | 110 |
| 92.0 | H2S | 107 |
| 92.60 | CH3OH | 124 |
| 92.69 | CH3OH | 124 |
| 92.87 | NH3 | 110 |
| 94 | D2O | 106 |
| 94.62 | D2O | 106 |
| 95.5 | CH2F2 | 118 |
| 95.763 | HE | 1 |
| 95.8 | C2H6O2 | 132 |
| 96.4 | H2S | 107 |
| 96.401 | HCN | 103 |
| 96.522394 | CH3OH | 124 |
| 97.19 | PH3 | 111 |
| 97.30 | PH3 | 111 |
| 97.48 | CH3OH | 124 |
| 98 | D2O | 106 |
| 98.693 | HCN | 103 |
| 99 | D2O | 106 |
| 99.00 | D2O | 105 |
| 99.5 | CH3NH2 | 131 |
| 100.6 | CH3OD | 128 |
| 101.257 | HCN | 103 |
| 101.9 | H2CO | 115 |
| 102.6 | CD3OH | 126 |
| 102.62 | PH3 | 111 |
| 103.3 | H2S | 107 |
| 103.33 | D2O | 105 |
| 103.48081 | CH3OH* | 125 |
| 104 | CH3OD | 128 |
| 104 | CH3NH2 | 131 |
| 104.4 | PH3 ? | 111 |
| 105.5 | CH2F2 | 118 |
| 106.0 | NE | 3 |
| 106.04 | PH3 | 111 |
| 106.05 | PH3 | 111 |
| 106.09 | PH3 | 111 |
| 106.23 | PH3 | 111 |
| 107.72019 | D2O | 105 |
| 107.91 | D2O | 105 |
| 108.8 | H2S | 107 |
| 108.88 | D2O | 105 |
| 109 | CH2DOH | 128 |
| 109.1 | C2H6O2 | 132 |
| 109.3 | CH2F2 | 118 |
| 109.7 | PH3 ? | 111 |
| 110 | CH3OD | 128 |
| 110.240 | HCN15 | 103 |
| 110.49 | D2O | 105 |
| 111.74 | D2O | 105 |
| 111.9 | N[15]H | 111 |
| 112 | CLO2 | 138 |
| 112.066 | HCN | 103 |
| 112.3 | CD3OH | 126 |
| 112.58 | D2O | 106 |
| 112.98 | NH3 | 110 |
| 113.311 | HCN15 | 103 |
| 114 | D2O | 106 |
| 114.29 | NH3 | 110 |
| 115.32 | H2O | 105 |
| 115.5 | CH3NH2 | 131 |
| 115.82324 | CH3OH* | 125 |
| 116 | D2O | 106 |
| 116.132 | HCN | 103 |
| 116.8 | H2S | 107 |
| 116.88 | PH3 | 111 |
| 116.88 | PH3 | 111 |
| 117 | CH3OD | 128 |
| 117.01 | PH3 | 111 |
| 117.1 | C2H6O2 | 132 |
| 117.7 | CH2F2 | 118 |
| 117.95 | CH3OH | 124 |
| 118 | C2H6O2 | 132 |
| 118 | CH3NH2 | 131 |
| 118.01314 | CH3OH* | 125 |
| 118.59104 | H2O | 105 |
| 118.83409 | CH3OH | 124 |
| 118.9 | C2H6O2 | 132 |

| WAVELENGTH [MICROMETER] | ACTIVE MEDIUM | PAGE | WAVELENGTH [MICROMETER] | ACTIVE MEDIUM | PAGE |
|---|---|---|---|---|---|
| 119 | CD3OD | 128 | 141.06 | SO2 | 106 |
| 119 | D2O | 106 | 142 | D2O | 106 |
| 119.02 | NH3 | 110 | 142.00 | SO2 | 106 |
| 119.6 | H2CO | 115 | 143 | CH3NH2 | 131 |
| 120.08 | H2O | 105 | 144.0 | CD3OH | 126 |
| 121 | O3 | 85 | 145.66171 | CH3OD | 128 |
| 121.45 | PH3 | 111 | 145.88 | PH3 | 111 |
| 121.7 | CH2F2 | 118 | 146.07 | PH3 | 111 |
| 122.4 | CH2F2 | 118 | 146.07 | PH3 | 111 |
| 122.4 | CH2F2 | 118 | 146.09739 | CH3OH* | 125 |
| 122.8 | H2CO | 115 | 146.34 | PH3 | 111 |
| 123 | OCS | 102 | 147 | CH3NH2 | 131 |
| 124.6 | NE | 3 | 147.04 | NH3 | 110 |
| 125 | CH2DOH | 128 | 147.15 | NH3 | 110 |
| 125.8 | C2H6O2 | 132 | 147.2 | NH3 | 110 |
| 125.9 | H2CO | 115 | 148 | C3H2O | 138 |
| 126 | CH3NH2 | 131 | 148.5 | CH3NH2 | 131 |
| 126.1 | NE | 3 | 148.59041 | CH3OH* | 125 |
| 126.164 | HCN | 103 | 149.27226 | CH3OH* | 125 |
| 126.2 | H2S | 107 | 149.94 | SO2 | 106 |
| 126.5 | HF | 55 | 150 | CD3OD | 128 |
| 128.629 | HCN | 103 | 151 | CH2DOH | 128 |
| 128.7 | CD3OH | 126 | 151.08 | SO2 | 106 |
| 129.1 | H2S | 107 | 151.35 | SO2 | 106 |
| 129.5497 | CH3OH | 124 | 151.35 | CH3OH | 124 |
| 129.78 | PH3 | 111 | 151.49 | NH3 | 110 |
| 129.98 | PH3 | 111 | 151.5 | NH3 | 110 |
| 129.98 | PH3 | 111 | 152.07568 | CH3OH* | 125 |
| 130.14 | PH3 | 111 | 153 | CH3NH2 | 131 |
| 130.60 | H2S | 107 | 155.07 | PH3 | 111 |
| 130.838 | HCN | 103 | 155.1 | H2CO | 115 |
| 132 | OCS | 102 | 155.17 | NH3 | 110 |
| 132 | C2H6O2 | 132 | 155.28 | NH3 | 110 |
| 132 | C2H6O2 | 132 | 156 | C3H2O | 138 |
| 132.6 | NE | 3 | 156.34 | PH3 | 111 |
| 133.1196 | CH3OH | 124 | 157.6 | H2CO | 115 |
| 134 | CH3NH2 | 131 | 157.92848 | CH3OH* | 126 |
| 134 | CH3NH2 | 131 | 158 | CD3OH | 126 |
| 134 | CH3OD | 128 | 158.5 | CH2F2 | 118 |
| 134.0 | CH2F2 | 118 | 158.9 | CH2F2 | 118 |
| 134.7 | CH3OD | 128 | 159 | CH3NH2 | 131 |
| 134.8 | CH3OD | 128 | 159.2 | CH3OH | 124 |
| 134.932 | HCN | 103 | 159.5 | H2CO | 115 |
| 135 | C2H6O2 | 132 | 162 | CH3OH | 124 |
| 135.3 | CH2F2 | 118 | 162.4 | H2S | 107 |
| 135.3 | D2S | 107 | 163.03353 | CH3OH | 124 |
| 135.3 | H2S | 107 | 163.61 | O3 | 85 |
| 135.95 | PH3 | 111 | 163.8 | H2CO | 115 |
| 136 | CH3OD | 128 | 163.9 | CH3OH | 124 |
| 136.71 | PH3 | 111 | 164 | CH3OH | 124 |
| 138.768 | HCN15 | 103 | 164 | C2H6O2 | 132 |
| 139 | CH3NH2 | 131 | 164 | C2H6O2 | 132 |
| 139.83 | SO2 | 106 | 164 | C2H6O2 | 132 |
| 140.6 | H2S | 107 | 164 | CH3NH2 | 131 |
| 140.82 | SO2 | 106 | 164 | CH2DOH | 128 |
| 140.85 | PH3 | 111 | 164.3 | CH3OH | 124 |
| 141 | CH3NH2 | 131 | 164.5076 | CH3OH | 124 |

| WAVELENGTH [MICROMETER] | ACTIVE MEDIUM | PAGE |
|---|---|---|
| 164.77 | CH3OH | 124 |
| 164.7832 | CH3OH | 124 |
| 165 | CHD2OH | 129 |
| 165 | CD3OD | 128 |
| 165.150 | HCN15 | 103 |
| 165.8 | CH2F2 | 118 |
| 165.9 | CH2F2 | 118 |
| 166 | CH3NH2 | 131 |
| 166.6 | CH2F2 | 118 |
| 166.6 | CH2F2 | 118 |
| 166.73 | PH3 | 111 |
| 166.79 | PH3 | 111 |
| 166.84 | PH3 | 111 |
| 166.87 | PH3 | 111 |
| 167 | CH2DOH | 128 |
| 168 | CHD2OH | 129 |
| 168 | CH3NH2 | 131 |
| 169 | C2H6O2 | 133 |
| 170.08 | D2O | 105 |
| 170.2 | H2CO | 115 |
| 170.57638 | CH3OH | 124 |
| 171 | C2H6O2 | 133 |
| 171 | CH2DOH | 128 |
| 171.3 | CH3OH | 124 |
| 171.5 | O3 | 85 |
| 171.67 | D2O | 105 |
| 175 | CH3NH2 | 131 |
| 176 | CH3NH2 | 131 |
| 176 | CLO2 | 138 |
| 177 | CH3NH2 | 131 |
| 179 | CD3OH | 126 |
| 179 | CHD2OH | 129 |
| 180 | CH3NH2 | 131 |
| 180.54 | PH3 | 112 |
| 181.789 | DCN | 104 |
| 182 | PH3 ? | 112 |
| 182.4 | CD3OH | 126 |
| 183 | CH3NH2 | 131 |
| 183.2 | HDS | 107 |
| 184 | CD3OD | 128 |
| 184 | CD3OH | 126 |
| 184.3 | CH2F2 | 118 |
| 184.4 | H2CO | 115 |
| 185 | C2H6O2 | 133 |
| 185 | C2H6O2 | 133 |
| 185.5 | CH3OH | 124 |
| 186.03 | CH3OH | 124 |
| 186.25 | PH3 | 112 |
| 187.56 | PH3 | 112 |
| 189 | C2H6O2 | 133 |
| 189 | C2H6O2 | 133 |
| 189.9490 | DCN | 104 |
| 190.0080 | DCN | 104 |
| 190.3 | CH3F | 120 |
| 190.3209 | CH3OH | 124 |
| 191.2 | CH3OH | 124 |
| 191.57 | CH3OH | 124 |
| 191.58 | CH3OH | 124 |
| 191.63 | CH3OH | 124 |
| 191.8 | CH2F2 | 118 |
| 191.9 | CD3OH | 126 |
| 192 | C2H6O2 | 133 |
| 192.78 | CH3F | 120 |
| 192.80 | SO2 | 106 |
| 192.9 | H2S | 107 |
| 193.2 | CH3OH | 124 |
| 193.9 | CH2F2 | 118 |
| 194 | CH3NH2 | 131 |
| 194.01 | CH3OH | 124 |
| 194.47 | PH3 | 112 |
| 194.5 | CH2F2 | 118 |
| 194.70 | PH3 | 112 |
| 194.7027 | DCN | 104 |
| 194.7644 | DCN | 104 |
| 194.89 | PH3 | 112 |
| 195 | HDCO | 115 |
| 195.0 | CH3F | 120 |
| 195.18 | PH3 | 112 |
| 196 | HDCO | 115 |
| 196 | CH3F | 120 |
| 196 | CLO2 | 138 |
| 197 | C2H6O2 | 133 |
| 198 | CH3NH2 | 131 |
| 198.0 | CH3NH2 | 131 |
| 198.8 | CH3OH | 124 |
| 199.14 | CH3F | 120 |
| 200 | C2H6O2 | 133 |
| 200.3 | CH3F | 120 |
| 201 | CH3NH2 | 131 |
| 201 | CD3OH | 126 |
| 201.059 | HCN | 103 |
| 202.4 | CH3OH | 125 |
| 202.5 | CH2F2 | 118 |
| 203.63578 | CH3OH* | 126 |
| 204 | CLO2 | 138 |
| 204.3872 | DCN | 104 |
| 205.3 | CH3OH | 125 |
| 206.53 | SO2 | 106 |
| 206.60 | C2H5F | 133 |
| 206.90 | CH3OH | 125 |
| 207 | CH2DOH | 128 |
| 208 | CH3NH2 | 131 |
| 208.41205 | CH3OH* | 126 |
| 209.89 | CH3OH | 125 |
| 211.001 | HCN | 103 |
| 211.25 | CH3OH | 125 |
| 212.8 | CH3OD | 128 |
| 214.35 | CH3OH | 125 |
| 214.5 | CH2F2 | 118 |
| 215.27 | SO2 | 106 |
| 215.3 | CH3F | 120 |
| 215.37244 | CH3OD | 128 |
| 216 | CLO2 | 138 |
| 216.12 | HE | 1 |

| WAVELENGTH [MICROMETER] | ACTIVE MEDIUM | PAGE |
|---|---|---|
| 216.44 | NH3 | 110 |
| 217.1 | C2H5F | 133 |
| 218.0 | CH3NH2 | 131 |
| 218.22 | CH3OH | 125 |
| 218.5 | D2O | 105 |
| 218.9 | N[15]H | 111 |
| 219 | CH3NH2 | 131 |
| 219.9 | CD3OH | 126 |
| 220.230 | H2O | 105 |
| 222 | CD3OH | 126 |
| 222.949 | HCN | 103 |
| 223 | CD3OH | 126 |
| 223.07 | PH3 | 112 |
| 223.5 | CH3OH | 125 |
| 223.91 | NH3 | 110 |
| 224 | CD3CL | 121 |
| 225 | CH3OD | 128 |
| 225.07 | NH3 | 110 |
| 225.3 | H2S | 107 |
| 225.39 | NH3 | 110 |
| 226.9 | C2H5F | 133 |
| 227.15 | CH3CL | 120 |
| 227.6 | CH2F2 | 119 |
| 229 | CD3OD | 128 |
| 229.1 | CH3OD | 128 |
| 229.39 | HCOOH | 116 |
| 230.1 | CH2F2 | 119 |
| 231 | C2H6O2 | 133 |
| 232 | CD3OH | 126 |
| 232.85 | CH3OH | 125 |
| 232.93906 | CH3OH | 125 |
| 233 | D2CO | 115 |
| 235.7 | CH2F2 | 119 |
| 236 | CD3OH | 126 |
| 236.25 | CH3CL | 120 |
| 236.5 | CH2F2 | 119 |
| 237.6 | CH3OH | 125 |
| 238 | CH2DOH | 128 |
| 238 | CHD2OH | 129 |
| 238 | CH3OD | 128 |
| 238.3 | CD3OH | 126 |
| 238.52268 | CH3OH* | 126 |
| 239 | D2O | 106 |
| 240 | C2H6O2 | 133 |
| 240.98 | CH3CL | 120 |
| 242.4727 | CH3OH | 125 |
| 242.79 | CH3OH | 125 |
| 243 | CH3NH2 | 131 |
| 245 | CD3CL | 121 |
| 245 | D2CO | 115 |
| 245.04 | CH3BR | 122 |
| 246 | CH3OH | 125 |
| 246 | CD3CL | 121 |
| 249 | CD3CL | 121 |
| 249 | CD2CL2 | 119 |
| 250 | C2H6O2 | 133 |
| 250 | CH2DOH | 128 |
| 250.4 | CH3CL | 120 |
| 250.78129 | CH3OH | 125 |
| 251.0 | CH3NH2 | 131 |
| 251.13983 | CH3OH | 125 |
| 251.56 | CH3OH | 125 |
| 251.91 | CH3F | 120 |
| 252 | C2H6O2 | 133 |
| 253.2 | CD3OH | 126 |
| 253.6 | CH3OH | 125 |
| 254 | CD2CL2 | 119 |
| 254 | CH3CL | 120 |
| 254.1 | CH3OH | 125 |
| 254.80 | HCOOH | 116 |
| 255 | CD3OD | 128 |
| 255.9 | CH2F2 | 119 |
| 256.61 | NH3 | 110 |
| 258 | CH2CL2 | 119 |
| 258.7 | CD3OH | 126 |
| 260 | CHD2OH | 129 |
| 260 | H13COOH | 118 |
| 261.03 | CH3CL | 120 |
| 261.7 | CH2F2 | 119 |
| 262 | C2H6O2 | 133 |
| 263 | D2O | 106 |
| 263.43 | NH3 | 110 |
| 263.43 | NH3 | 110 |
| 263.7 | CH3OH | 125 |
| 264.05 | CH3BR | 122 |
| 264.6 | CH3OH | 125 |
| 264.7 | C2H5F | 133 |
| 266 | CD3OH | 126 |
| 266.2 | CD3OH | 126 |
| 267 | CD3OH | 126 |
| 267 | CH3NH2 | 131 |
| 267.4432 | CH3OH | 125 |
| 268 | CH3NH2 | 131 |
| 268 | CD3OH | 126 |
| 268.57203 | CH3OH* | 126 |
| 270.0 | CH2F2 | 119 |
| 270.6 | C2H3CN | 137 |
| 271.29 | CH3CL | 120 |
| 272 | CD3I | 123 |
| 272 | CH2DOH | 128 |
| 272.2 | CH2F2 | 119 |
| 273.7 | CH3CL | 120 |
| 275.00 | CH3CL | 121 |
| 275.09 | CH3CL | 121 |
| 276 | D2O | 106 |
| 276 | CD3OH | 126 |
| 277 | CD3OH | 126 |
| 277 | C2H6O2 | 133 |
| 278 | CD3OH | 126 |
| 278.61 | HCOOH | 116 |
| 278.8 | CH3OH | 125 |
| 279 | D2CO | 115 |
| 279.81 | CH3BR | 122 |
| 280 | CH3NC | 130 |

| WAVELENGTH [MICROMETER] | ACTIVE MEDIUM | PAGE |
|---|---|---|
| 280.21826 | CH3OH* | 126 |
| 280.23974 | CH3OH* | 126 |
| 280.5 | NH3 | 110 |
| 280.96 | CH3OH | 125 |
| 281.18 | CH3CN | 129 |
| 281.35 | NH3 | 110 |
| 281.48 | NH3 | 110 |
| 281.67 | CH3CL | 121 |
| 281.96 | CH3CN | 129 |
| 282.3 | C2H5F | 133 |
| 283 | C2H3BR | 136 |
| 284 | HCN | 103 |
| 285 | CD3OH | 126 |
| 286.6 | CD3OH | 126 |
| 286.79 | CH3CL | 121 |
| 286.88 | CH3CN | 129 |
| 287.4 | CD3OH | 126 |
| 287.7 | CH2F2 | 119 |
| 288 | CD3CL | 121 |
| 288 | CD3CL | 121 |
| 288 | CH3NC | 130 |
| 288 | C2H6O2 | 133 |
| 288 | CH3NH2 | 131 |
| 288.5 | CH2CF2 | 132 |
| 289.4 | CH2F2 | 119 |
| 290 | C2H6O2 | 133 |
| 290.0 | CD3OH | 126 |
| 290.4 | NH3 | 110 |
| 290.62 | CH3OH | 125 |
| 290.9 | NH3 | 110 |
| 291.2 | NH3 | 110 |
| 291.27 | CD3CL | 121 |
| 291.35 | NH3 | 110 |
| 291.95 | NH3 | 110 |
| 292.2 | CH3OH | 125 |
| 292.5 | CH3OH | 125 |
| 293.78 | CH3OH | 125 |
| 293.9 | CH2F2 | 119 |
| 294.28 | CH3BR | 122 |
| 294.81097 | CH3OD | 128 |
| 295 | CH2DOH | 128 |
| 296 | CH2DOH | 128 |
| 297 | CD3OH | 126 |
| 297 | CD3OH | 126 |
| 298.2 | CH2F2 | 119 |
| 299 | CD3OD | 128 |
| 299 | CD3OH | 126 |
| 299 | C2H6O2 | 133 |
| 301 | CD3I | 123 |
| 301.2 | NH3 | 110 |
| 301.9943 | CH3OH | 125 |
| 302.08 | HCOOH | 116 |
| 302.2781 | HCOOH | 116 |
| 303.54 | CH3CN | 129 |
| 305 | DCOOD | 118 |
| 305.72610 | CH3OD | 128 |
| 306.28 | NH3 | 110 |
| 307.65 | CH3CL | 121 |
| 308 | CH2DOH | 128 |
| 308 | FCN | 139 |
| 309 | CD3OH | 126 |
| 309.23 | HCOOH | 116 |
| 309.7140 | HCN | 103 |
| 310 | CD3OH | 127 |
| 310.8870 | HCN | 103 |
| 311.07 | CH3BR | 122 |
| 311.10 | CH3BR | 122 |
| 311.20 | CH3BR | 122 |
| 311.21 | CH3BR | 122 |
| 311.45 | HCOOH | 116 |
| 311.75 | NH3 | 110 |
| 312 | CD3OD | 128 |
| 312 | CD3OD | 128 |
| 313 | H13COOH | 118 |
| 314 | CH3NH2 | 131 |
| 318 | CD3CL | 121 |
| 319.48 | HCOOH | 116 |
| 321 | CD3OH | 127 |
| 322 | CH2DOH | 128 |
| 326.5 | CH2F2 | 119 |
| 330.1 | CH3OD | 128 |
| 330.2 | C2H5F | 133 |
| 332.6033 | CH3OH* | 126 |
| 332.86 | CH3BR | 122 |
| 333.15 | CH3BR | 122 |
| 333.96 | CH3CL | 121 |
| 334.82 | HCOOH | 116 |
| 334.91 | HCOOH | 116 |
| 335.1831 | HCN | 103 |
| 336 | HCOOH | 116 |
| 336 | CD3OH | 127 |
| 336 | CD3OH | 127 |
| 336 | C3H2O | 138 |
| 336.3 | HCOOH | 116 |
| 336.5578 | HCN | 103 |
| 336.7 | C2H5F | 133 |
| 338.9637 | CH3OH* | 126 |
| 339 | CD3OD | 128 |
| 342 | CD2CL2 | 119 |
| 342.74 | HCOOH | 116 |
| 344 | C2H6O2 | 133 |
| 346 | CD3OH | 127 |
| 346 | CHD2OH | 129 |
| 346.32 | CH3CN | 129 |
| 347 | CH3NH2 | 131 |
| 349 | DCOOD | 118 |
| 349.34 | CH3CL | 121 |
| 350 | CD3OH | 127 |
| 350.20 | H2O | 105 |
| 351 | CD3OH | 127 |
| 351 | CD3OH | 127 |
| 352 | CD3OH | 127 |
| 352 | CD3OH | 127 |
| 352.75 | CH3BR | 122 |

| WAVELENGTH [MICROMETER] | ACTIVE MEDIUM | PAGE |
|---|---|---|
| 353 | CD3OH | 127 |
| 354 | CD3OD | 128 |
| 354 | CH3CL | 121 |
| 355 | CHD2OH | 129 |
| 355.2 | CH2F2 | 119 |
| 356 | C2H3BR | 136 |
| 358 | C2H6O2 | 133 |
| 358.5 | D2O | 106 |
| 359.81 | HCOOH | 116 |
| 362.1 | C2H5F | 133 |
| 363 | CHD2OH | 129 |
| 363 | CH2DOH | 129 |
| 364.5 | CH3CL | 121 |
| 368 | HCOOH | 116 |
| 368.0 | C2H3CL | 135 |
| 369.11368 | CH3OH | 125 |
| 370 | C2H3BR | 136 |
| 370 | CD3OH | 127 |
| 370 | CD3OH | 127 |
| 372.5283 | HCN | 103 |
| 372.68 | CH3F | 120 |
| 372.87 | CH3CN | 129 |
| 374 | CH2DOH | 129 |
| 375 | C2H6O | 137 |
| 375.0 | CH2CF2 | 132 |
| 375.9 | N[15]H | 111 |
| 376.0 | C2H5F | 133 |
| 377.45 | CH3I | 122 |
| 378.0 | C2H5F | 133 |
| 378.57 | CH3CL | 121 |
| 379 | C2H3F3 | 134 |
| 380.02 | CH3BR | 122 |
| 380.71 | CH3CN | 129 |
| 381 | DCOOD | 116 |
| 381.8 | CH2F2 | 119 |
| 382.9 | CH2F2 | 119 |
| 383.28 | CD3CL | 121 |
| 384 | C3H6O3 | 116 |
| 385 | D2O | 106 |
| 385 | CD3OH | 127 |
| 386 | CD3OH | 127 |
| 386.20 | CH3OH | 125 |
| 386.41 | CH3CN | 129 |
| 387.31 | CH3CN | 129 |
| 388 | C2H6O2 | 133 |
| 388 | NH3 | 110 |
| 388 | HCOOH | 116 |
| 388.39 | CH3CN | 129 |
| 390 | CD3I | 123 |
| 390.1 | CH3OH | 125 |
| 390.53 | CH3I | 122 |
| 392 | HCOOH | 116 |
| 392.06871 | CH3OH | 125 |
| 392.48 | CH3I | 122 |
| 393.6311 | HCOOH | 116 |
| 393.6311 | HCOOH | 116 |
| 394.2 | HCOOH | 116 |
| 394.7 | CH2F2 | 119 |
| 396 | HCOOH | 116 |
| 396 | CH2DOH | 129 |
| 396 | C2H3BR | 136 |
| 396 | C2H5OH | 135 |
| 397.51 | CH3F | 120 |
| 397.6 | CH3CL | 121 |
| 398 | CD3OH | 127 |
| 401 | HCOOH | 116 |
| 403 | HCOOH | 116 |
| 404 | C2H5F | 133 |
| 404 | C2H5F | 133 |
| 404 | CH3NC | 130 |
| 404.1 | HCOOH | 116 |
| 404.69 | NH3 | 110 |
| 405.0 | C2H5F | 133 |
| 405.50 | C2H5F | 133 |
| 405.5848 | HCOOH | 116 |
| 405.75 | HCOOH | 116 |
| 406 | CD3OD | 128 |
| 406.0 | HCOOH | 116 |
| 407 | CD3OH | 127 |
| 407.72 | CH3BR | 122 |
| 409 | CD3OH | 127 |
| 410 | CD3OH | 127 |
| 411 | C2H3BR | 136 |
| 412 | CD3OH | 127 |
| 413 | HCOOH | 117 |
| 414 | HCOOH | 117 |
| 414 | CD3OD | 128 |
| 414.98 | CH3BR | 122 |
| 415 | CH2CF2 | 132 |
| 415 | C2H6O2 | 133 |
| 416 | C2H3BR | 136 |
| 416.5224 | CH3OH | 125 |
| 417.1 | CH3OD | 128 |
| 417.8 | CH3OH | 125 |
| 418.1 | CH2F2 | 119 |
| 418.31 | CH3BR | 122 |
| 418.51 | HCOOH | 117 |
| 418.6 | HCOOH | 117 |
| 419 | CH3F | 120 |
| 419 | C2H3BR | 136 |
| 419.0 | CD3OH | 127 |
| 419.55 | HCOOH | 117 |
| 420.0 | HCOOH | 117 |
| 420.26 | HCOOH | 117 |
| 421 | HCOOH | 117 |
| 421 | CD3OH | 127 |
| 421.0 | HCOOH | 117 |
| 422 | CD3OH | 127 |
| 422.14 | CH3CN | 129 |
| 422.78 | CH3BR | 122 |
| 424 | C2H3CL | 135 |
| 424 | C2H3BR | 136 |
| 426 | CHD2OH | 129 |
| 427 | C2H3BR | 136 |

| WAVELENGTH [MICROMETER] | ACTIVE MEDIUM | PAGE |
|---|---|---|
| 427.04 | CH3CN | 129 |
| 427.89 | CH3CCH | 130 |
| 428 | HCOOH | 117 |
| 428.87 | CH3CCH | 130 |
| 430.55 | CH3CN | 129 |
| 432.1093 | HCOOH | 117 |
| 432.4 | CH2F2 | 119 |
| 432.6313 | HCOOH | 117 |
| 432.6325 | HCOOH | 117 |
| 433 | C3H6O3 | 116 |
| 433.10 | HCOOH | 117 |
| 433.1038 | CD3I | 123 |
| 434.9 | CH2F2 | 119 |
| 435 | HCOOH | 117 |
| 435 | CD3OH | 127 |
| 437.70 | HCOOH | 117 |
| 438 | HCOOH | 117 |
| 438.5 | C2H3BR | 136 |
| 441 | HCOOH | 117 |
| 441.15 | CH3CN | 129 |
| 443.26 | CD3CL | 121 |
| 443.5 | C2H3BR | 136 |
| 444.3862 | CD3I | 123 |
| 445 | C2H3BR | 136 |
| 445 | C2H3CL | 135 |
| 445.21 | HCOOH | 117 |
| 445.81 | HCOOH | 117 |
| 445.8971 | HCOOH | 117 |
| 446.5054 | HCOOH | 117 |
| 446.75 | HCOOH | 117 |
| 446.8730 | HCOOH | 117 |
| 447.0 | HCOOH | 117 |
| 447.1424 | CH3I | 122 |
| 447.58 | HCOOH | 117 |
| 448.5335 | H13COOH | 118 |
| 449.79 | CD3CL | 121 |
| 451.9 | CH3OH | 125 |
| 451.903 | CH3F | 120 |
| 451.924 | CH3F | 120 |
| 453.41 | CH3CN | 129 |
| 455 | CD3OH | 127 |
| 457.25 | CH3I | 122 |
| 458 | C2H4F2 | 134 |
| 458.0 | CH2CF2 | 132 |
| 458.43 | HCOOH | 117 |
| 458.5229 | HCOOH | 117 |
| 458.6 | HCOOH | 117 |
| 459.18 | CH3I | 122 |
| 460 | C3H6O3 | 116 |
| 460.51 | HCOOH | 117 |
| 460.5619 | CD3I | 123 |
| 461 | C2H6O | 137 |
| 461.20 | CH3CL | 121 |
| 461.3848 | CH3OH* | 126 |
| 462 | HCOOD | 118 |
| 462.92 | C2H5F | 133 |
| 464 | C2H4F2 | 134 |
| 464.3 | CH2CF2 | 132 |
| 464.5 | CH2F2 | 119 |
| 464.76 | CD3CL | 121 |
| 466.25 | CH3CN | 129 |
| 466 | CH2DOH | 129 |
| 469 | CD2CL2 | 119 |
| 469.02330 | CH3OH | 125 |
| 470 | CH3OH | 125 |
| 471 | CH3OH | 125 |
| 472 | CD3OH | 127 |
| 477.87 | CH3I | 123 |
| 480 | CD3OH | 127 |
| 480 | H13COOH | 118 |
| 480 | C2H6O | 137 |
| 480.01 | CH3CN | 129 |
| 480.31 | CD3CL | 121 |
| 482.96 | C2H3BR | 136 |
| 483 | CHD2OH | 129 |
| 483 | CD3OH | 127 |
| 483 | CD3OH | 127 |
| 486 | C2H5F | 133 |
| 486.1 | CH3OH | 125 |
| 487 | C2H3CL | 135 |
| 487.2260 | CD3I | 123 |
| 488.88 | CH3CCH | 130 |
| 489 | C2H3CN | 137 |
| 490.08 | C2H3BR | 136 |
| 490.3909 | CD3I | 123 |
| 492 | HCOOH | 117 |
| 492 | DCOOD | 118 |
| 492 | C2H6O | 137 |
| 493.28 | HCOOH | 117 |
| 494 | CH3F | 120 |
| 494.74 | CH3CN | 129 |
| 495 | CD3OD | 128 |
| 495 | C2H6O | 137 |
| 495 | CD3OH | 127 |
| 496 | HCOOH | 117 |
| 496.072 | CH3F | 120 |
| 496.1009 | CH3F | 120 |
| 498.0 | CD3OH | 127 |
| 502.2 | C2H5F | 133 |
| 503 | C2H3CN | 137 |
| 503.6 | CH2F2 | 119 |
| 506 | C2H3BR | 136 |
| 507.7 | C2H3CL | 135 |
| 508 | CD3OH | 127 |
| 508.37 | CH3I | 123 |
| 508.48 | CH3BR | 122 |
| 509 | HCCF | 138 |
| 510.16 | CH3CN | 129 |
| 511.3 | CH2F2 | 119 |
| 511.90 | CH3CL | 121 |
| 512 | C3H6O3 | 116 |
| 512.88 | HCOOH | 117 |
| 513 | CHD2OH | 129 |
| 513.2 | HCOOH | 117 |

| WAVELENGTH [MICROMETER] | ACTIVE MEDIUM | PAGE |
|---|---|---|
| 515.1690 | HCOOH | 117 |
| 516 | C3H2O | 138 |
| 516.77 | CH3CCH | 130 |
| 517 | CD3OH | 127 |
| 517.33 | CH3I | 123 |
| 518.83 | HCOOH | 117 |
| 519 | C2H5F | 134 |
| 519 | C2H3CL | 135 |
| 519.30 | CD3CL | 121 |
| 520 | CD2CL2 | 119 |
| 520 | C2H6O | 137 |
| 523.4061 | CD3I | 123 |
| 525.32 | CH3I | 123 |
| 526 | DCOOD | 118 |
| 528.49 | C2H3BR | 136 |
| 529.28 | CH3I | 123 |
| 530 | HCOOH | 117 |
| 530 | HCOOH | 117 |
| 531.06 | CH3BR | 122 |
| 531.08 | CH3CCH | 130 |
| 532 | CH2CF2 | 132 |
| 532 | C2H3CL | 135 |
| 533 | C2H4F2 | 134 |
| 533.6773 | HCOOH | 117 |
| 533.6773 | HCOOH | 117 |
| 534.5 | HCOOH | 117 |
| 534.8 | HCOOH | 117 |
| 538 | C2H3CL | 135 |
| 538.2 | HCN | 103 |
| 540 | CD3I | 123 |
| 540.8 | CH2F2 | 119 |
| 540.9 | C2H5F | 134 |
| 541.113 | CH3F | 120 |
| 541.147 | CH3F | 120 |
| 542.99 | CH3I | 123 |
| 545.21 | CH3BR | 122 |
| 545.39 | CH3BR | 122 |
| 545.4 | HCN | 103 |
| 550.0 | C2H3CN | 137 |
| 551 | CD3OH | 127 |
| 553 | CD3OH | 127 |
| 553 | CD3OH | 127 |
| 553.69 | C2H3BR | 136 |
| 554 | CD3OH | 127 |
| 554.4 | CH2CF2 | 132 |
| 556.8755 | CD3I | 123 |
| 561.41 | CH3CN | 129 |
| 563.13 | CH3CCH | 130 |
| 564.68 | CH3BR | 122 |
| 566.44 | CH3CCH | 130 |
| 567.5 | CH2F2 | 119 |
| 568 | CH2CF2 | 132 |
| 568.81 | CH3CL | 121 |
| 569 | DCOOD | 118 |
| 569.4773 | CD3I | 123 |
| 570.56864 | CH3OH | 125 |
| 574 | C2H3CL | 135 |
| 574.4 | C2H3CN | 137 |
| 576.17 | CH3I | 123 |
| 577 | HCOOH | 117 |
| 578 | C2H3CN | 137 |
| 578.90 | CH3I | 123 |
| 580.3872 | HCOOH | 117 |
| 580.52 | HCOOH | 117 |
| 582.0 | HCOOH | 117 |
| 583 | CD3OH | 127 |
| 583 | CD3OH | 127 |
| 583.77 | CH3CCH | 130 |
| 583.87 | CH3I | 123 |
| 584.0 | C2H3CN | 137 |
| 585.72 | CH3BR | 122 |
| 586.6 | C2H3CN | 137 |
| 588.1 | CH2F2 | 119 |
| 593.32 | C2H5F | 134 |
| 594.72 | C2H3BR | 136 |
| 595 | CH3F | 120 |
| 599 | CD3OH | 127 |
| 599.5499 | CD3I | 123 |
| 603 | C2H3CL | 135 |
| 603.06 | CH3OH | 125 |
| 614.1098 | CD3I | 123 |
| 614.92 | CH3OH | 125 |
| 616 | CH2DOH | 129 |
| 618.44 | C2H3BR | 136 |
| 619 | C3H6O3 | 116 |
| 620.4 | C2H5F | 134 |
| 623 | C2H3CN | 137 |
| 624.09 | C2H3BR | 136 |
| 627.34 | CH3OH | 125 |
| 631 | CD2CL2 | 119 |
| 631 | C2H3CN | 137 |
| 631.93 | CH3BR | 122 |
| 632.00 | CH3BR | 122 |
| 634.4 | C2H3CL | 135 |
| 635.35 | C2H3BR | 136 |
| 636 | C2H3CL | 135 |
| 639.73 | CH3I | 123 |
| 640 | CD3I | 123 |
| 642.5 | CH2F2 | 119 |
| 644 | CD3I | 123 |
| 646 | CD3OH | 127 |
| 646 | C2H3BR | 136 |
| 647.89 | CH3CCH | 130 |
| 648 | CD3OH | 127 |
| 649.42 | C2H3BR | 136 |
| 649.59 | CH3CCH | 130 |
| 652.68 | CH3CN | 129 |
| 657.2 | CH2F2 | 119 |
| 658.53 | CH3BR | 122 |
| 660.5822 | CD3I | 123 |
| 660.70 | CH3BR | 122 |
| 663.3 | CH2CF2 | 132 |
| 667.2322 | CD3I | 123 |
| 669.5308 | HCOOH | 117 |

| WAVELENGTH [MICROMETER] | ACTIVE MEDIUM | PAGE |
|---|---|---|
| 670.0 | HCOOH | 117 |
| 670.0940 | CD3I | 123 |
| 670.1143 | CD3I | 123 |
| 670.99 | CH3I | 123 |
| 675.29 | CH3CCH | 130 |
| 676 | HCN | 103 |
| 680 | C3H6O3 | 116 |
| 680 | CD3OH | 127 |
| 680.54 | C2H3BR | 136 |
| 685 | CD3OH | 127 |
| 690 | C13D3I | 123 |
| 691.1292 | CD3I | 123 |
| 693.13 | C2H3BR | 136 |
| 694.17 | CH3OH | 125 |
| 695 | CD3OH | 127 |
| 695 | CH3OH | 125 |
| 696 | C3H6O3 | 116 |
| 696 | C2H6O2 | 133 |
| 698.55 | CD3CL | 121 |
| 699 | C2H3CL | 135 |
| 699.42258 | CH3OH | 125 |
| 702 | CD3OH | 127 |
| 703 | CD3OH | 127 |
| 704.53 | CH3CN | 129 |
| 707 | C2H3CL | 135 |
| 707.22 | C2H3BR | 136 |
| 711 | CD3OH | 127 |
| 712 | C3H6O3 | 116 |
| 712 | C2H3BR | 136 |
| 713.72 | CH3CN | 129 |
| 715.40 | CH3BR | 122 |
| 719.30 | CH3I | 123 |
| 722 | CD3OH | 127 |
| 722 | C2H3CN | 137 |
| 724.13 | C2H3BR | 136 |
| 725.1 | CH2F2 | 119 |
| 730.3234 | CD3I | 123 |
| 733.5739 | D2CO | 115 |
| 734.2624 | CD3I | 123 |
| 735.12 | CD3CL | 121 |
| 738 | C2H3CN | 137 |
| 741.11 | C2H3BR | 136 |
| 741.62 | CH3CN | 130 |
| 742.5723 | HCOOH | 117 |
| 743.0 | HCOOH | 117 |
| 744.0503 | HCOOH | 117 |
| 745 | CD3I | 123 |
| 745 | CD3OH | 127 |
| 745.0 | HCOOH | 117 |
| 749.29 | CH3BR | 122 |
| 749.36 | CH3BR | 122 |
| 750 | C3H6O3 | 116 |
| 752.6807 | D2CO | 115 |
| 755 | C2H4F2 | 134 |
| 757.41 | CH3CCH | 130 |
| 760 | CD3OH | 127 |
| 761 | HCOOH | 117 |
| 764.1 | CH2CF2 | 132 |
| 773.5 | HCN | 103 |
| 774 | CD3OH | 127 |
| 775 | C2H3CN | 137 |
| 780.13 | C2H3BR | 136 |
| 784.26 | C2H3BR | 136 |
| 785 | HCOOH | 117 |
| 786.1617 | HCOOH | 117 |
| 786.48 | CD3I | 123 |
| 786.9192 | H13COOH | 118 |
| 790 | DCOOD | 118 |
| 793 | C2H3CN | 137 |
| 798.55 | CH3CCH | 130 |
| 806 | C13D3I | 123 |
| 815 | C3H6O3 | 116 |
| 823.4 | CFBR | 113 |
| 826.94 | C2H3BR | 136 |
| 828 | C2H3CN | 137 |
| 828 | C2H3CL | 135 |
| 829 | CD2CL2 | 119 |
| 831.13 | CH3BR | 122 |
| 851.9 | C2H5F | 134 |
| 853.43 | C2H3BR | 136 |
| 854.41 | CH3CN | 130 |
| 862.0 | CD3OH | 127 |
| 869 | CD3OD | 128 |
| 870.80 | CH3CL | 121 |
| 883.59 | CD3CL | 121 |
| 884 | CH2CF2 | 132 |
| 885.2 | CFBR | 113 |
| 890 | C3H6O3 | 116 |
| 890.0 | CH2CF2 | 132 |
| 890.1 | CH2CF2 | 132 |
| 891 | C3H6O3 | 116 |
| 895 | CD3I | 123 |
| 900 | C2H5CL | 135 |
| 900.13 | C2H3BR | 136 |
| 910 | C2H3CN | 137 |
| 918.6101 | CD3I | 123 |
| 925 | HCOOD | 118 |
| 925.52 | CH3BR | 122 |
| 935 | C2H3CL | 135 |
| 936.15 | C2H3BR | 136 |
| 937 | DCOOD | 118 |
| 940 | C2H3CN | 137 |
| 943.22 | C2H3BR | 136 |
| 943.97 | CH3CL | 121 |
| 948.9247 | C3H6O3 | 116 |
| 953.8799 | CD3I | 123 |
| 958.25 | CH3CL | 121 |
| 963.48 | C2H3BR | 136 |
| 964 | CH3I | 123 |
| 966 | CH3CL | 121 |
| 968 | CD3OH | 127 |
| 981.7094 | CD3I | 123 |
| 985.85 | C2H3BR | 136 |
| 989.19 | C2H3BR | 136 |

| WAVELENGTH [MICROMETER] | ACTIVE MEDIUM | PAGE |
|---|---|---|
| 990.0 | CH2CF2 | 132 |
| 990.15 | CH3BR | 122 |
| 990.63 | C2H3BR | 136 |
| 995 | C2H3CL | 135 |
| 1005.3476 | CD3I | 123 |
| 1013 | C2H5F | 134 |
| 1014.89 | CH3CN | 130 |
| 1016.33 | CH3CN | 130 |
| 1020 | CH2CF2 | 132 |
| 1028 | HCCF | 138 |
| 1030.3782 | H13COOH | 118 |
| 1041 | C2H3CL | 135 |
| 1063.29 | CH3I | 123 |
| 1069 | C2H5F | 134 |
| 1086.89 | CH3CN | 130 |
| 1097.11 | CH3CCH | 130 |
| 1099.5441 | CD3I | 123 |
| 1100 | CD3OH | 127 |
| 1146 | CD3OH | 127 |
| 1146.83 | CH3CN | 130 |
| 1156 | C2H3CN | 137 |
| 1174.87 | CH3CCH | 131 |
| 1184 | C2H3CN | 137 |
| 1221.79 | CH3F | 120 |
| 1239.47 | CD3CL | 121 |
| 1247.59 | C2H3BR | 136 |
| 1253.738 | CH3I | 123 |
| 1290 | CD3OH | 127 |
| 1310.38 | CH3BR | 122 |
| 1350 | C2H5CL | 135 |
| 1351.78 | CH3CN | 130 |
| 1363.88 | C2H3BR | 136 |
| 1394.06 | C2H3BR | 136 |
| 1400 | C2H5CL | 135 |
| 1446.1 | CH2F2 | 119 |
| 1546 | C2H5F | 134 |
| 1549.5048 | CD3I | 123 |
| 1572.64 | CH3BR | 122 |
| 1614.88 | C2H3BR | 136 |
| 1720 | C2H5CL | 135 |
| 1814.37 | CH3CN | 130 |
| 1866.87 | CH3CL | 121 |
| 1899.889 | C2H3BR | 136 |
| 1965.34 | CH3BR | 122 |
| 1990.75 | CD3CL | 121 |

# References


[001] BOCKASTEN,K., T.LUNDHOLM, AND O.ANDRADE:
LASER LINES IN ATOMIC AND MOLECULAR HYDROGEN.
J.OPT.SOC.AM.56,1260-1261(1966).

[002] ABRAMS,R.L., AND G.J.WOLGA:
NEAR-INFRARED LASER TRANSITIONS IN PURE HELIUM.
IEEE J.QUANT.ELECTRON. QE-3,368(1967).

[003] BENNETT,W.R.,JR.:
INVERSION MECHANISM IN GAS LASER.
APPL.OPT.,SUPPL. ON CHEMICAL LASERS 14-20(1965).

[004] BENNETT,W.R.,JR.:
GASEOUS OPTICAL MASERS.
APPL.OPT. SUPPL. ON OPTICAL MASERS 1,24-61(1962).

[005] ISAEV,A.A., P.I.ISCHENKO, AND G.G.PETRASH:
SUPER-RADIANCE AT TRANSITIONS TERMINATING AT METASTABLE
LEVELS OF HELIUM AND THALLIUM.
JETP LETT.6,118-121(1967).

[006] BROCHARD,J., AND S.LIBERMAN:
EMISSION STIMULEE DE NOUVELLES TRANSITIONS
INFRAROUGES DE L HELIUM ET DU NEON.
COMPT.REND.260,6827-6829(1965).

[007] MATHIAS,L.E.S., A.CROCKER, AND M.S.WILLS:
PULSED LASER EMISSION FROM HELIUM AT 95 MICROMETERS.
IEEE J.QUANT.ELECTRON QE-3,170(1967).

[008] PIXTON,R.M. AND G.R.FOWLES:
VISIBLE LASER OSCILLATIONS IN HELIUM AT 7065A.
PHYS.LETT.29A,654-655(1969).

[009] LEVINE,J.S. AND A.JAVAN:
FAR-INFRARED CONTINUOUS-WAVE OSCILLATION IN PURE HELIUM.
APPL.PHYS.LETT.14,348-349(1969).

[010] DEZENBERG,G.J.:
NEW UNIDENTIFIED HIGH-GAIN OSCILLATION AT 486.1 AND 434.0 NM
IN THE PRESENCE OF NEON.
IEEE J.QUANT.ELECTRON. QE-7 10,491-493(1971).

[011] BROCHARD,J., J.-F.LESPRIT, AND L.LIBERMAN:
MEASUREMENT OF THE ISOTOPIC SEPARATION OF TWO INFRARED
LASER LINES OF HE I.
COMPT.REND.A.SC.270B,9,600-602(1970).

[012] CHEO,P.K. AND H.G.COOPER:
ULTRAVIOLET ION-LASER TRANSITIONS BETWEEN 2300A
AND 4000A.
J.APPL.PHYS.36,1862-1865(1965).

[013] BRIDGES,W.B., AND A.N.CHESTER:
VISIBLE AND UV LASER OSCILLATION AT 118 WAVELENGTHS
IN IONIZED NEON,ARGON,KRYPTON,XENON,
OXYGEN AND OTHER GASES.
APPL.OPT.4,573-580(1965).

[014] DANA,L., P.LAURES, AND R.ROCHEROLLES:
RAIES LASER ULTRAVIOLETTES DANS LE NEON,
L ARGON ET LE XENON.
COMPT.REND.260,481-484(1965).

[015] BRIDGES,W.B., R.J.FREIBERG, AND A.S.HALSTED:
NEW CONTINUOUS UV ION TRANSITIONS IN NEON,
ARGON AND KRYPTON
IEEE J.QUANT.ELECTRON. QE-3,339(1967).

[016] CLUNIE,D.M., R.S.A.THORN, AND K.E.TREZISE:
ASYMETRIC VISIBLE SUPER-RADIANT EMISSION FROM
A PULSED NEON DISCHARGE.
PHYS.LETT.14,28-29(1965).

[017] LEONARD,D.A., R.A.NEAL, AND E.T.GERRY:
OBERVATION OF A SUPER-RADIANT SELF TERMINATING
GREEN LASER TRANSITION IN NEON.
APPL.PHYS.LETT.7,175(1965).

[018] ROSENBERGER D.:
LASERUEBERGAENGE UND SUPERSTRAHLUNG BEI 6143A UND
5944 A IN EINER GEPULSTEN NEON-ENTLADUNG.
PHYS.LETT.13,228-229(1964).

[019] WHITE,A.D., AND J.D.RIGDEN:
THE EFFECT OF SUPER-RADIANCE AT 3,39 MICROMETERS
ON THE VISIBLE TRANSITIONS IN THE HE-NE MASER.
APPL.PHYS.LETT.2,211-212(1963).

[020] HEARD,H.G., AND J.PETERSEN:
SUPER-RADIANT YELLOW AND ORANGE LASER
TRANSITIONS IN PURE NEON.
PROC.IEEE 52,1285(1964).

[021] RIGDEN,J.D., AND A.D.WHITE:
SIMULTANEOUS GAS-MASER ACTION IN THE
VISIBLE AND INFRARED.
PROC.IRE 50,2366-2367(1962).

[022] BLOOM,A.L.:
OBSERVATION OF NEW VISIBLE GAS-LASER TRANSITIONS
BY REMOVAL OF DOMINANCE.
APPL.PHYS.LETT.2,101-102(1963).

[023] ZITTER R.N.:
2S-2P AND 3P-2S TRANSITIONS OF NEON IN A LASER
TEN METERS LONG.
J.APPL.PHYS.35,3073-3071(1964).

[024] MCFARLANE,R.A., C.K.N.PATEL, W.R.BENNETT,JR., AND W.L.FAUST:
NEW HE-NE OPTICAL-MASER TRANSITIONS.
PROC.IEEE 50,2111-2112(1962).

[025] JAVAN,A., W.R.BENNETT,JR., AND D.R.HERRIOTT:
POPULATION INVERSION AND CONTINUOUS OPTICAL-MASER
OSCILLATION IN A GAS DISCHARGE CONTAINING
A HE-NE MIXTURE.
PHYS.REV.LETT.6,106-110(1961).

[026] RIGDEN,J.D., AND A.D.WHITE:
THE INTERACTION OF VISIBLE AND INFRARED MASER
TRANSITIONS IN THE HELIUM-NEON SYSTEM.
QUANTUM ELECTRONICS I,ED.BY P.GRIVET,
N.BLOEMBERGEN,NEW YORK,
COLUMBIA UNIVERSITY PRESS 1964,S.499-505.
PROC.IEEE 51,943-945(1963).

[027] BENNETT,W.R.,JR., AND J.W.KNUTSON,JR.:
SIMULTANEOUS LASER OSCILLATION ON THE NEON
DOUBLET AT 1.1523 MICRONS.
PROC.IEEE 52,861(1964).

[028] ROSENBERGER,D.:
SCHWINGVERHALTEN UND WECHSELWIRKUNG DER
0.63 MICROMETERS UND 3.39 MICROMETERS OSZILLATIONEN
BEI EINEM HE-NE-LASER MIT KLEINEM SPIEGELABSTAND.
PHYS.LETT.8,187-189(1964).

[029] DER AGOBIAN,R., I.L.OTTO, R.ECHARD, AND R.CAGNARD:
EMISSION STIMULEE DE NOUVELLES TRANSITIONS
INFRAROUGES DU NEON.
COMPT.REND.257,3844-3847(1963).

[030] CAGNARD,R., R.DER AGOBIAN, R.ECHARD, AND I.L.OTTO:
L'EMISSION STIMULEE DE QUELQUES TRANSITIONS
INFRAROUGES DE L'HELIUM ET DU NEON.
COMPT.REND.257,1044-1047(1963).

[031] MCFARLANE,R.A., W.L.FAUST, AND C.K.N.PATEL:
OSCILLATIONS ON F-D TRANSITIONS IN NEON
IN A GAS OPTICAL MASER.
PROC.IEEE 51,468(1963).

[032] DER AGOBIAN,R., I.L.OTTO, R.CAGNARD, AND R.ECHARD:
NEW NE LASER TRANSITIONS IN THE NEAR INFRARED.
J.APPL.PHYS.35,2787(1964).

[033] FAUST,W.L., R.A.MCFARLANE, C.K.N.PATEL, AND C.G.B.GARRETT:
NOBLE GAS OPTICAL-MASER LINES AT WAVELENGTHS
BETWEEN 2 AND 35 MICRONS.
PHYS.REV.133A,1476-1486(1964).

[034] GERRITSEN H.J., AND P.V.GOEDERTIER:
A GASEOUS(HE-NE)CASCADE LASER.
APPL.PHYS.LETT.4,20(1964).

[035] GRUDZINSKY,R., M.PAILLETTE, AND I.DECRELLE:
ETUDE DE TRANSITIONS LASER COUPLEES
DANS MELANGE HELIUM-NEON.
COMPT.REND.258,1452-1454(1964).

[036] PATEL,C.K.N., W.L.FAUST, R.A.MFARLANE, AND C.G.B.GARRETT:
LASER ACTION UP TO 57.355 MICRONS IN
GASEOUS DISCHARGES(NE,HE-NE).
APPL.PHYS.LETT.4,18-19(1964).

[037] PATEL,C.K.N., W.L.FAUST, R.A.MCFARLANE, AND C.G.B.GARRETT:
CW OPTICAL-MASER ACTION UP TO 133 MICRONS
(0.133 MM)IN NEON DISCHARGES.
PROC.IEEE 52,713(1964).

[038] MCFARLANE,R.A., W.L.FAUST, C.K.N.PATEL, AND C.G.B.GARRETT:
NEON GAS-MASER LINES AT 68.329 MICRONS
AND 85.047 MICRONS.
PROC.IEEE 52,318(1964).

[039] MCFARLANE,R.A.:
OPTICAL-MASER OSCILLATION ON ISO-ELECTRONIC
TRANSITIONS IN AR III AND CL II.
APPL.OPT.3,1196(1964).

[040] BRIDGES,W.B., AND A.N.CHESTER:
SPECTROSCOPY OF ION LASERS.
IEEE J.QUANT.ELECTRON. QE-1,66-84(1965).

[041] BRIDGES,W.B.:
LASER OSCILLATION IN SINGLY IONIZED ARGON
IN THE VISIBLE SPECTRUM.
APPL.PHYS.LETT.4,128-130(1964).

[042] CONVERT,G., M.ARMAND, AND P.MARTINOT-LAGARDE:
TRANSITIONS LASER VISIBLES DANS L'ARGON IONISE.
COMPT.REND.258,4467-4469(1964).

[043] BENNETT,W.R.,JR.,J.W.KNUTSON, JR.,G.N.MERCER, AND J.L.DETCH:
SUPERRADIANCE,EXCITATION MECHANISMS,AND QUASI-CW
OSCILLATION IN THE VISIBLE AR+ LASER.
APPL.PHYS.LETT.4,180-182(1964).

[044] SINCLAIR,D.C.:
NEAR-INFRARED OSCILLATION IN PULSED NOBLE-GAS-ION LASERS.
J.OPT.SOC.AM.55,571-572(1965).

[045] HORRIGAN,F.A., S.H.KOOZEKANANI, AND R.A.PAANANEN:
INFRARED LASER ACTION AND LIFETIMES IN ARGON II.
APPL.PHYS.LETT.6,41-43(1965).

[046] LIBERMAN,S.:
EMISSION STIMULEE DE NOUVELLES TRANSITIONS INFRAROUGES
DE L'ARGON,DU KRYPTON ET DU XENON.
COMPT.REND.261,2601-2604(1965).

[047] PAANANEN,R.:
CONTINUOUSLY OPERATED ULTRAVIOLET LASERS.
APPL.PHYS.LETT.9,34-35(1966).

[048] BRIDGES,W.B., AND A.S.HALSTEAD:
NEW CW LASER TRANSITIONS IN ARGON,KRYPTON,AND XENON.
IEEE J.QUANT.ELECTRON. QE-2,84(1966).

[049] LABUDA,E.F., AND A.M.JOHNSON:
THRESHOLD PROPERTIES OF CONTINUOUS-DUTY RARE-GAS
ION-LASER TRANSITIONS.
IEEE J.QUANT.ELECTRON. QE-2,700-701(1966).

[050] GORDON E.I., AND E.F.LABUDA:
CONTINUOUS VISIBLE LASER ACTION IN SINGLY IONIZED
ARGON,KRYPTON AND XENON.
APPL.PHYS.LETT.4,178-180(1964).

[051] BOCKASTEN,K., T.LUNDHOLM, AND O.ANDRADE:
NEW NEAR-INFRARED LASER LINES IN ARGON I.
PHYS.LETT.22,145-146(1966).

[052] LAURES,P., L.DANA, AND C.FRAPARD:
NOUVELLES TRANSITIONS LASER DANS LE DOMAINE
0.42-0.52 MICROMETRE OBTENUES A PARTIR DU SPECTRE
DU KRYPTON IONISE.
COMPT.REND.258,6363-6365(1964).

[053] DANA,L., AND P.LAURES:
STIMULATED EMISSION IN KRYPTON AND XENON IONS
BY COLLISIONS WITH METASTABLE ATOMS.
PROC.IEEE 53,78-79(1965).

[054] BRIDGES,W.B.:
LASER ACTION IN SINGLY IONIZED KRYPTON AND XENON.
PROC.IEEE 52,843-844(1964).

[055] DER AGOBIAN,R., R.L.OTTO, R.CAGNARD,
J.BARTHELEMY, AND R.ECHARD:
EMISSION STIMULEE EN REGIME PERMANENT DANS LE SPECTRE
VISIBLE DU KRYPTON IONISE.
COMPT.REND.260,6327-6329(1965).

[056] NEUSEL,R.H.:
A NEW KRYPTON LASER OSCILLATION AT 5016.4A.
IEEE J.QUANT.ELECTRON. QE-2,106(1966).

[057] NEUSEL,R.H.:
NEW LASER OSCILLATIONS IN XENON AND KRYPTON.
IEEE J.QUANT.ELECTRON. QE-2,758(1966).

[058] COTTRELL,T.H.E., D.C.SINCLAIR, AND J.M.FORSYTH:
NEW LASER WAVELENGTHS IN KRYPTON.
IEEE J.QUANT.ELECTRON. QE-2,703(1966).

[059] ROSENBERGER,D.:
SUPERSTRAHLUNG IN GEPULSTEN ARGON-,KRYPTON-UND
XENON-ENTLADUNGEN.
PHYS.LETT.14,32(1965).

[060] FABRIKANT,V.A.:
DISSERTATION,LEBEDEV INSTITUTE,
ACADEMY OF SCIENCES, USSR (1939).

[061] NEUSEL,R.H.:
NEW LASER OSCILLATIONS IN KRYPTON AND XENON.
IEEE J.QUANT.ELECTRON. QE-2,334(1966).

[062] LAURES,P., L.DANA, AND C.FRAPARD:
NOUVELLES RAIES LASER VISIBLES DANS LE XENON IONISE.
COMPT.REND.259,745-747(1964).

[063] NEUSEL,R.H.:
A NEW XENON LASER OSCILLATION AT 5401A.
IEEE J.QUANT.ELECTRON. QE-2,70(1966).

[064] PATEL,C.K.N., W.L.FAUST, AND R.A.MCFARLANE:
HIGH-GAIN GASEOUS(XE-HE)OPTICAL MASERS.
APPL.PHYS.LETT.1,84-85(1962).

[065] WALTER,W.T., AND S.M.JARRETT:
STRONG 3.27 MICRON LASER OSCILLATION IN XENON.
APPL.OPT.3,789(1964).

[066] FAUST,W.L., R.A.MCFARLANE, C.K.N.PATEL, AND C.G.B.GARRETT:
GAS-MASER SPECTROSCOPY IN THE INFRARED.
APPL.PHYS.LETT.1,85-88(1962).

[067] PAANANEN,R.A., AND D.L.BOBROFF:
VERY HIGH GAIN GASEOUS(XE-HE)OPTICAL MASER AT 3.5 MICROMETER
APPL.PHYS.LETT.2,99-100(1963).

[068] PETROV,Y.N., AND A.M.PROKHOROV:
75 MICROMETERS QUANTUM GENERATOR.
JETP LETT.1,24-25(1965).

[069] CHEO,P.K., AND H.G.COOPER:
UV AND VISIBLE LASER OSCILLATIONS IN FLUORINE,
PHOSPHORUS AND CHLORINE.
APPL.PHYS.LETT.7,202-204(1966).

[070] PALENIUS,H.P.:
THE IDENTIFICATION OF SOME SI AND CL LASER LINES
OBSERVED BY CHEO AND COOPER.
APPL.PHYS.LETT.8,82-83(1966).

[071] ZAROWIN,C.B.:
NEW VISIBLE CW LASER LINES IN SINGLY IONIZED CHLORINE.
APPL.PHYS.LETT.9,241-242(1966).

[072] FOWLES,G.R., J.A.ZURYK, AND R.C.JENSEN:
INFRARED LASER LINES IN ARSENIC VAPOR.
IEEE J.QUANT.ELECTRON. QE-10,849(1974).

[073] PAANANEN,R.A., AND F.A.HORRIGAN:
NEAR-INFRARED LASERING IN NE/CL2 AND HE/CL2.
PROC.IEEE 52,1261(1964).

[074] JARRETT,S.M., J.NUNEZ, AND G.GOULD:
LASER OSCILLATION IN ATOMIC CL AND I IN THE HCL
AND HI GAS DISCHARGES.
APPL.PHYS.LETT.8,150-151(1966).

[075] PAANANEN,R.A., C.L.TANG, AND F.A.HORRIGAN:
LASER ACTION IN CL2 AND HE/CL2.
APPL.PHYS.LETT.3,154(1963).

[076] BOCKASTEN,K.:
ON THE CLASSIFICATION OF LASER LINES IN CHLORINE AND IODINE.
APPL.PHYS.LETT.4,118(1964).

[077] KEEFFE,W.M., AND W.J.GRAHAM:
LASER OSCILLATION IN THE VISIBLE SPECTRUM OF SINGLY
IONIZED PURE BROMINE VAPOR.
APPL.PHYS.LETT.7,263-264(1965).

[078] BELL,E.W., A.L.BLOOM, AND J.P.GOLDSBOROUGH:
VISIBLE LASER TRANSITIONS IN IONIZED SELENIUM,
ARSENIC,AND BROMINE.
IEEE J.QUANT.ELECTRON. QE-1,400(1965).

[079] KEEFFE,W.M., AND W.J.GRAHAM:
OBSERVATION OF NEW BR+ LASER TRANSITIONS.
PHYS.LETT.20,643(1966).

[080] JARRETT,S.M., J.NUNEZ, AND G.GOULD:
INFRARED LASER OSCILLATION IN HBR AND HI GAS DISCHARGES.
APPL.PHYS.LETT.7,294-296(1965).

[081] JENSEN,R.C., AND G.R.FOWLES:
NEW LASER TRANSITIONS IN IODINE-INERT-GAS MIXTURES.
PROC.IEEE 52,1350(1964).

[082] FOWLES,G.R., AND R.C.JENSEN:
VISIBLE LASER TRANSITIONS IN THE SPECTRUM
OF SINGLY IONIZED IODINE.
PROC.IEEE 52,851-852(1964).

[083] WILLET,C.S., AND O.S.HEAVENS:
LASER TRANSITION AT 651.6 NM IN IONIZED IODINE.
OPTICA ACTA 13,271-273(1966).

[084] KASPAR,J.V.V., AND G.C.PIMENTEL:
ATOMIC IODINE PHOTODISSOCIATION LASER.
APPL.PHYS.LETT.5,231-233(1964).

[085] MCFARLANE,R.A.:
LASER OSCILLATION ON VISIBLE AND ULTRAVIOLET TRANSITIONS
OF SINGLY AND MULTIPLY IONIZED OXYGEN,CARBON,AND NITROGEN.
APPL.PHYS.LETT.5,91-93(1964).

[086] MCFARLANE,R.A.:
STIMULATED-EMISSION SPECTROSCOPY OF SOME DIATOMIC MOLECULES.
IN: PHYSICS OF QUANTUM ELECTRONICS.
MCGRAW HILL, NEW YORK 1969)PP.655-663.

[087] COOPER,H.G., AND P.K.CHEO:
ION-LASER OSCILLATIONS IN SULFUR.
IN: PHYSICS OF QUANTUM ELECTRONICS
(MCGRAW HILL, NEW YORK 1969)PP.690-697.

[088] PATEL,C.K.N., R.A.MCFARLANE, AND W.C.FAUST:
OPTICAL-MASER ACTION IN C,N,O,S AND BR ON DISSOCIATION
OF DIATOMIC OR POLYATOMIC MOLECULES.
PHYS.REV.133A,1244(1964).

[089] BELL,W.E., A.L.BLOOM, AND J.P.GOLDSBOROUGH:
NEW LASER TRANSITIONS IN ANTIMONY AND TELLURIUM.
IEEE J.QUANT.ELECTRON. QE-2,154(1966).

[090] ALLEN,R.B., R.B.STARNES, AND A.A.DOUGAL:
A NEW PULSED ION LASER TRANSITION IN NITROGEN AT 3995A.
IEEE J.QUANT.ELECTRON. QE-2,334(1966).

[091] SHIMAZU,M., AND Y.SUZAKI:
LASER OSCILLATIONS IN SILICON TETRACHLORIDE VAPOR.
JAPAN J.APPL.PHYS.4,819(1965).

[092] SILFVAST,W.T., G.R.FOWLES, AND B.D.HOPKINS:
LASER ACTION IN SINGLY IONIZED SN,PB,IN,CD,AND ZN.
APPL.PHYS.LETT.8,318-319(1966).

[093] COOPER,H.G., AND P.K.CHEO:
LASER TRANSITIONS IN B II BR II AND SN.
IEEE J.QUANT.ELECTRON. QE-2,785(1966).

[094] CARR,W.C., AND R.W.GROW:
A NEW LASER LINE IN TIN USING STANIC CHLORIDE VAPOR.
PROC.IEEE 55,1198(1967).

[095] FOWLES G.R., AND W.T.SILFVAST:
HIGH-GAIN LASER TRANSITION IN LEAD VAPOR.
APPL.PHYS.LETT.6,236-237(1965).

[096] FOWLES,G.R., AND W.T.SILFVAST:
LASER ACTION IN THE IONIC SPECTRA OF ZINC AND CADMIUM.
IEEE J.QUANT.ELECTRON. QE-1,131(1965).

[097] GERRITSEN,H.J., AND P.V.GOEDERTIER:
BLUE GAS LASER USING HG2+.
J.APPL.PHYS.35,3060-3061(1964).

[098] BELL,W.E.:
VISIBLE LASER TRANSITIONS IN HG+.
APPL.PHYS.LETT.4,34-35(1964).

[099] CONVERT,G., M.ARMAND, AND P.MARTINOT-LAGARDE:
EFFET LASER DANS DES MELANGES MERCURE-GAZ RARES.
COMPT.REND.258,3259-3260(1964).

[100] BLOOM,A.L., W.E.BELL, AND F.O.LOPEZ:
LASER SPECTROSCOPY OF A PULSED MERCURY-HELIUM DISCHARGE.
PHYS.REV.135A,578-579(1964).

[101] HERNQVIST,K.G., AND J.R.FENDLEY:
MODERN OPTICS.PROCEEDINGS OF A SYMPOSIUM.NEW YORK,(MAR,1967)
J.FOX,EDITOR.NEW YORK,POLYTECH.PRESS AND INTERSCIENCE PUBL.
(MICROWAVE RESEARCH INST.,SYMPOSIA SERIES,17,383-387(1968).)

[102] CORTI,M.:
PULSED NEON LASER AT 5401A WITH SUBNANOSECOND EMISSION.
OPT.COMMUN.4,373-376(1972).

[103] BROCHARD,J. ET AL:
MEASUREMENT OF THE ISOTOPE SHIFT OF SIX INFRARED ARGON
LASER LINES.
COMPT.REND.A.SC.265,467-470(1967).

[104] BOCKASTEN,K., AND O.ANDRADE:
IDENTIFICATION OF HIGH-GAIN LASER LINES IN ARGON.
NATURE 215,382(1967).

[105] JOHNSON,A.M., AND C.E.WEBB:
NEW CW-LASER WAVELENGTH IN KR II.
IEEE J.QUANT.ELECTRON. QE-3,369(1967).

[106] BRIDGES,W.B., A.N.CHESTER, A.S.HALSTED, AND J.V.PARKER:
ION-LASER PLASMAS.
PROC.IEEE 59,724-737(1971).

[107] JENNINGS,W.C., J.H.NOON, AND E.H.HOLT:
COMPARISON OF HOLLOW-CATHODE AND CONVENTIONAL
ION LASERS.
REV.SCI.INSTR.41,322-326(1970).

[108] WOOD,O.R., E.G.BURKHARDT, M.A.POLLACK, AND T.J.BRIDGES:
HIGH-PRESSURE LASER ACTION IN 13 GASES
WITH TRANSVERSE EXITATION.
APPL.PHYS.LETT.18,112-115(1971).

[109] MARANTZ,H., R.J.RUDKO, AND C.L.TANG:
THE SINGLY IONIZED KRYPTON LASER.
IEEE J.QUANT.ELECTRON. QE-5,38-44(1969).

[110] KOVACS,M.A. AND ULTEE,C.J.:
VISIBLE LASER ACTION IN FLUORINE I.
APPL.PHYS.LETT.17,39-40(1970).

[111] KOVALCHUK,V.M., AND G.G.PETRASH:
NEW GENERATION LINES OF A PULSED IODINE-VAPOR LASER.
JETP LETT.4,144-146(1966).

[112] ANDRADE,O., M.GALLARDO, AND K.BOCKASTEN:
HIGH-GAIN LASER LINES IN NOBLE GASES.
ARKIV FOER FYSIK 37,354-355(1968).

[113] HOFFMANN,V., AND P.TOSCHEK:
NEW LASER EMISSION FROM IONIZED XENON.
IEEE J.QUANT.ELECTRON. QE-6,757(1970).

[114] TELL,B.,R.J.MARTIN, AND D.MACNAIR:
CW LASER OSCILLATION IN IONIZED XENON AT 9697A.
IEEE J.QUANT.ELECTRON. QE-3,96(1967).

[115] HUGHES,W.H., J.SHANNON, AND R.HUNTER:
126.1 NM MOLECULAR ARGON LASER.
APPL.PHYS.LETT.24,488-490(1974).

[116] TANG,R.:
HIGH-REPETITION RATE XENON LASER WITH TRANSVERSE EXCITATION.
IEEE J.QUANT.ELECTRON. QE-8,166-169(1972).

[117] GENSEL,P.,K.HOHLA,AND K.L.KOMPA:
ENERGY STORAGE OF CF3J PHOTODISSOCIATION LASER.
APPL.PHYS.LETT.18,48-49(1971).

[118] JEFFERS,W.Q., AND C.E.WISWALL:
LASER ACTION IN ATOMIC FLUORINE BASED ON
COLLISIONAL DISSOCIATION OF HF.
APPL.PHYS.LETT.17,444-447(1970).

[119] KIM,K.H., R.A.PAANANEN, P.L.HANST, AND T.F.DEUTSCH:
IODINE INFRARED LASER.
IEEE J.QUANT.ELECTRON. QE-4,908(1968).

[120] BOCKASTEN,K., M.GARAVAGLIA, B.A.LENGYEL, AND T.LUNDHOLM:
LASER LINES IN HG I.
J.OPT.SOC.AM.55,1051-1053(1965).

[121] PAANANEN,R.A., C.L.TANG, F.A.HORRIGAN, AND H.STATZ:
OPTICAL-MASER ACTION IN HE-HG RF DISCHARGES.
J.APPL.PHYS.34,3148(1963).

[122] RIGDEN,J.D., AND A.D.WHITE:
OPTICAL-MASER ACTION IN IODINE AND MERCURY DISCHARGES.
NATURE 198,774(1963).

[123] WALTER,W.I., N.SOLIMENE, M.PILTCH, AND G.GOULD:
EFFICIENT PULSED GAS DISCHARGE LASERS.
IEEE J.QUANT.ELECTRON. QE-2,474-479(1966).

[124] WALTER,W.T. M.PLITCH, N.SOLIMENE, AND G.GOULD:
PULSED LASER ACTION IN ATOMIC COPPER VAPOR.
BULL.AM.PHYS.SOC.11,113(1966).

[125] PILTCH,M., W.T.WALTER, N.SOLIMENE, AND G.GOULD:
PULSED LASER TRANSITIONS IN MANGANESE VAPOR.
APPL.PHYS.LETT.7,309-310(1965).

[126] BRIDGES,W.B. AND A.M.CHESTER:
IONIZED GAS LASERS.
IN HANDBOOK OF LASERS (CRC PRESS,CLEVELAND, OHIO,1971).

[127] BIRNBAUM,M., A.W.TUCKER, J.A.GELBWACHS, AND C.L.FINCHER:
NEW OII 6649A LASER LINE.
IEEE J.QUANT.ELECTRON. QE-7,208(1971).

[128] POWELL,F.X., AND N.I.DJEU:
CW ATOMIC OXYGEN LASER AT 4.56 MICROMETER.
IEEE J.QUANT.ELECTRON. QE-7,176(1971).

[129] FLYNN,G.W., M.S.FELD, AND B.J.FELDMAN:
NEW INFRARED-LASER TRANSITIONS AND G VALUES IN
ATOMIC OXYGEN.
BULL.AM.PHYS.SOC.12,15(1967).

[130] TUNITZKI,L.N. AND E.M.CHERKASOV:
PURE-OXYGEN LASER.
SOV.PHYS.-TECH.PHYS.13,993-994(1968).

[131] PATEL,C.K.N., R.A.MCFARLANE, AND W.L.FAUST:
OPTICAL-MASER ACTION IN C,N,O,S AND BR ON DISSOCIATION
OF DIATOMIC OR POLYATOMIC MOLECULES.
PHYS.REV.133A,1244(1964).

[132] BENNETT,W.R.,JR., W.L.FAUST, R.A.MCFARLANE, AND C.K.N.PATEL:
DISSOCIATIVE EXCITATION TRANSFER AND OPTICAL MASER
OSCILLATION IN NE-O2 AND AR-O2 DISCHARGES.
PHYS.REV.LETT.8,470-474(1963).

[133] HUEBNER,G., AND C.WITTIG:
SOME NEW INFRARED LASER TRANSITIONS IN ATOMIC
OXYGEN AND SULFUR.
J.OPT.SOC.AM.61,415-416(1971).

[134] SILFVAST,W.T., AND M.B.KLEIN:
CW-LASER ACTION ON 24 VISIBLE WAVELENGTHS IN SE II.
APPL.PHYS.LETT.17,400-403(1970).

[135] KLEIN,M.B., AND W.T.SILFVAST:
NEW CW-LASER TRANSITIONS IN SE II.
APPL.PHYS.LETT.18,482-485(1971).

[136] WEBB,C.E.:
NEW PULSED LASER TRANSITIONS IN TE II.
IEEE J.QUANT.ELECTRON. QE-4,426-427(1968).

[137] SILFVAST,W.T., AND J.S.DEECH:
SIX,DB/CM SINGLE-PASS GAIN AT 7229A IN LEAD VAPOR.
APPL.PHYS.LETT.11,97(1967).

[138] ISAYEV,A.A., AND G.G.PETRASH:
NEW GENERATION AND SUPERRADIANCE LINES OF LEAD VAPOR.
JETP LETT.10,119-121(1969).

[139] SILFVAST,W.T.:
NEW CW METAL-VAPOR TRANSITION IN CD,SN AND ZN.
APPL.PHYS.LETT.15,23-25(1969).

[140] COLLINS,G.J.:
PROPERTIES OF HE/NE/ZN LASER.
J.APPL.PHYS.42,3812-3815(1971).

[141] SUGAWARA,Y., Y.TOKIWA, AND T.IIJIMA:
NEW CW LASER OSCILLATIONS IN CD-HE AND ZN-HE
HOLLOW CATHODE LASERS.
JAPAN.J.APPL.PHYS.9,1537(1970).

[142] KARABUT,E.K., V.S.MIKHALEVSKII, V.F.PAPAKIN, AND M.F.SEM:
CONTINOUS GENERATION OF COHERENT RADIATION IN A DISCHARGE
IN ZN AND CD VAPORS OBTAINED BY CATHODE SPUTTERING.
SOV.PHYS.-TECH.PHYS.14,1447-1448(1970).

[143] RISEBERG L.A., AND L.D.SCHEARER:
ON THE EXCITATION MECHANISM OF THE HE-ZN LASER.
IEEE J.QUANT.ELECTRON. QE-7,40-41(1971).

[144] JENSEN,R.C., G.J.COLLINS, AND W.R.BENNETT,JR.:
LOW-NOISE CW HOLLOW-CATHODE ZINC-ION LASER.
APPL.PHYS.LETT.18,50-51(1971).

[145] SILFVAST,W.T.:
EFFICIENT CW LASER OSCILLATION AT 4416A IN CD(II).
APPL.PHYS.LETT.13,169-171(1968).

[146] BLOOM,A.L., AND J.P.GOLDSBOROUGH:
NEW CW LASER TRANSITIONS IN CADMIUM AND ZINC ION.
IEEE J.QUANT.ELECTRON. QE-6,164(1970).

[147] SCHUEBEL,W.K.:
NEW CW CD-VAPOR LASER TRANSITIONS IN A
HOLLOW-CATHODE STRUCTURE.
APPL.PHYS.LETT.16,470-472(1970).

[148] SILFVAST,W.T., AND L.H.SZETO:
SIMPLIFIED LOW-NOISE HE-CD LASER WITH SEGMENTED BORE.
APPL.PHYS.LETT.19,445-447(1971).

[149] GOLDSBOROUGH,J.P.:
CONTINUOUS LASER OSCILLATION AT 3250A IN CADMIUM ION.
IEEE J.QUANT.ELECTRON. QE-5,133(1969).

[150] BYER R.L., W.E.BELL, E.HODGES, AND A.L.BLOOM:
LASER EMISSION IN IONIZED MERCURY: ISOTOPE SHIFT, LINE WIDTH
AND PRECISE WAVELENGTH.
J.OPT.SOC.AM.55,1598(1965).

[151] GOLDSBOROUGH,J.P., AND A.L.BLOOM:
NEAR-INFRARED OPERATING CHARACTERISTICS OF THE MERCURY-
ION LASER.
IEEE J.QUANT.ELECTRON. QE-5,459-460(1969).

[152] DEECH,J.S., AND J.H.SANDERS:
NEW SELF-TERMINATING LASER TRANSITIONS IN CALCIUM
STRONTIUM.
IEEE J.QUANT.ELECTRON. QE-4,474(1968).

[153] CAHUZAC,PH.:
NEW INFRARED LASER LINES IN BARIUM VAPOR.
PHYS.LETT.32A,150-151(1970).

[154] KEYDAN,V.F., AND V.S.MIKHALEVSKIY:
PULSE GENERATION IN BISMUTH VAPORS.
ZHURNAL PRIKLADNOY SPEKTROSKOPII 9,713(1968).

[155] BUKOWSKIY,B.L., L.G.VASIL'YEVA, L.A.SAKAYEVA,
YU.F.TOMASHEVSKIY, A.K.TOROPOV, AND YU.A.FEDOROV:
AN INSTALLATION FOR RELATIVE MEASUREMENTS OF
LASER WAVELENGTHS IN A WIDE SPECTRAL RANGE.
INSTR.AND EXP.TECH.17,831-833(1974).

[156] ISAEV,A.A., AND G.G.PETRASH:
PULSED SUPERRADIANCE AT THE GREEN LINE OF THALLIUM
IN TLI-VAPOR.
SOV.PHYS.JETP LETT.7,156-158(1968).

[157] WALTER,W.T.:
40-KW PULSED COPPER LASER.
PRESENTED AT THE ANN. MEETING,AM.PHYS.SOC.,NEW YORK,
(1967)30.1.-3.2.(PAPER FE13).

[158] SOROKIN,P.P., AND J.R.LANKARD:
INFRARED LASERS RESULTING FROM PHOTODISSOCIATION
OF CS2 AND RB2.
J.CHEM.PHYS.51,2929-2931(1969).

[159] CAHUZAC,PH.:
INFRARED LASER EMISSION FROM RARE-EARTH VAPORS.
PHYS.LETT.31A,541-542(1970).

[160] CAHUZAC,PH.:
EMISSION LASER INFRAROUGES DANS LES VAPEURS DE THULIUM
ET D'YTTERBIUM.
PHYS.LETT.27A,473-474(1968).

[161] WAYNANT,R.W., J.D.SHIPMAN,JR., R.C.ELTON, AND A.W.ALI:
LASER EMISSION IN THE VACUUM ULTRAVIOLET
FROM MOLECULAR HYDROGEN.
PROC.IEEE 52,679-684(1971).

[162] HODGSON,R.T., AND R.W.DREYFUS:
ELECTRON-BEAM EXCITATION OF VACUUM-ULTRAVIOLET
HYDROGEN LASER.
PHYS.LETT.38A,213-214(1972).

[163] PIXTON,R.H., AND G.R.FOWLES:
NEW LASER OSCILLATION IN H2 AT 7525A.
IEEE J.QUANT.ELECTRON. QE-5,478-479(1969).

[164] WAYNANT,R.W., AND A.W.ALI:
EXPERIMENTAL OBSERVATIONS AND CALCULATED BAND STRENGTHS
FOR THE D2 LYMAN-BAND LASER.
J.APPL.PHYS.42,3406-3408(1971).

[165] PIPER,J.A., G.J.COLLINS, AND C.E.WEBB:
CW LASER OSCILLATION IN SINGLY IONIZED IODINE.
APPL.PHYS.LETT.21,203-205(1972).

[166] BREDERLOW,G., K.J.WITTE, E.FILL, K.HOHLA, AND R.VOLK:
THE ASTERIX III PULSED HIGH-POWER IODINE LASER.
IEEE J.QUANT.ELECTRON. QE-12,152-155(1976).

[167] GALLARDO,M., M.GARAVAGLIA, A.A.TAGLIAFERRI,
AND E.GALLEGO LLUESMA:
ABOUT UNIDENTIFIED IONIZED XE LASER LINES.
IEEE J.QUANT.ELECTRON. QE-6,745-747(1970).

[168] HODGES,D.T., AND C.L.TANG:
NEW CW ION-LASER TRANSITIONS IN ARGON,KRYPTON AND XENON.
IEEE J.QUANT.ELECTRON. QE-6,757-758(1970).

[169] JEFFERS,W.Q.:
SINGLE-WAVELENGTH OPERATION OF A PULSED WATER-VAPOR LASER.
APPL.PHYS.LETT.11,178-180(1967).

[170] HARTMANN,B., B.KLEMAN, AND G.SPANGSTEDT:
WATER-VAPOR LASER LINES IN THE 7 MICRON REGION.
IEEE J.QUANT.ELECTRON. QE-4,296(1968).

[171] JEFFERS,W.Q., AND F.D.COLEMAN:
THE FAR-INFRARED STIMULATED-EMISSION SPECTRUM OF D2O.
PROC.IEEE 55,1222-1223(1967).

[172] KASUYA,T., A.MINOH, AND K.SHIMODA:
A NEW LASER EMISSION FROM DEUTERIUM OXIDE VAPOR.
J.PHYS.SOC.JAPAN 25,1201(1968).

[173] DYUBKO,S.F.:
A SUBMILLIMETER CW D2O-LASER.
SOV.PHYS.-TECH.PHYS.15,1216(1971).

[174] PUERTA,J., W.HERMANN, G.BOURAUEL, AND W.URBAN:
EXTENDED SPECTRAL DISTRBUTION OF LASING TRANSITIONS
IN A LIQUID COOLED CO-LASER.
APPL.PHYSICS (SUBMITTED).
ALSO PRIVATE COMMUNICATION ON CALCULATIONS OF CO-LASER
LINES USING THE IMPROVED DUNHAM COEFICIENTS OF H.KILDAL,
R.S.ENG, AND A.H.ROSS IN: J.MOL.SPECTROSC.53,479-488(1974).

[175] WOOD,O.R. AND T.Y.CHANG:
TRANSVERSE-DISCHARGE HYDROGEN HALIDE LASERS.
APPL.PHYS.LETT.20,77-79(1972).

[176] SUCHARD,S.N., R.W.F.GROSS, AND J.S.WHITTIER:
TIME-RESOLVED SPECTROSCOPY OF A FLASH-INITIATED
H2-F2 LASER.
APPL.PHYS.LETT.19,411-413(1971).

[177] COOL,T.A., AND R.R.STEPHENS:
EFFICIENT, PURELY CHEMICAL CW-LASER OPERATION.
APPL.PHYS.LETT.16,55-58(1970).

[178] GERBER,R.A., E.L.PATTERSON, L.S.BLAIR, AND R.N.GREINER:
MULTI-KILOJOULE HF LASER USING INTENSE-ELECTRON-BEAM
INITIATION OF H2-F2 MIXTURES.
APPL.PHYS.LETT.25,281-283(1974).

[179] CHESTER,A.N.:
CITED IN'LASER FOCUS,DECEMBER(1972)18-22.

[180] CHESTER,A.N., AND L.D.HESS:
STUDY OF THE HF CHEMICAL LASER BY PULSE-
DELAY MEASUREMENTS.
IEEE J.QUANT.ELECTRON. QE-8,1-13(1972).

[181] DEUTSCH,T.F.:
MOLECULAR LASER ACTION IN HYDROGEN AND DEUTERIUM HALIDES.
APPL.PHYS.LETT.10,234-236(1967).

[182] SUCHARD,S.N., AND G.C.PIMENTEL:
DEUTERIUM FLUORIDE VIBRATIONAL-OVERTONE CHEMICAL LASER.
APPL.PHYS.LETT.18,530-531(1971).

[183] DOLGOV-SAVEL'EV,G.G., V.F.ZHAROV,
YU.S.NEGANOV, AND G.M.CHUMAK:
VIBRATIONAL-ROTATIONAL TRANSITIONS
IN AN $H_2+F_2$ CHEMICAL LASER.
SOV.PHYS.JETP 34,34-37(1972).

[184] TIFFANY,W.B., R.TARG, AND J.D.FOSTER:
KILOWATT $CO_2$-GASTRANSPORT LASER.
APPL.PHYS.LETT.15,91-95(1969).

[185] GERRY,E.T.:
GASDYNAMIC LASERS.
IEEE SPECTRUM,NOV.51-58(1970).

[186] COOL,T.A.:
THE TRANSFER CHEMICAL LASER: A REVIEW OF RECENT RESEARCH.
IEEE J.QUANT.ELECTRON. QE-8,72-83(1973).

[187] WISWALL,C.E., D.P.AMES, AND T.J.MENNE:
CHEMICAL LASER DEVICE BIBLIOGRAPHY.
IEEE J.QUANT.ELECTRON. QE-9,181-188(1973).

[188] BEAULIEU A.J.:
TRANSVERSELY EXCITED ATMOSPHERIC-PRESSURE $CO_2$ LASERS.
APPL.PHYS.LETT.16,504-505(1970).

[189] SEGUIN,H.J., K.MANES, AND J.TULIP:
SIMPLE, INEXPENSIVE LABORATORY-QUALITY ROGOWSKI TEA-LASER.
REV.SCI.INSTR.43,1134-1139(1972).

[190] ROBINSON,A.M., AND D.C.JOHNSON:
A CARBON DIOXIDE LASER BIBLIOGRAPHIE,1964-1969
IEEE J.QUANT.ELECTRON. QE-6,590-605(1970).

[191] SIDDOWAY,J.C.:
CALCULATED AND OBSERVED LASER TRANSITIONS USING C(14)O(16)2.
J.APPL.PHYS.39,4854-4855(1968).

[192] WALLACE,S.C., AND R.W.DREYFUS:
CONTINUOUSLY TUNABLE XENON LASER AT 1720A.
APPL.PHYS.LETT.25,489-500(1974).

[193] ERNST,G.J., AND W.J.WITTEMAN:
TRANSITION SELECTION WITH ADJUSTABLE OUTCOUPLING
FOR A LASER DEVICE APPLIED TO CO2.
IEEE J.QUANT.ELECTRON. QE-7,484-488(1971).

[194] SADIE,F.G., P.A.BUEGER, AND O.G.MALAN:
INVESTIGATIONS ON THE CS2-O2 CHEMICAL LASER.
Z.NATURFORSCH.27A,1260-1263(1972).

[195] LIN,M.C., AND S.H.BAUER:
A CHEMICAL CO LASER.
CHEM.PHYS.LETT.7,223-225(1970).

[196] SUART,R.D., P.H.DAWSON, AND G.H.KIMBELL:
CS2/O2 CHEMICAL LASERS'
CHEMISTRY AND PERFORMANCE CHARACTERISTICS.
J.APPL.PHYS.43,1022-1032(1972).

[197] WEST,G.A. AND M.J.BERRY:
CN PHOTODISSOCIATION AND PREDISSOCIATION CHEMICAL LASERS:
MOLECULAR ELECTRONIC AND VIBRATIONAL EMISSIONS.
J.CHEM.PHYS.61,4700-4716(1974).

[198] BRUNET,H., AND M.MABRU:
ELECTRICAL CO-MIXING GAS-DYNAMIC LASER.
APPL.PHYS.LETT.21,432-433(1972).

[199] COHN,D.B.:
CO TEA LASER AT 77K.
APPL.PHYS.LETT.21,343-345(1972).

[200] SHARP,L.E., AND A.T.WETHERELL:
HIGH-POWER PULSED HCN LASER.
APPL.OPT.11,1737-1741(1972).

[201] MATHIAS,L.E.S., AND A.CROCKER:
STIMULATED EMISSION IN THE FAR-INFRARED FROM WATER
VAPOUR AND DEUTERIUM OXIDE DISCHARGES.
PHYS.LETT.13,35(1964).

[202] WITTEMAN,J.W., AND R.BLEEKRODE:
PULSED AND CONTINUOUS MOLECULAR FAR-INFRARED GAS LASER.
PHYS.LETT.13,126-127(1964).

[203] MUELLER,W.M., AND G.T.FLESHER:
CONTINUOUS-WAVE SUBMILLIMETER OSCILLATION IN H2O,
D2O AND CH3CN.
APPL.PHYS.LETT.8,217-218(1966).

[204] FLESHER,G.T., AND W.M.MUELLER:
SUBMILLIMETER GAS LASER.
PROC.IEEE 54,543-546(1966).

[205] DEUTSCH,T.F.:
LASER EMISSION FROM HF ROTATIONAL TRANSITIONS.
APPL.PHYS.LETT.11,18-20(1967).

[206] KASPAR,J.V.V., AND G.C.PIMENTEL:
HCL CHEMICAL LASER.
PHYS.REV.LETT.14,352-354(1965).

[207] HOCKER,L.O., AND A.JAVAN:
ABSOLUTE FREQUENCY MEASUREMENTS ON NEW
CW HCN SUBMILLIMETER LASER LINES.
PHYS.LETT.25A,489-490(1967).

[208] LIDE,JR.D.R., AND A.G.MAKI:
ON THE EXPLANTION OF THE SO-CALLED CN LASER.
APPL.PHYS.LETT.11,62-64(1967).

[209] POLLACK,M.A.:
LASER ACTION IN OPTICALLY PUMPED CN.
APPL.PHYS.LETT.9,230-232(1966).

[210] BARRY,J.D., W.E.BONEY, AND J.E.BRANDELIK:
CO(2-1)LASER TRANSITIONS FROM HE-AIR-CH4.
IEEE J.QUANT.ELECTRON. QE-7,461(1971).

[211] FLORIN,A.E., AND R.J.JENSEN:
PULSED-LASER OSCILLATION AT 0.7311 MICROMETER
FROM F ATOMS.
IEEE J.QUANT.ELECTRON. QE-7,472(1971).

[212] GREGG,D.W., AND S.J.THOMAS:
ANALYSIS OF THE CS2-O2 CHEMICAL LASER
SHOWING NEW LINES AND SELECTIVE EXCITATION.
J.APPL.PHYS.39,4399-4404(1968).

[213] MATHIAS,L.E.S., A.CROCKER, AND M.S.WILLS:
SPECTROSCOPIC MEASUREMENTS ON THE LASER EMISSION FROM
DISCHARGES IN COMPOUNDS OF HYDROGEN,CARBON + NITROGEN.
IEEE J.QUANT.ELECTRON. QE-4,205-208(1968).

[214] HENRY,A.:
EMISSION LASER DE L'OXYDE DE CARBONE
DANS LE SPECTRE VISIBLE.
COMPT.REND.261,1495-1497(1965).

[215] MATHIAS,L.E.S., AND J.T.PARKER:
VISIBLE LASER OSCILLATIONS FROM CARBON MONOXIDE.
PHYS.LETT.7,194(1963).

[216] PATEL,C.K.N.:
VIBRATIONAL-ROTATIONAL LASER ACTION
IN CARBON MONOXIDE.
PHYS.REV.141,71-83(1966).

[217] POLLACK,M.A.:
LASER OSCILLATION IN CHEMICALLY FORMED CO.
APPL.PHYS.LETT.8,237-238(1966).

[218] MCFARLANE,R.A., AND J.A.HOWE:
STIMULATED EMISSION IN THE SYSTEM CO/CO2.
PHYS.LETT.19,208-210(1965).

[219] PATEL,C.K.N.:
CW LASER ON VIBRATIONAL-ROTATIONAL TRANSITIONS OF CO.
APPL.PHYS.LETT.7,246-247(1965).

[220] SCHIFFNER,G.:
CALCULATION OF ACCURATE CO2-LASER TRANSITION
FREQUENCIES AND THEIR STANDARD DEVIATIONS.
OPTO-ELECTRONICS 4,215-223(1972).

[221] FLYNN,G.W., L.O.HOCKER, A.JAVAN, M.A.KOVACS, AND C.K.RHODES:
PROGRESS AND APPLICATIONS OF Q-SWITCHING
TECHNIQUES USING MOLECULAR-GAS LASERS.
IEEE J.QUANT.ELECTRON. QE-2,378-381(1966).

[222] HOCKER,L.O., M.A.KOVACS, C.K.RHODES, G.W.FLYNN, AND A.JAVAN:
VIBRATIONAL RELAXATION MEASUREMENTS IN CO2
USING AN INDUCED-FLUORESCENCE TECHNIQUE.
PHYS.REV.LETT.17,233-235(1966).

[223] LAURES,P., ET X.ZIEGLER:
LASERS MOLECULAIRES DE GRANDE PUISSANCE EN FONCIONNEMENT
CONTINUE ET EN IMPULSIONS.
J.CHEM.PHYS.64,100-106(1967).

[224] PATEL,C.K.N.:
CONTINUOUS-WAVE LASER ACTION ON
VIBRATIONAL-ROTATIONAL TRANSITIONS OF CO2.
PHYS.REV.136A,1187-1193(1964).

[225] HOWE,J.A.:
EFFECT OF FOREIGN GASES ON THE CO2 LASER
R-BRANCH TRANSITIONS.
APPL.PHYS.LETT.7,21-22(1965).

[226] PATEL,C.K.N.:
SELECTIVE EXCITATION THROUGH VIBRATIONAL ENERGY
TRANSFER AND OPTICAL-MASER ACTION IN N2-CO2.
PHYS.REV.LETT.13,617(1964).

[227] DEUTSCH,T.F.:
OSC MOLECULAR LASER.
APPL.PHYS.LETT.8,334-335(1966).

[228] MAKI,A.G.:
INTERPRETATION OF THE CS2 LASER TRANSITIONS.
APPL.PHYS.LETT.11,204-205(1967).

[229] LIDE,D.R.,JR.:
INTERPRETATION OF THE FAR-INFRARED
LASER OSCILLATION IN AMMONIA.
PHYS.LETT.24A,599-600(1967).

[230] MATHIAS,L.E.S., A.CROCKER, AND M.S.WILLS:
LASER OSCILLATIONS AT WAVELENGTHS BETWEEN 21 AND
32 MICROMETER FROM A PULSED DISCHARGE THROUGH AMMONIA.
PHYS.LETT.14,33-34(1965).

[231] KASUYA,T., AND D.R.LIDE,JR.:
MEASUREMENTS ON THE MOLECULAR NITROGEN PULSED LASER.
APPL.OPT.6,69-70(1967).

[232] SHIPMAN,J.D., AND A.C.KOLB:
A HIGH-POWER PULSED NITROGEN LASER.
IEEE J.QUANT.ELECTRON. QE-2,298(1966).

[233] DEUTSCH,T.F.:
NO MOLECULAR LASER.
APPL.PHYS.LETT.9,295-297(1966).

[234] HOWE,J.A.:
R-BRANCH LASER ACTION IN N2O.
PHYS.LETT.17,252-253(1965).

[235] PATEL,C.K.N.:
CW LASER ACTION IN N2O(N2-N2O SYSTEM).
APPL.PHYS.LETT.6,12-13(1965).

[236] MATHIAS,L.E.S., A.CROCKER, AND M.S.WILLS:
LASER OSCILLATIONS FROM NITROUS OXIDE
AT WAVELENGTHS AROUND 10,9 MICROMETERS.
PHYS.LETT.13,303-304(1964).

[237] SADIE,F.G, P.A.BUEGER, AND O.G.MALAN:
CONTINUOUS-WAVE OVERTONE BANDS IN A CS2/O2 CHEMICAL LASER.
J.APPL.PHYS.43,2906-2907(1972).

[238] HODGSON R.T.:
VACUUM-ULTRAVIOLET LASING ACTION
OBSERVED IN CO:1800-2000A.
J.CHEM.PHYS.55,5378-5379(1971).

[239] CHANG,T.Y., T.J.BRIDGES, AND E.G.BURKHARDT:
CW SUBMILLIMETER LASER ACTION IN OPTICALLY PUMPED
METHYL FLUORIDE,METHYL ALCOHOL,AND VINYL CHLORIDE GASES.
APPL.PHYS.LETT.17,249-251(1970).

[240] AKITT,D.P., AND C.F.WITTIG:
LASER EMISSION IN AMMONIA
J.APPL.PHYS.40,902-903(1969).

[241] CHANG,T.Y., T.J.BRIDGES, AND E.G.BURKHARDT:
CW LASER ACTION AT 81,5 AND 263,4 MICROMETER
IN OPTICALLY PUMPED AMMONIA GAS.
APPL.PHYS.LETT.17,357-358(1970).

[242] SKRIBANOWITZ,N., I.P.HERMAN, AND M.S.FELD:
LASER OSCILLATION AND ANISOTROPIC GAIN IN THE
1-0 VIBRATIONAL BAND OF OPTICALLY PUMPED HF GAS.
APPL.PHYS.LETT.21,466-470(1972).

[243] SKRIBANOWITZ,N., I.P.HERMAN, R.M.OSGOOD,JR.,
M.S.FELD, AND A.JAVAN:
ANISOTROPIC ULTRAHIGH GAIN EMISSION OBSERVED IN
ROTATIONAL TRANSITIONS IN OPTICALLY PUMPED HF GAS.
APPL.PHYS.LETT.20,428-431(1972).

[244] ZHUKOV,V.V., E.L.LATUSH, V.S.MIKHALEVSKII, AND M.F.SEM:
NEW LASER TRANSITIONS IN THE SPECTRUM OF
TIN AND POPULATION-INVERSION MECHANISM.
SOV.J.QUANT.ELECTRON.5,468-469(1975).

[245] ANDRADE,O., M.GALLARDO, AND K.BOCKASTEN:
NEW LASER LINES IN A PULSED N2 LASER.
APPL.OPT.6,2006(1967).

[246] MASSONE,C.A., M.GARAVAGLIA, M.GALLARDO,
J.A.E.CALATRONI, AND A.A.TAGLIAFERRI:
INVESTIGATION OF A PULSED MOLECULAR-
NITROGEN LASER AT LOW TEMPERATURES.
APPL.OPT.11,1317-1328(1972).

[247] KASLIN,V.M., AND G.G.PETRASH:
ROTATIONAL STRUCTURE OF ULTRAVIOLET
GENERATION OF MOLECULAR NITROGEN.
JETP LETT.3,55-57(1966).

[248] KASLIN,V.M., AND G.G.PETRASH:
EFFECT OF TEMPERATURE ON THE PROPERTIES OF PULSED
LASER ACTION ON ELECTRON TRANSITIONS IN DIATOMIC MOLECULES.
SOV.PHYS.KETP 27,561-567(1968).

[249] HODGSON,R.T., AND R.W.DREYFUS:
ELECTRON-BEAM EXCITATION OF THE NITROGEN LASER.
APPL.PHYS.LETT.20,195-196(1972).

[250] MASSONE,C.A., M.GARAVAGLIA, AND M.GALLARDO:
NEW IR LASER LINES IN A N2 PULSED DISCHARGE.
IEEE J.QUANT.ELECTRON. QE-5,553(1969).

[251] RAU,D.R., L.O.HOCKER, A.JAVAN, AND K.KNABLE:
SPECTROSCOPIC STUDIES OF 4.3 MICROMETER TRANSIENT
LASER OSCILLATION IN CO2.
J.MOL.SPECTROSC.25,410-411(1968).

[252] FRAPPARD,C., P.LAURES, M.ROULOT,
X.ZIEGLER, AND N.LEGAY-SOMMAIRE:
MISE EN EVIDENCE DE 85 OSCILLATIONS LASER NOUVELLES
SUR TROIS TRANSITIONS VIBRATIONELLES DE L'ANHYDRIDE
CARBONIQUE.
COMPT.REND.B 262,543-546(1966).

[253] HARTMANN,B., AND B.KLEMAN:
LASER LINES FROM CO2 IN THE 11-18 MICRON REGION.
CAN.J.PHYS.44,1609-1612(1966).

[254] GALLARDO,M., C.A.MASSONE, AND M.GARAVAGLIA:
SUPERRADIANT AND LASER SPECTROSCOPY
IN THE SECOND POSITIVE SYSTEM OF N2.
APPL.OPT.7,2418(1968).

[255] MAKI,A.G.:
ASSIGNMENT OF SOME DCN AND HCN LASER LINES.
APPL.PHYS.LETT.12,122-123(1968).

[256] MATHIAS,L.E.S., A.CROCKER, AND M.S.WILLS:
LASER OSCILLATIONS AT SUBMILLIMETER WAVELENGTHS FROM A PULSED GAS DISCHARGES IN COMPOUNDS OF HYDROGEN, CARBON AND NITROGEN.
ELECTRONICS LETT.1,45-46(1965).

[257] STEFFEN,H., AND F.K.KNEUBUEHL:
RESONATOR INTERFEROMETRY OF PULSED SUBMILLIMETER-WAVE LASERS.
IEEE J.QUANT.ELECTRON. QE-4,992-1008(1968).

[258] SHELTON,C.F., AND F.T.BYRNE:
LASER EMISSION NEAR 8 MICRONS FROM H2-C2H2-HE MIXTURE.
APPL.PHYS.LETT.17,436-437(1970).

[259] CALLEAR,A.B., AND H.E.VAN DEN BERGH:
AN HYDROXYL RADICAL INFRARED LASER.
CHEM.PHYS.LETT.8,17-18(1971).

[260] WAUCHOP,T.S., H.I.SCHIFF, AND K.H.WELGE:
PULSED DISCHARGE INFRARED OH-LASER.
REV.SCI.INSTR.45,653-655(1974).

[261] WILLET,C.S.:
NEW LASER OSCILLATIONS IN SINGLY IONIZED IODINE.
IEEE J.QUANT.ELECTRON. QE-3,33(1967).

[262] DJEU,N., AND F.X.POWELL:
MORE INFRARED LASER TRANSITIONS IN ATOMIC IODINE.
IEEE J.QUANT.ELECTRON. QE-7,537-538(1971).

[263] SOROKIN,P.P., AND J.R.LANKARD:
INFRARED LASERS RESULTING FROM GIANT-PULSE LASER EXCITATION OF ALKALI METAL MOLECULES.
J.CHEM.PHYS.54,2184-2190(1971).

[264] BYER,R.L., R.L.HERBST, AND H.KILDAL:
OPTICALLY PUMPED MOLECULAR IODINE VAPOR-PHASE LASER.
APPL.PHYS.LETT.20,463-466(1972).

[265] POLLACK,M.A.:
MOLECULAR-LASER ACTION IN NITRIC OXIDE BY PHOTODISSOCIATION OF NOCL.
APPL.PHYS.LETT.9,94-96(1966).

[266] GIULIANO,C.R., AND L.D.HESS:
CHEMICAL REVERSIBILITY AND SOLAR EXCITATION RATES OF THE NITROSYL CHLORIDE PHOTODISSOCIATIVE LASER.
J.APPL.PHYS.38,4451-4454(1967).

[267] HASSLER,J.C., AND P.D.COLEMANN:
FAR-INFRARED LASING IN H2S,OCS AND SO2.
APPL.PHYS.LETT.14,135-136(1969).

[268] LEGAY-SOMMAIRE,N.:
INTERPRETATION AND MECHANISMS OF THE CS2-N2 LASER.
APPL.PHYS.LETT.12,34-35(1968).

[269] MOELLER,G., AND J.D.RIGDEN:
OBSERVATION OF LASER ACTION IN THE R BRANCH OF
CO2 AND N2O VIBRATIONAL SPECTRA.
APPL.PHYS.LETT.8,69-70(1966).

[270] DJEU,N., AND G.J.WOLGA:
OBSERVATION OF NEW LASER TRANSITIONS IN N2O.
IEEE J.QUANT.ELECTRON. QE-5,50(1969).

[271] CHANG,T.Y., AND J.D.MC GEE:
MILLIMETER AND SUBMILLIMETER-WAVE LASER ACTION IN
SYMMETRIC-TOP MOLECULES OPTICALLY PUMPED VIA PARALLEL
ABSORPTION BANDS.
APPL.PHYS.LETT.19,103-105(1971).

[272] HUBNER,G., J.C.HASSLER, P.D.COLEMAN, AND G.STEENBECKELIERS:
ASSIGNMENT OF THE FAR-INFRARED SO2 LASER LINES.
APPL.PHYS.LETT.18,511-513(1971).

[273] HARD,T.M.:
SULFUR DIOXIDE SUBMILLIMETER LASER.
APPL.PHYS.LETT.14,130(1969).

[274] COLEMAN,P.D.:
FAR-INFRARED MOLECULAR-LASER RESEARCH.
NATL.TECH.INFO.SERVICE,AD-732946, 62 PP(1971).

[275] POURTER,G.L., AND G.BALOG:
NEW INFRARED LASER LINE IN OCS AND
NEW METHOD FOR C-ATOM LASING.
IEEE J.QUANT.ELECTRON. QE-8,917-918(1972).

[276] BURAK,J., Y.NOTER, A.M.RONN, AND A.SZOEKE:
TEA CHEMICAL LASERS FROM H2+CL2 AND H2+BR2.
CHEM.PHYS.LETT.13,322-324(1972).

[277] AIREY,J.R.:
A NEW PULSED IR CHEMICAL LASER.
IEEE J.QUANT.ELECTRON. QE-3,208(1967).

[278] NAEGELI,D.W., AND C.J.ULTEE:
A CW HCL CHEMICAL LASER.
CHEM.PHYS.LETT.6,121-122(1970).

[279] GLAZE,J.A., J.FINZI, AND W.F.KRUPKE:
A TRANSVERSE-FLOW CW HCL CHEMICAL LASER.
APPL.PHYS.LETT.18,173-175(1971).

[280] DEUTSCH,T.F.:
NEW INFRARED LASER TRANSITIONS IN HCL,
HBR,DCL AND DBR.
IEEE J.QUANT.ELECTRON. QE-3,419-421(1967).

[281] HENRY,A., F.BOURCIN, I.ARDITI, R.CHARNEAU, AND J.MENARD:
EFFECT LASER PAR REACTION CHIMIQUE DE
L HYDROGENE SUR DU CHLORE OU DU CHLORURE
DE NITROSYLE.
COMPT.REND.267 B,6161(1968).

[282] AKITT,D.P., AND J.J.YARDLEY:
FAR-INFRARED LASER EMISSION IN GAS DISCHARGES
CONTAINING BORON TRIHALIDES.
IEEE J.QUANT.ELECTRON. QE-6,113-116(1970).

[283] TIBILOV,A.S., AND A.M.SHUKHTIN:
INVESTIGATION OF GENERATION OF RADIATION IN
THE NA-HE MIXTURE.
OPT.SPECTROSC.25,221-224(1968).

[284] WALTER,W.T.:
METAL-VAPOR LASERS.
IEEE J.QUANT.ELECTRON. QE-4,355(1968).

[285] ASMUS,J.F., AND N.K.MONCUR:
PULSE BROADENING IN A MHD COPPER-VAPOR LASER.
APPL.PHYS.LETT.13,384-385(1968).

[286] FOWLES,G.R., W.T.SILFVAST, AND R.C.JENSEN:
LASER ACTION IN IONIZED SULFUR AND PHOSPHORUS.
IEEE J.QUANT.ELECTRON. QE-1,183-184(1965).

[287] CORNELL,P.H., AND G.C.PIMENTEL:
HYDROGEN-CHLORINE EXPLOSION LASER II.DCL.
J.CHEM.PHYS.49,1379-1386(1968).

[288] SCHENCK,P., AND H.METCALF:
LOW-COST NITROGEN-LASER DESIGN FOR DYE-LASER PUMPING.
APPL.OPT.12,183-186(1973).

[289] HODGSON,R.T. IN:
MOLECULAR GAS LASERS.
HANDBOOK OF LASERS,P.320-321.
CRC PRESS,CLEVELAND-OHIO(1971)

[290] HODGSON,R.T., AND R.W.DREYFUS:
VACUUM-UV LASER ACTION OBSERVED IN H2 WERNER BANDS:
1161-1240A.
PHYS.REV.LETT.28,536-539(1972).

[291] WAYNANT,R.W.:
OBSERVATIONS OF GAIN BY STIMULATED EMISSION IN
THE WERNER BAND OF MOLECULAR HYDROGEN.
PHYS.REV.LETT.28,533-535(1972).

[292] HODGSON,R.T.:
VACUUM-ULTRAVIOLET LASER ACTION OBSERVED IN THE
LYMAN BANDS OF MOLECULAR HYDROGEN.
PHYS.REV.LETT.25,494-497(1970).

[293] WAYNANT,R.W.:
VACUUM-ULTRAVIOLET LASER EMISSION FROM C IV.
APPL.PHYS.LETT.22,419-420(1973).

[294] SOSNOWSKI,T.P.:
CATHAPHORESIS IN THE HELIUM-CADMIUM LASER DISCHARGE TUBE.
J.APPL.PHYS.43,5138-5144(1969).

[295] COLLINS,G.J., R.C.JENSEN, AND W.R.BENNETT,JR.:
CHARGE-EXCHANGE EXCITATION IN THE HE-CD LASER.
APPL.PHYS.LETT.19,125-128(1971).

[296] AHMED,S.A, AND A.J.CAMPILLO:
HE-NE-CD LASER WITH TWO-COLOR OUTPUT.
PROC.IEEE 57,2084-2085(1969).

[297] HERNQVIST,K.G.:
HE-CD LASERS USING RECIRCULATION GEOMETRY.
IEEE J.QUANT.ELECTRON. QE-8,740-743(1972).

[298] BROWN,F., E.SILVER, C.L.CHASE, K.J.BUTTON, AND B.LAX:
10-W METHYL FLUORIDE LASER AT 496 MICRONS.
IEEE J.QUANT.ELECTRON. QE-8,499-500(1972).

[299] HODGSON,D.T.:
CW LASER OSCILLATION IN SINGLY IONIZED MAGNESIUM.
APPL.PHYS.LETT.18,454-456(1971).

[300] CAHUZAC,P.:
NEW INFRARED LASER LINES IN MG VAPOR.
IEEE J.QUANT.ELECTRON. QE-8,500(1972).

[301] OKAJIMA,S., AND A.MURAI:
FAR-INFRARED LASER EMISSION FROM H2CO
IN A LARGE GAS-LASER TUBE.
IEEE J.QUANT.ELECTRON. QE-8,677-678(1972).

[302] CHANG,T.Y., AND O.R.WOOD:
OPTICALLY PUMPED N2O-LASER.
APPL.PHYS.LETT.22,93-94(1973).

[303] DAUGER, A.B., AND O.M.STAFSUDD:
OBSERVATION OF CW LASER ACTION IN CHLORINE, ARGON
AND HELIUM GAS MIXTURES.
IEEE J.QUANT.ELECTRON. QE-6,572-574(1970).

[304] TARG,R.:
PULSED NITROGEN LASER AT HIGH REPETITION RATE.
IEEE J.QUANT.ELECTRON. QE-8,726-728(1972).

[305] SILFVAST,W.T., AND M.B.KLEIN:
CW LASER ACTION ON 31 TRANSITIONS IN TELLURIUM VAPOR.
APPL.PHYS.LETT.20,501-504(1972).

[306] DUCAS,T.W., L.D.GEOFFRION, R.M.OSGOOD,JR., AND A.JAVAN:
OBSERVATION OF LASER OSCILLATION IN PURE ROTATIONAL
TRANSITION OF OH AND OD FREE RADICALS.
APPL.PHYS.LETT.21,42-44(1972).

[307] PIPER,J.A., AND C.E.WEBB:
CONTINUOUS-WAVE LASER OSCILLATION IN SINGLY IONIZED ARSENIC.
J.PHYS.B:ATOM.MOL.PHYS.6,L116-L120(1973).

[308] ULTEE,C.J.:
INFRARED LASER EMISSION FROM DISCHARGES THROUGH GASEOUS
SULFUR COMPOUNDS.
J.APPL.PHYS.44,1406(1973).

[309] UL TEE,C.J.:
COMPACT PULSED DEUTERIUM FLUORIDE LASER.
IEEE J.QUANT.ELECTRON. QE-6,820(1972).

[310] ULTEE,C.J.:
COMPACT PULSED HF LASERS.
REV.SCI.INSTR.42,1174-1176(1971).

[311] ISAYEV,A.A., M.A.KAZARYAN, AND G.G.PETRASH:
EFFECTIVE PULSED COPPER-VAPOR LASER WITH HIGH AVERAGE GENERATION POWER.
JETP LETT.16,27-29(1972).

[312] DYUBKO,S.F., V.A.SVICH, AND L.D.FESENKO:
SUBMILLIMETER-BAND GAS LASER PUMPED BY A CO2-LASER.
JETP LETT.16,418-419(1972).

[313] LIS,L.:
CHARACTERISTICS OF 3S2-3P1 (LAMBDA=4218NM)
LASER ACTION IN NEON.
ACTA PHYS..POL.43A,453-459(1973).

[314] HASHINO,Y., ET AL.:
LASER OSCILLATION OF O4+ IN Z-PINCH DISCHARGE.
JAPAN.J.APPL.PHYS.12,470(1973).

[315] DJEU,N.:
CW SINGLE-LINE CO-LASER ON THE V=1-0 BAND.
APPL.PHYS.LETT.23,309-310(1973).

[316] FETTERMANN,H.R., H.R.SCHLOSSBERG, AND J.WALDMAN:
SUBMILLIMETER LASERS OPTICALLY PUMPED OFF RESONANCE.
OPT.COMMUN.6,156-159(1972).

[317] WAGNER,R.J., A.I.ZELANO, AND L.H.NGAI:
NEW SUBMILLIMETER LASER LINES IN
OPTICALLY PUMPED GAS MOLECULES.
OPT.COMMUN.8,46-47(1973).

[318] HOFFMANN,V. AND P.TOSCHEK:
ON THE IONIC ASSIGNMENT OF XENON LASER LINES.
J.OPT.SOC.AM.66,152-154(1976).

[319] MAYER,S.W., D.TAYLOR, AND M.A.KWOK:
HF CHEMICAL LASING AT HIGHER VIBRATIONAL LEVELS.
APPL.PHYS.LETT.23,434-436(1973).

[320] FOWLES,G.R., J.A.ZORYK, AND R.C.JENSEN:
INFRARED LASER LINES IN NEUTRAL ATOMIC PHOSPHORUS.
IEEE J.QUANT.ELECTRON. QE-10,394-395(1974).

[321] FETTERMAN,H.R., H.R.SCHLOSSBERG, AND C.D.PARKER:
CW SUBMILLIMETER LASER GENERATION IN
OPTICALLY PUMPED STARK-TUNED NH3.
APPL.PHYS.LETT.23,684-686(1973).

[322] LINFORD,G.J.:
NEW PULSED LASER LINES IN KRYPTON.
IEEE J.QUANT.ELECTRON. QE-9,610-611(1973).

[323] LINFORD,G.J.:
NEW PULSED AND CW LASER LINES IN THE
HEAVY NOBLE GASES.
IEEE J.QUANT.ELECTRON. QE-9,611-612(1973).

[324] LINFORD,G.J.:
HIGH-GAIN NEUTRAL LASER LINES IN PULSED
NOBLE-GAS DISCHARGES.
IEEE J.QUANT.ELECTRON. QE-8,477-482(1972).

[325] DREYFUS,R.W., AND R.T.HODGSON:
MOLECULAR-HYDROGEN LASER:1098-1613A.
PHYS.REV.A 9,2635-2647(1974).

[326] SANDERS,J.H., AND J.E.THOMSON:
NEW HIGH-GAIN LASER TRANSITIONS IN NEON.
J.PHYS.B:ATOM.MOL.PHYS.6,2177-2183(1973).

[327] KNYAZEV,I., N.LETKOHOV, AND V.G.MOVSHEV:
EFFECTIVE AND PRACTICABLE HYDROGEN VUV LASERS.
VIII. IQEC PAPER M3 (1974)

[328] WEAVER,L.A., AND E.W.SUCOV:
SUPERRADIANT EMISSION AT 5108, 5700 AND 5782A IN
PULSED COPPER DISCHARGES.
IEEE J.QUANT.ELECTRON. QE-10,140-147(1974).

[329] PETERSEN,A.B., AND M.BIRNBAUM:
THE SINGLY IONIZED CARBON LASER AT 6783, 6578 AND 5145A.
IEEE J.QUANT.ELECTRON. QE-10,468(1974).

[330] IIJIMA,T., AND Y.SUGAWARA:
NEW CW LASER OSCILLATION IN HE-ZN HOLLOW-CATHODE LASER.
J.APPL.PHYS.45,5091-5092(1974).

[331] DOWNEY,G.D., AND D.W.ROBINSON:
A SINGLE LINE, FAR-INFRARED WATER LASER.
CHEM.PHYS.LETT.24,108-110(1974).

[332] GULLBERG,K., B.HARTMANN, AND B.KLEMAN:
SUBMILLIMETER EMISSION FROM OPTICALLY PUMPED 14NH3.
PHYSICA SCRIPTA 8,177-182(1973).

[333] BROWN,F., S.KRONHEIM, AND E.SILVER:
TUNABLE FAR INFRARED METHYL FLUORIDE LASER
USING TRANSVERSE OPTICAL PUMPING.
APPL.PHYS.LETT.25,394-396(1974).

[334] DYUBKO,S.F., V.A.SVICH, AND L.D.FESENKO:
SUBMILLIMETER LASER USING FORMIC ACID VAPOR
PUMPED WITH CARBON DIOXIDE LASER.
SOV.J.QUANT.ELECTRON.3,446(1974).

[335] PLANT,T.K., P.D.COLEMAN, AND T.A.DETEMPLE:
NEW OPTICALLY PUMPED FAR-INFRARED LASERS.
IEEE J.QUANT.ELECTRON. QE-962-963(1973).

[336] ROSS,A.H.M., R.S.ENG, AND H.KILDAL:
HETERODYNE MEASUREMENTS OF [12]C[16]O,[13]C[18]O,
AND [13]C[18]O LASER FREQUENCIES:
MASS DEPENDENCE OF DUNHAM COEFFICIENT.
OPT.COMMUN.12,433-438(1974).

[337] ROH,W.B., AND K.N.RAO:
CO LASER SPECTRA.
J.MOL.SPECTROSC.49,317-321(1974).

[338] FREED,CH., A.H.M.ROSS, AND R.G.O.DONNELL:
DETERMINATION OF LASER-LINE FREQUENCIES AND
VIBRATIONAL-ROTATIONAL CONSTANTS OF THE
[12]C[18]O2,[13]C[18]O2 AND [13]C[18]O2 ISOTOPES
FROM MEASUREMENTS OF CW BEAT FREQUENCIES WITH
FAST HGCDTE PHOTODIODES AND MICROWAVE-FREQUENCY COUNTERS.
J.MOL.SPECTROSC.49,439-453(1974).

[339] CHE JEN CHEN:
MANGANESE LASER USING MANGANESE CLORIDE AS LASANT.
APPL.PHYS.LETT.24,499-500(1974).

[340] DUXBURY,G. AND H.HERMAN:
OPTICALLY PUMPED MILLIMETER LASERS.
J.PHYS.B:ATOM.MOL.PHYS.11,935-949(1978).

[341] JENNINGS,D.A., K.M.EVENSON, AND J.J.JIMENEZ:
NEW CO2 PUMPED CW FAR-INFRARED LASER LINES.
IEEE J.QUANT.ELECTRON. QE-11,637(1975).

[342] SUTTON,D.G., L.GALVAN, AND S.N.SUCHARD:
NEW LASER OSCILLATION IN THE OXYGEN ATOM.
IEEE J.QUANT.ELECTRON. QE-11,92(1975).

[343] SUTTON,D.G., L.GALVAN, P.R.VALENZULLA, AND S.N.SUCHARD:
ATOMIC LASER ACTION IN RARE-GAS SF6-MIXTURES.
IEEE J.QUANT.ELECTRON. QE-11,54-57(1975).

[344] BROWN,F., P.D.HISLOP, AND J.O.TARPINIAN:
A HIGH-POWER NARROW-LINEWIDTH LINEARLY PUMPED
CH3F OSCILLATOR-AMPLIFIER SYSTEM.
IEEE J.QUANT.ELECTRON. QE-13,445-446(1977).

[345] ANDERSON,R.S., L.SPRINGER, B.G.BRICKS, AND T.W.KARRAS:
A DISCHARGE-HEATED COPPER-VAPOR LASER.
IEEE J.QUANT.ELECTRON. QE-11,172-174(1975).

[346] RADFORD,H.E.:
NEW CW LINES FROM A SUBMILLIMETER-WAVEGUIDE LASER.
IEEE J.QUANT.ELECTRON. QE-11,213-214(1975).

[347] HODGES,D.T., R.D.REEL, AND D.H.BARKER:
LOW-THRESHOLD CW SUBMILLIMETER-AND MILLIMETER-
WAVE LASER ACTION IN CO2-LASER-PUMPED C2H4F2,C2H2F2,AND CH3O
IEEE J.QUANT.ELECTRON. QE-9,1159-1160(1973).

[348] COHN,D.R., T.FUSE, K.J.BUTTON, B.LAX, AND Z.DRUZDOWICZ:
DEVELOPMENT OF AN EFFICIENT 9-KW 496-MICROMETER
CH3F LASER OSCILLATOR.
APPL.PHYS.LETT.27,280-282(1975).

[349] IZATT,J.R., B.L.BEAN, AND G.F.CAUDLE:
ONE WATT,FAR-INFRARED CH3OH LASER.
OPT.COMMUN.14,385-387(1975).

[350] BRAU,C.A., AND J.J.EWING:
354-NM LASER ACTION ON XEF.
APPL.PHYS.LETT.27,435-437(1975).

[351] EWING,J.J., AND C.A.BRAU:
LASER ACTION ON THE 2SIGMA+1/2-2SIGMA+1/2
BANDS OF KRF AND XECL.
APPL.PHYS.LETT.27,350-352(1975).

[352] TISONE,G.C., A.K.HAYS, AND J.M.HOFFMAN:
100 MW, 248.4 NM, KRF LASER EXCITED BY
AN ELECTRON BEAM.
OPT.COMMUN.15,188-189(1975).

[353] SEARLES,S.K., AND G.A.HART:
STIMULATED EMISSION AT 281.8 NM FROM XEBR.
APPL.PHYS.LETT.27,243-245(1975).

[354] MANGANO,J.A., AND J.H.JACOB:
ELECTRON-BEAM-CONTROLLED DISCHARGE
PUMPING OF THE KRF LASER.
APPL.PHYS.LETT.27,495-498(1975).

[355] HOFF,P.W., J.C.SWINGLE, AND C.K.RHODES:
DEMONSTRATION OF TEMPORAL COHERENCE,
SPATIAL COHERENCE, AND THRESHOLD EFFECTS
IN THE MOLECULAR XENON LASER.
OPT.COMMUN.8,128-131(1973).

[356] HUGHES,W.H., J.SHANNON, AND R.HUNTER:
EFFICIENT HIGH-ENERGY-DENSITY MOLECULAR
XENON LASER.
APPL.PHYS.LETT.25,85-87(1974).

[357] HUGHES,W.H., J.SHANNON, A.KOLB, E.AULT, AND M.BHAUMIK:
HIGH-POWER ULTRAVIOLET LASER RADIATION FROM
MOLECULAR XENON.
APPL.PHYS.LETT.23,385-387(1973).

[358] GERARDO,J.B., AND A.W.JOHNSON:
HIGH PRESSURE XENON LASER AT 2730A.
IEEE J.QUANT.ELECTRON. QE-9,748-755(1973).

[359] HOFF,P.W., J.C.SWINGLE, AND C.K.RHODES:
OBSERVATION OF STIMULATED EMISSION FROM HIGH-PRESSURE
KRYPTON AND ARGON/XENON MIXTURES.
APPL.PHYS.LETT.23,245-246(1973).

[360] BARON,K.U., AND B.STADLER:
HALLOW-CATHODE-EXCITED LASER TRANSITIONS IN
CALCIUM, STRONTIUM AND BARIUM.
PRESENTED AT IX.IQEC, AMSTERDAM (1976), PAPER R9.

[361] BARON,K.U., AND B.STADLER:
NEW VISIBLE LASER TRANSITIONS IN BAI AND BAII.
IEEE J.QUANT.ELECTRON. QE-11,852-853(1975).

[362] MCNEIL,J.R., G.J.COLLINS, K.B.PERSSON, AND D.L.FRANZEN:
CW LASER ACTION IN CU II.
APPL.PHYS.LETT.27,595-598(1975).

[363] MCNEIL,J.R., G.J.COLLINS, K.B.PERSSON, AND D.L.FRANZEN:
ULTRAVIOLET LASER ACTION FROM CU II IN THE 2500-A REGION.
APPL.PHYS.LETT.28,207-209(1976).

[364] DJEU,N., AND R.BURNHAM:
OPTICALLY PUMPED CW HG LASER AT 546.1 NM.
APPL.PHYS.LETT.25,350-351(1974).

[365] ISHCHENKO,V.N., V.N.LISITSIN, A.M.RAZHEV,
V.N.STARINSKY, AND P.L.CHAPOVSKY:
THE N2+ LASER.
OPT.COMMUN.13,231-234(1975).

[366] PIPER,J.A., AND C.E.WEBB:
POWER LIMITATIONS OF THE CW HE-HG LASER
OPT.COMMUN.13,122-125(1975).

[367] PIPER,J.A., AND C.WEBB:
HIGH-CURRENT CHARACTERISTICS OF THE CONTINUOUS-WAVE
HOLLOW CATHODE HE-J2 LASER.
IEEE J.QUANT.ELECTRON. QE-12,21-25(1976).

[368] HATTORI,S., H.KANO, K.TOKUTOME, G.J.COLLINS, AND T.GOTO:
CW IODINE-ION LASER IN A POSITIVE-COLUMN DISCHARGE.
IEEE J.QUANT.ELECTRON. QE-10,530-531(1974).

[369] CHANG,T.Y., AND J.D.MCGEE:
MILLIMETER AND SUBMILLIMETER-WAVE LASER ACTION
IN SYMMETRIC-TOP MOLECULES OPTICALLY PUMPED VIA
PERPENDICULAR ABSORPTION BANDS.
IEEE J.QUANT.ELECTRON. QE-12,62-65(1976).

[370] KARLOV,N.V., YU.B.KONEV, YU.N.PETROV, A.M.PROKHOROV,
AND O.M.STEL'MAKH:
LASER BASED ON BORON TRICHLORIDE.
JETP LETT.8,12-14(1968).

[371] MARLING,J.B.:
ULTRAVIOLET ION LASER PERFORMANCE AND SPECTROSCOPY-PART I:
NEW STRONG NOBLE-GAS TRANSITIONS BELOW 2500 A.
IEEE J.QUANT.ELECTRON. QE-11,822-834(1975).

[372] DETEMPLE,T.A., AND E.J.DANIELEWICZ:
CONTINUOUS-WAVE CH3F WAVEGUIDE LASER AT 496 MICROMETER:
THEORY AND EXPERIMENT.
IEEE J.QUANT.ELECTRON. QE-12,40-47(1976).

[373] AULT,E.R., R.S.BRADFORD,JR., AND M.L.BHAUMIK:
HIGH-POWER XENON FLUORIDE LASER.
APPL.PHYS.LETT.27,413-415(1975).

[374] AKIRTAVA,O.S., V.L.DZHIKIYA, AND M.OLEINIK:
LASER UTILIZING CU I TRANSITIONS IN COPPER HALIDE VAPORS.
SOV.J.QUANT.ELECTRON.5,1001-1002(1976).

[375] BARCH,W.E., H.R.FETTERMANN, AND H.R.SCHLOSSBERG:
OPTICALLY PUMPED 15.90 MICROMETER SF6 LASER.
OPT.COMMUN.15,358-360(1975).

[376] KWOK,M.A., R.R.GIEDT, AND R.W.T.GROSS:
COMPARISON OF HF AND DF CONTINUOUS
CHEMICAL LASERS: II.SPECTROSCOPY.
APPL.PHYS.LETT.16,386-387(1970).

[377] NELSON,L.Y., C.H.FISHER, AND S.R.BYRON:
HIGH-PRESSURE CS2 ELECTRIC DISCHARGE LASER.
APPL.PHYS.LETT.25,517-520(1974).

[378] COLLINS,C.B. A.J. CUNNINGHAM, AND M.STOCKTON:
A NITROGEN-ION LASER PUMPED BY CHARGE TRANSFER.
APPL.PHYS.LETT.25,344-345(1974).

[379] SEARLES,S.K.D.:
SUPERFLUORESCENT LASER EMISSION FROM ELECTRON-BEAM-PUMPED
AR-N2 MIXTURES.
APPL.PHYS.LETT.25,735-737(1974).

[380] JOHNS,J.W.C., A.R.W.MCKELLAR, AND A.D.WEITZ:
WAVELENGTH MEASUREMENTS OF C[13]O[16] TRANSITIONS.
J.MOL.SPECTROSC.51,539-545(1974).

[381] BENEDICT,W.S., M.A.POLLACK, AND W.J.TOMLINSON,III:
THE WATER-VAPOR LASER.
IEEE J.QUANT.ELECTRON. QE-5,108-124(1969).

[382] PETERSEN,F.R., K.M.EVENSON, D.A.JENNINGS,
J.S.WELLS, K.GOTO, AND J.J.JIMENES:
FAR-INFRARED FREQUENCY SYNTHESIS WITH STABILIZED CO2-LASERS:
ACCURATE MEASUREMENTS OF THE WATER-VAPOR AND
METHYL ALCOHOL FREQUENCIES.
IEEE J.QUANT.ELECTRON. QE-11,838-843(1975).
CORRECTION TO THE ABOVE PAPER:
IEEE J.QUANT.ELECTRON, QE-12,86-87(1976).

[383] KRAMER,G. AND C.O.WEISS:
FREQUENCIES OF SOME OPTICALLY PUMPED SUBMILLMETER
LASER LINES,
APPL.PHYS.LETT.10,187-188(1976).

[384] POWELL,H.T., J.R.MURRAY, AND C.K.RHODES:
LASER OSCILLATION ON THE GREEN BANDS OF XEO AND KRO.
APPL.PHYS.LETT.25,730-732(1974).

[385] WHITFORD,B.G., K.J.SIEMSEN, H.D.RICCIUS, AND G.R.HANES:
ABSOLUTE FREQUENCY MEASUREMENTS
OF N2O LASER TRANSITIONS.
OPT.COMMUN.14,70-74(1975).

[386] ISAEV,A.A., M.A.KAZARYAN, S.V.MARKOVA, AND G.G.PETRASH:
INVESTIGATION OF PULSE INFRARED STIMULATED EMISSION
FROM BARIUM VAPOR.
SOV.J.QUANT.ELECTRON.5,285-287(1975).

[387] KLIMKIN,V.M.:
INVESTIGATION OF AN YTTERBIUM VAPOR LASER.
SOV.J.QUANT.ELECTRON.5,326-329(1975).

[388] BELLAND,P., D.VERON, AND L.B.WHITBOURN:
MODE STUDY, BEAM CHARACTERISTICS AND OUTPUT POWER OF
A CW 337 MICROMETER HCN WAVEGUIDE LASER.
J.PHYS.D:APPL.PHYS.8,2113-2122(1975).

[389] NEWMAN,L.A., AND T.A.DETEMPLE:
HIGH-PRESSURE INFRARED AR-XE LASER SYSTEM:
IONIZER-SUSTAINER MODE OF OPERATION.
APPL.PHYS.LETT.27,678-680(1975).

[390] EWING,J.J. AND C.A.BRAU:
LASER ACTION ON THE 342-NM MOLECULAR IODINE BAND.
APPL.PHYS.LETT.27,557-559(1975).

[391] BRADFORD,R.S.,JR., E.R.AULT, AND M.L.BHAUMIK:
HIGH-POWER J2 LASER IN THE 342-NM BAND SYSTEM.
APPL.PHYS.LETT.27,546-548(1975)

[392] EWING,J.J., J.H.JACOB, J.A.MANGANO, AND H.A.BROWN:
DISCHARGE PUMPING OF THE BR2* LASER.
APPL.PHYS.LETT.28,656-658(1976).

[393] HOFFMANN,J.M., A.K.HAYS, AND G.C.TISONE:
HIGH-POWER UV NOBLE-GAS-HALIDE LASER.
APPL.PHYS.LETT.28,538-539(1976).

[394] CHOU,MAU SONG AND T.A.COOL:
LASER OPERATION BY DISSOCIATION OF METAL COMPLEXES:
NEW TRANSITIONS IN AS, BI, GA, HG, IN, PB, SB, AND TL.
J.APPL.PHYS.47,1055-1061(1976).

[395] BLANEY,T.G., D.J.E.KNIGHT, AND E.K.MURRAY:
FREQUENCY MEASUREMENTS OF SOME OPTICALLY-PUMPED
LASER LINES IN CH3OD.
OPT.COMMUN.25,176-178(1978).

[396] BIGIO,I.J. AND R.F.BEGLEY:
HIGH-POWER VISIBLE LASER ACTION IN NEUTRAL ATOMIC FLUORINE.
APPL.PHYS.LETT.28,263-264(1976).

[397] SCHLOSSBERG,H.R. AND H.R.FETTERMAN:
OPTICALLY PUMPED VIBRATIONAL TRANSITION
LASER IN OCS.
APPL.PHYS.LETT.26,316-318(1975).

[398] ARTUSY,M., N.HOLMES, AND A.E.SIEGMAN:
DC-EXCITED AND SEALED-OFF OPERATION OF THE
OPTICALLY PUMPED 546.1-NM HG LASER.
APPL.PHYS.LETT.28,133-134(1976).

[399] SADIE,F.G., P.A.BUGER, AND O.G.MALAN:
A NEW CONTINUOUS-WAVE CO CHEMICAL LASER
FROM THE OXIDATION OF ACETYLENE.
Z.FUER NATURF.28A,309-310(1973).

[400] TARASENKO,U.F., A.I.FEDOROV, AND YU.I.BYCHOV:
HIGH-POWER NITROGEN LASER.
SOV.J.QUANT.ELECTRON.4,674(1974).

[401] NELSON,L.Y., G.J.MULLANEY, AND S.R.BYRON:
SUPERFLUORESCENCE IN N2 AND H2 ELECTRON-BEAM-
STABILIZED DISCHARGES.
APPL.PHYS.LETT.25,79-80(1973).

[402] DUXBURY,G., T.J.GAMBLE, AND H.HERMAN:
ASSIGNMENTS OF OPTICALLY PUMPED LASERLINES
OF 1,1-DIFLUOROETHYLENE.
IEEE TRANS.ON MICROWAVE THEORY+TECH.,MTT.22,1108-1109(1974).

[403] ZHUKOV,V.V., V.G.IL'YUSHKO, E.L.LATUSH, AND M.F.SEM:
PULSE STIMULATED EMISSION FROM BERYLLIUM VAPOR.
SOV.J.QUANT.ELECTRON.5,757-760(1975).

[404] DUNN,M.H. AND J.N.ROSS:
THE ARGON ION LASER.
PROG.QUANT.ELECTRON.4,233-269(1976).

[405] DESSERT,R.A.:
VISIBLE CW AND LONG-PULSE LASING IN S II AND S III.
GOVT.REPT.ANNOUNCE,76,NO.1(1967),AD-A017217/1GA,#,ARAC.

[406] CHAPOVSKY,P.L., V.N.LISITSYN, AND A.R.SOROKIN:
HIGH-PRESSURE GAS LASERS ON AR I, XE I, AND KR I TRANSITIONS
OPT.COMMUN.26,33-36(1976).

[407] SUTTON,D.G., L.GALVAN,JR., AND S.N.SUCHARD:
TWO-ELECTRON LASER TRANSITION IN SN(1).
GOVT.REPT.ANNOUNCE,75,NO.12(1975),AD-A008160/4GA,#,ARAC.

[408] MURRAY,J., J.C.SWINGLE, AND C.E.TURNER,JR.:
LASER OSCILLATION ON THE 292-NM BAND SYSTEM OF BR2*.
APPL.PHYS.LETT.28,530-531(1976).

[409] MCNEIL,J.R., W.L.JOHNSON, G.J.COLLINS, AND K.B.PERSSON:
ULTRAVIOLET LASER ACTION IN HE-AG AND NE-AG MIXTURES.
APPL.PHYS.LETT.29,172-174(1976).

[410] GRISCHKOWSKY,D., P.P.SOROKIN, AND J.R.LANKARD:
AN ATOMIC RYDBERG STATE 16-MICRON LASER.
IEEE J.QUANT.ELECTRON. QE-13,392-396(1977).

[411] REID,J. AND K.SIEMSEN:
NEW CO2 LASER BANDS IN THE 9 TO 11 MICRON WAVELENGTH REGION.
APPL.PHYS.LETT.29,250-251(1976).

[412] HORIUCHI,Y. AND A.MURAI:
FAR-INFRARED LASER OSCILLATIONS FROM H2CO.
IEEE J.QUANT.ELECTRON. QE-12,547-549(1976).

[413] KON,S., E.HAGIWARA, T.YANO, AND H.HIROSE:
FAR-INFRARED LASER ACTION IN OPTICALLY PUMPED CD3OD.
JAP.J.APPL.PHYS.14,731-732(1975).

[414] KON,S., T.HAGIWARA, AND H.HIROSE:
FAR-INFRARED LASER ACTION IN OPTICALLY PUMPED CH3OD.
JAP.J.APPL.PHYS.14,1861-1862(1975).

[415] MURRAY,J.R. AND H.T.POWELL:
KRCL LASER OSCILLATION AT 222 NM.
APPL.PHYS.LETT.29,252-253(1976).

[416] DYUBKO,S.F., L.D.FESENKO, U.I.BASKAKOW, AND V.A.SVICH:
USE OF THE MOLECULES CD3I, CH3I AND CD3CL AS ACTIVE
MEDIA FOR OPTICALLY PUMPED SUBMILLIMETER LASERS.
ZH.PRIKL.SPECTROSC.23,317-320(1975).

[417] BUCHWALD,M.I., C.R.JONES, M.R.FETTERMAN,
AND M.R.SCHLOSSBERG:
DIRECT OPTICALLY PUMPED MULTIWAVELENGTH CO2 LASER.
APPL.PHYS.LETT.29,300-302(1976).

[418] PETERSEN,A.B., C.WITTIG, AND S.R.LEONE:
ELECTRONIC-TO-VIBRATIONAL PUMPED CO2 LASER
OPERATING AT 4.3, 10.6, AND 14.1 MICROMETER.
J.APPL.PHYS.47,1051-1054(1976).

[419] MANUCCIA,T.J., J.A.STREGACK, N.W.HARRIS, AND B.L.WEXLER:
14 AND 16 MICRON GASDYNAMIC CO2 LASERS.
APPL.PHYS.LETT.29,360-362(1976).

[420] PUMMER,H., D.PROCH, U.SCHMAILZL, AND K.L.KOMPA:
THE GENERATION OF PARTIAL AND TOTAL VIBRATIONAL
INVERSION IN COLLIDING MOLECULAR SYSTEMS
INITATED BY IR-LASER ABSORPTION.
OPT.COMMUN.19,273-278(1976).

[421] PUMMER,M.:
INVESTIGATIONS ON THE SPECTRUM OF A PULSED
DISCHARGE-INITIATED HF-LASER.
IPP REPORT IV/31, GARCHING (1972).

[422] GENSEL,P., K.L.KOMPA, AND J.WANNER:
IF5-H2 HYDROGEN FLUORIDE CHEMICAL
LASER INVOLVING A CHAIN REACTION.
CHEM.PHYS.LETT.5,179-180(1970).

[423] PROCH,D. AND J.WANNER:
TABLES OF VIBRATIONAL ROTATIONAL TRANSITIONS
IN DIATOMIC MOLECULES PERTINENT TO CHEMICAL LASERS.
IPP REPORT IV/17, GARCHING (1971).

[424] JOHNSON,W.L., J.R.MCNEIL, AND G.J.COLLINS:
CW LASER ACTION IN THE BLUE-GREEN SPECTRAL
REGION FROM AG II.
APPL.PHYS.LETT.29,101-102(1976).

[425] OSGOOD,R.M.,JR.:
OPTICALLY PUMPED 16 MICROMETER CO2 LASER.
APPL.PHYS.LETT.28,342-345(1976).

[426] EDEN,J.G. AND S.K.SEARLES:
LASER EMISSION IN KRCL AT 222 NM.
PRESENTED AT IX. IQEC, AMSTERDAM(1976), PAPER S6.

[427] MARLING,J.B. AND D.B.LANG:
VACUUM ULTRAVIOLET LASING FROM HIGHLY IONIZED
NOBLE GASES.
PROC.III.COLL.ON ELECTRON.TRANS.LASERS,
SNOWMASS-IN-ASPEN, COLORADO, SEPT.7-10 (1976).
APPL.PHYS.LETT.31,181-184(1977).

[428] TURNER,R. AND R.A.MURPHY:
THE FAR INFRARED HELIUM LASER.
INFRARED PHYS.16,197-200(1976).

[429] DEYOUNG,J.D., W.E.WELLS, AND G.H.MILEY:
LASING IN A TERNARY MIXTURE OF HE-NE-O2
AT PRESSURES UP TO 200 TORR.
J.APPL.PHYS.47,1477-1478(1976).

[430] SUTTON,D.G.:
NEW LASER OSCILLATIONS IN THE N ATOM QUARTET MANIFOLD.
IEEE J.QUANT.ELECTRON. QE-12,315-316(1976).

[431] BURKHARD,P., H.R.LUETHI, AND W.SEELIG:
QUASI-CW LASER ACTION FROM HG-III LINES.
OPT.COMMUN.18,485-487(1976).

[432] OLSON,R.A., P.BLETZINGER, AND A.GARSCADDEN:
NEW PULSED XE-NEUTRAL LASER LINE.
IEEE J.QUANT.ELECRTRON. QE-12,316-317(1976).

[433] ANDERSON,R.S., B.G.BRICKS, T.W.KARRAS, AND L.W.SPRINGER:
DISCHARGE-HEATED LEAD VAPOR LASER.
IEEE J.QUANT.ELECTRON. QE-12,313-315(1976).

[434] HENESIAN,M.A., R.L.HERBST, AND R.L.BYER:
OPTICALLY PUMPED SUPERFLUORESCENT NA2 MOLECULAR LASER.
J.APPL.PHYS.47,1515-1518(1976).

[435] ITOH,H., H.UCHIKI, AND M.MATSUOKA:
STIMULATED EMISSION FROM MOLECULAR SODIUM.
OPT.COMMUN.18,271-273(1976).

[436] PLANT,T.K., L.A.NEWMAN, E.J.DANIELEWICZ,
T.A.DETEMPLE, AND P.D.COLEMAN:
HIGH POWER OPTICALLY PUMPED FAR INFRARED LASERS.
IEEE TRANS. ON MICROW.THEORY AND TECH. MTT-22,988-990(1974).

[437] KEILMANN,F., R.L.SHEFFIELD, J.R.R.LEITE, M.S.FELD, AND A.JAVAN:
OPTICALL PUMPING AND TUNABLE LASER SPECTROSCOPY OF THE NUE-2 BAND OF D2O.
APPL.PHYS.LETT.26,19-22(1975).

[438] PIPER,J.A.:
SIMULTANOUS CW LASER OSCILLATION ON TRANSITIONS OF CD+ AND I+ IN A HOLLOW CATHODE HE-CDI2 DISCHARGE.
OPT.COMMUN.19,189-192(1976).

[439] EVANS,D.E., L.E.SHARP, W.A.PEEBLES, AND G.TAYLOR:
FAR-INFRARED SUPER-RADIANT LASER ACTION IN HEAVY WATER.
OPT.COMMUN.18,479-484(1976).

[440] FUSS,W. AND K.HOHLA:
A CLOSED CYCLE IODINE LASER.
OPT.COMMUN.18,427-430(1976).

[441] DYUBKO,S.F., M.N.EFIMENKO, V.A.SVICH, AND L.D.FESENKO:
STIMULATED EMISSION OF RADIATION FROM OPTICALLY PUMPED VINYL BROMIDE MOLECULES.
SOV.J.QUANT.ELECTRON.6,600-601(1976).

[442] REID,R.D., W.L.JOHNSON, J.R.MCNEIL, AND G.J.COLLINS:
NEW INFRARED LASER TRANSITIONS IN AG II.
IEEE J.QUANT.ELECTRON. QE-12,778-779(1976).

[443] DAIBER,J.W. AND H.M.THOMPSON:
PERFORMANCE OF A LARGE, CW, PREEXCITED CO SUPERSONIC LASER.
IEEE J.QUANT.ELECTRON. QE-13,10-17(1977).

[444] QUICK,C.R.,JR., C.WITTIG, AND J.B.LAUDENSLAGER:
ELECTRONIC TRANSITION CN LASER.
OPT.COMMUN.18,268-270(1976).

[445] BOSCHER,J.:
EXPERIMENTELLE UND THEORETISCHE UNTERSUCHUNGEN DES ANREGUNGSMECHANISMUS DER LASERLINIE O-I 6446 AE IN GEPULSTEN SAUERSTOFF-ENTLADUNGEN BEI HOHEN E/P WERTEN.
DISS.-TU.BERLIN, D83(1977).

[446] FRY,S.M.:
OPTICALLY PUMPED MULTILINE NH3 LASER.
OPT.COMMUN.19,320-324(1976).

[447] GERSTENBERGER,D.C., R.D.REID, AND G.J.COLLINS:
HOLLOW-CATHODE ALUMINUM ION LASER.
APPL.PHYS.LETT.30,466-468(1977).

[448] WAYNANT,R.W.:
A DISCHARGE-PUMPED ARCL SUPERFLUORESCENT LASER AT 175.0 NM.
APPL.PHYS.LETT.30,234-235(1977).

[449] WAYNANT,R.W.:
J.OPT.SOC.AM.67,574(1977).

[450] KUDRYAVTSEV,YU.A. AND N.P.KUZMINA:
EXCIMER ULTRAVIOLET GAS DISCHARGE XEF, XECL, KRF LASERS.
KVANT.ELECTR.4,220-222(1977).

[451] KUDRYAVTSEV,YU.A. AND N.P.KUZMINA:
EXCIMER GAS-DISCHARGE TUNABLE ARF LASER.
APPL.PHYS.13,107-108(1977).

[452] WELLEGEHAUSEN,B., S.SHADIN, D.FRIEDE, AND H.WELLING:
CONTINUOUS LASER OSCILLATION IN DIATOMIC MOLECULAR SODIUM.
APPL.PHYS.13,97-99(1977).

[453] SMITH,D.S. AND H.D.RICCIUS:
OBSERVATION OF NEW HELIUM-NEON LASER TRANSITIONS
NEAR 1.49 MICROMETERS.
IEEE J.QUANT.ELECTRON. QE-13,366(1977).

[454] BLAU,E.J., B.F.HOCHHEIMER, J.T.MASSEY, AND A.G.SCHULZ:
IDENTIFICATION OF LASING ENERGY LEVELS
BY SPECTROSCOPIC TECHNIQUES.
J.APPL.PHYS.34,703(1963).

[455] DOWNEY,G.D., D.W.ROBINSON, AND J.H.SMITH:
A PURE-ROTATIONAL COLLISIONALLY PUMPED OH LASER.
J.CHEM.PHYS.66,1685-1688(1977).

[456] NELSON,L.Y., C.H.FISHER, S.J.HOVERSON, AND S.R.BYRON:
ELECTRON-BEAM-CONTROLLED DISCHARGE EXCITATION
OF A CO-C2H2 ENERGY TRANSFER LASER.
APPL.PHYS.LETT.30,192-195(1977).

[457] LUETHI,H.R., W.SEELIG, AND J.STEINGER:
POWER ENHANCEMENT OF CONTINUOUS ULTRAVIOLET LASERS.
APPL.PHYS.LETT.31,670-672(1977).

[458] HENNINGSEN,J.O.:
ASSIGNMENT OF LASER LINES IN OPTICALLY PUMPED CH3OH.
IEEE J.QUANT.ELECTRON. QE-13,435-441(1977).

[459] BUSHNELL,A.H. AND M.GUNDERSEN:
NEW FAR-INFRARED LASER LINES IN N[15]H3.
IEEE J.QUANT.ELECTRON. QE-12,260-261(1976).

[460] HOCKER,L.O. AND T.B.PHI:
PRESSURE DEPENDENCE OF THE ATOMIC FLUORINE
TRANSITION INTENSITIES.
APPL.PHYS.LETT.29,493-494(1976).

[461] DANIELEWICZ,E.J., E.G.MALK, AND P.D.COLEMAN:
HIGH-POWER VIBRATION-ROTATION EMISSION FROM
N[14]H3 PUMPED OFF RESONANCE.
APPL.PHYS.LETT.29,557-559(1976).

[462] REID,R.D., J.R.MCNEIL, AND G.J.COLLINS:
NEW ION LASER TRANSITIONS IN HE-AU MIXTURES.
APPL.PHYS.LETT.29,666-668(1976).

[463] HODGES,D.T., F.B.FOOTE, AND R.D.REEL:
EFFICIENT HIGH-POWER OPERATION OF THE
CW FAR-INFRARED WAVEGUIDE LASER.
APPL.PHYS.LETT.29,662-664(1976).

[464] JACOBS,R.R., D.PROSNITZ, W.K.BISCHEL, AND CH.K.RHODES:
LASER GENERATION FROM 6 TO 35 MICRONS FOLLOWING
TWO-PHOTON EXCITATION OF AMMONIA.
APPL.PHYS.LETT.29,710-712(1976).

[465] CHANG,T.Y. AND J.D.MCGEE:
OFF-RESONANT INFRARED LASER ACTION IN NH3 AND C2H4
WITHOUT POPULATION INVERSION.
APPL.PHYS.LETT.29,725-727(1976).

[466] SIEMSEN,K. AND J.REID:
NEW N2O LASER BAND IN THE 10 MICRON WAVELENGTH REGION.
OPT.COMMUN.20,284-288(1977).

[467] BURNHAM,R.:
ATOMIC INDIUM PHOTODISSOCIATION LASER AT 451 NM.
APPL.PHYS.LETT.30,132-133(1977).

[468] TIEE,J.J. AND C.WITTIG:
OPTICALLY PUMPED MOLECULAR LASERS
IN THE 11-17 MICROMETER REGION.
J.APPL.PHYS.49,61-64(1978).

[469] CHOU,M.S. AND T.A.COOL:
LASER OPERATING BY DISSOCIATION OF METAL COMPLEXES. II.
NEW TRANSITIONS IN CD, FE, NI, SE, SN, TE, V, AND ZN.
J.APPL.PHYS.48,1551-1555(1977).

[470] LATUSH,E.L., V.S.MIKHALEVSKII, M.F.SEM,
G.N.TOMACHEV, AND V.YA.KHASILOV:
METAL-ION TRANSITION LASERS WITH TRANSVERSE HF EXCITATION.
JETP LETT.24,69-71(1976).

[471] SCHUEBEL,W.K.:
LASER ACTION IN AL II AND HE I IN A SLOT CATHODE DISCHARGE.
APPL.PHYS.LETT.30,152-154(1977).

[472] ROTHE,D.E. AND K.O.TAN:
HIGH-POWER N2+ LASER PUMPED BY CHARGE TRANSFER
IN A HIGH-PRESSURE PULSED GLOW DISCHARGE.
APPL.PHYS.LETT.30,152-154(1977).

[473] PETERSEN,A.B., L.W.BRAVERMAN, AND C.WITTIG:
H2O, NO, AND N2O INFRARED LASERS PUMPED DIRECTLY AND
INDIRECTLY BY ELECTRONIC-VIBRATIONAL ENERGY TRANSFER.
J.APPL.PHYS.48,230-233(1977).

[474] LOREE,T.R. AND R.C.SZE:
THE ATOMIC FLUORINE LASER: SPECTRAL PRESSURE DEPENDENCE.
OPT.COMMUN.21,255-257(1977).

[475] RICE,J.K., A.K.HAYS, AND J.R.WOODWORTH:
VUV EMISSIONS FROM MIXTURES OF F2 AND THE NOBLE GASES -
A MOLECULAR F2 LASER AT 1575 A.
APPL.PHYS.LETT.31,31-33(1977).

[476] MARKOVA,S.V. AND V.M.CHEREZOV:
INVESTIGATION OF PULSE STIMULATED EMISSION FROM GOLD VAPOR.
SOV.J.QUANT.ELECTRON.7,339-342(1977).

[477] BASOV,N.G., I.S.DATSKEVICH, V.S.ZUEV, L.D.MIKHLEV,
A.V.STARTSEV, AND A.P.SHIROKIKH:
OPTICALLY PUMPED ULTRAVIOLET MOLECULAR IODINE LASER.
SOV.J.QUANT.ELECTRON.7,352-353(1977).

[478] SIEMSEN,K. AND J.REID:
NEW N2O LASER BAND IN THE 10 MICRON WAVELENGTH REGION.
OPT.COMMUN.20,284-288(1977).

[479] SIEMSEN,K.J. AND B.G.WHITFORD:
HETERODYNE FREQUENCY MEASUREMENTS OF
CO2 LASER SEQUENCE-BAND TRANSITIONS.
OPT.COMMUN.22,11-16(1977).

[480] PARKS,J.H.:
LASER ACTION ON THE B2SIGMA+1/2 - X2SIGMA+1/2 BAND
OF HGBR AT 5018 A.
APPL.PHYS.LETT.31,297-300(1977).

[481] PARKS,J.H.:
LASER ACTION ON THE B2SIGMA+1/2 - X2SIGMA+1/2 BAND
OF HGCL AT 5576 A.
APPL.PHYS.LETT.31,192-194(1977).

[482] CUELLAR,E., J.H.PARKER, AND G.PIMENTEL:
ROTATIONAL CHEMICAL LASERS FROM HF ELIMINATION REACTIONS.
J.CHEM.PHYS.61,422-423(1974).

[483] ENG,R.S. AND D.L.SPEARS:
FREQUENCY STABILISATION AND ABSOLUTE FREQUENCY MEASUREMENTS
OF A CW HF/DF LASER.
APPL.PHYS.LETT.27,650-652(1975).

[484] HOCKER,L.O.:
NEW INFRARED TRANSITIONS IN NEUTRAL SULFUR.
J.APPL.PHYS.48,3127-3128(1977).

[485] CHAPOVSKY,P.L., S.A.KOCHUBEI, V.N.LISITSYN, AND A.M.RAZHEV:
EXCIMER ARF/XEF LASERS PROVIDING HIGH-POWER STIMULATED
RADIATION IN AR/XE AND F LINES.
APPL.PHYS.14,231-233(1977).

[486] ANDREWS,A.J., C.E.WEBB, R.C.TOBIN, AND R.G.DENNING:
A COPPER VAPOR LASER OPERATING AT ROOM TEMPERATURE.
OPT.COMMUN.22,272-274(1977).

[487] WHITFORD,B.G., K.J.SIEMSEN, AND J.REID:
HETERODYNE FREQUENCY MEASUREMENTS OF
CO2 LASER HOT-BAND TRANSITIONS.
OPT.COMMUN.22,261-264(1977).

[488] DANGOISSE,D., A.DELDALLE, J.P.SPLINGARD, AND J.BELLET:
CW OPTICALLY PUMPED LASER ACTION IN D2CO, HDCO AND (H2CO)3.
IEEE J.QUANT.ELECTRON. QE-13,731-732(1977).

[489] MCDERMOTT,W.E., N.R.PCHELKIN, D.J.BENARD, AND R.R.BOUSEK:
AN ELECTRIC TRANSITION CHEMICAL LASER.
APPL.PHYS.LETT.32,469-470(1978).

[490] SZE,R.C. AND P.B.SCOTT:
HIGH-ENERGY LASING OF XEBR IN AN ELECTRIC DISCHARGE.
APPL.PHYS.LETT.32,479-480(1978).

[491] ROKNI,M., J.A.MANGANO, J.H.JACOBS, AND J.C.HSIA:
RARE GAS FLUORIDE LASERS.
IEEE J.QUANT.ELECTRON. QE-14,464-481(1978).

[492] OSGOOD,R.M.,JR.:
LINE-TUNABLE OPTICALLY PUMPED 16 MICRON LASER.
APPL.PHYS.LETT.32,564-566(1978).

[493] DANIELEWICZ,E.J. AND C.O.WEISS:
FAR-INFRARED LASER LINES FROM CO2 LASER-PUMPED CD3OH.
IEEE J.QUANT.ELECTRON. QE-14,458-459(1978).

[494] DYUBKO,S.F., V.A.SVICH, AND L.D.FESENKO:
EXPERIMENTAL INVESTIGATION OF THE RADIATION SPECTRUM
OF A SUBMILLIMETER CD3OH MOLECULAR LASER.
RADIOFIZIKA,18,1434-1437(1975).

[495] DANIELEWICZ,E.J. AND P.D.COLEMAN:
ASSIGNMENTS OF THE HIGH POWER OPTICALLY PUMPED
CW LASER LINES OF CH3OH.
IEEE J.QUANT.ELECTRON.QE-13,435-441(1977).

[496] CHEN,C.J., A.M.BHANJI, AND G.R.RIJSSEL:
LONG-DURATION HIGH-EFFICIENCY OPERATION OF A PULSED COPPER
LASER UTILIZING COPPER BROMIDE AS LASANT.
APPL.PHYS.LETT.33,146-148(1978).

[497] HERNQUIST,K.G.:
ONTINUOUS LASER OSCILLITON AT 2703 A IN COPPER ION.
IEEE J.QUANT.ELECTRON. QE-13,929(1977).

[498] JANOSSY,M., L.CSILLAG, K.ROZSA, AND T.SALOMON:
NEAR INFRARED CW LASER OSCILLATION IN CUII.
PHYS.LETT.50A,13-14(1974).

[499] WELLEGEHAUSEN,B., K.H.STEPHAN, D.FRIEDL, AND H.WELLING:
OPTICALLY PUMPED CONTINUOUS I2 MOLECULAR LASER.
OPT.COMMUN.23,157-161(1977).

[500] LIPTON,K.S. AND J.P.NICHOLSON:
OBSERVATION OF NEW FIR SUPERRADIANT LINES FROM
OPTICALLY PUMPED D2O.
IEEE J.QUANT.ELECTRON. QE-13,811-812(1977).

[501] PRELAS,M.A., M.A.AKERMAN, F.P.BOODY, AND G.H.MILEY:
A DIRECT NUCLEAR PUMPED 1.45 MICROMETER ATOMIC CARBON LASER
IN MIXTURES OF HE-CO AND HE-CO2.
APPL.PHYS.LETT.31,428-430(1977).

[502] EDEN,J.G.:
GREEN HGCL (B2SIGMA+ - X2SIGMA+) LASER.
APPL.PHYS.LETT.31,448-450(1977).

[503] ROZSA,K., M.JANOSSY, L.CSILLAG, AND J.BERGOU:
CW CUII LASER IN A HOLLOW ANODE-CATHODE DISCHARGE.
OPT.COMMUN.23,162-164(1977).

[504] SCHIMITSCHEK,E.J., J.E.CELTO, AND J.A.TRIAS:
MERCURIC BROMIDE PHOTODISSOCIATION LASER.
APPL.PHYS.LETT.31,609-610(1977).

[505] MARLING,J.:
ULTRAVIOLET ION LASER PERFORMANCE AND SPECTROSCOPY
FOR SULFUR, FLUORINE, CHLORINE, AND BROMINE.
IEEE J.QUANT.ELECTRON. QE-14,4-6(1978).

[506] PETERSEN,A.B. AND C.WITTIG:
LINE TUNABLE CO2 LASER OPERATING IN THE REGION 2280-2360
CM-1 PUMPED BY ENERGY TRANSFER FROM BR(42P1/2).
J.APPL.PHYS.48,3665-3668(1977).

[507] REID,R.D., D.C.GERSTENBERGER, J.R.MCNEILL, AND G.J.COLLINS:
INVESTIGATIONS OF UNIDENTIFIED LASER TRANSITIONS IN AGII.
J.APPL.PHYS.48,3994(1977).

[508] PIPER,,J.A. AND M.BRANDT:
CW LASER OSCILLATION ON TRANSITIONS OF CD+ AND ZN+ IN
HE-CD-HALIDE AND HE-ZN-HALIDE DISCHARGES.
J.APPL.PHYS.48,4486-4494(1977).

[509] BERGMAN,R.C. AND J.W.RICH:
OVERTONE BANDS LASING AT 2.7-3.1 MICROMETER IN
ELECTRICALLY EXCITED CO.
APPL.PHYS.LETT.31,597-599(1977).

[510] KOFFEND,J.B. AND R.W.FIELD:
CW OPTICALLY PUMPED MOLECULAR IODINE LASER.
J.APPL.PHYS.48,4468-4472(1977).

[511] FRANZEN,D.L., B.L.DANIELSON, AND G.W.DAY:
A SIMPLE FIRST POSITIVE SYSTEM N2 LASER FOR
USE IN OPTICAL FIBER MEASUREMENTS.
IEEE J.QUANT.ELECTRON. QE-14,402-404(1978).

[512] MARKOVA,S.V., G.G.PETRASH, AND V.M.CHEREZOV:
PULSE STIMULATED EMISSION OF THE 472.2 NM LINE
OF THE BISMUTH ATOM.
SOV.J.QUANT.ELECTRON.7,657-658(1977).

[513] KLIMKIN,V.M. AND P.D.KOLBYSCHEVA:
TUNABLE SINGLE-FREQUENCY CALCIUM-HYDROGEN LASER
EMITTING AT 5.54 MICROMETER.
SOV.J.QUANT.ELECTRON.7,1037-1039(1977).

[514] DANIELEWICZ,E.J. AND C.O.WEISS:
FAR INFRARED EMISSION FROM 15NH3 OPTICALLY PUMPED
BY A CW SEQUENCE BAND CO2 LASER.
IEEE J.QUANT.ELECTRON. QE-14,222-223(1978).

[515] VASIL'EV,B.I., A.Z.GRASYUK, AND A.P.DYADKIN:
HIGH-POWER PULSE NH3 LASER PUMPED OPTICALLY BY
CO2 LASER RADIATION.
SOV.J.QUANT.ELECTRON.7,1027-1028(1977).

[516] FONTAINE,B. AND B.FORESTIER:
LONG PULSE ULTRAVIOLET LASER EMISSIONS IN AN AERODYNA-
MICALLY COOLED ELECTRON-BEAM EXCITED NE-NF3 MIXTURE.
OPT.COMMUN.26,243-247(1978).

[517] MALK,E.G., J.W.NIESEN, D.F.PARSONS, AND P.D.COLEMAN:
LASER EMISSION IN THE 63-223 MICROMETER REGION
FROM PH3 WITH LASER LINE ASSIGNMENTS.
IEEE J.QUANT.ELECTRON. QE-14,544-550(1978).

[518] WODARCZYK,F.J. AND H.R.SCHLOSSBERG:
AN OPTICALLY PUMPED MOLECULAR BROMINE LASER.
J.CHEM.PHYS.61,4476-4482(1977).

[519] LEONE,S.R. AND K.G.KOSNIK:
A TUNABLE VISIBLE AND ULTRAVIOLET LASER ON S2.
APPL.PHYS.LETT.30,346-348(1977).

[520] ZUEV,V.S., L.D.MIKHEEV, AND V.I.YALOVOI:
PHOTOCHEMICAL LASER UTILIZING THE 1SIGMA+G-3SIGMA+
VIBRONIC TRANSITION IN S2.
SOV.J.QUANT.ELECTRON.5,442-446(1975).

[521] HARTMANN,B., B.KLEMAN, AND O.STEINVALL:
QUASI-TUNABLE I2-LASER FOR ABSORPTION MEASUREMENTS
IN THE INFRARED.
OPT.COMMUN.21,33-36(1977).

[522] LLUESMA,E.G., A.A.TAGLIAFERRI, C.MASSONE,
M.GARAVAGLIA, AND M.GALLARDO:
COMMENTS ON THE IONIC ASSIGNMENT OF XENON
LASER LINES AT 3306 A.
EEE J.QUANT.ELECTRON. QE-13,809-810(1977).

[523] WEISS,C.O., M.GRINDA, AND K.SIEMSEN:
FIR LASER LINES OF CH3OH PUMPED BY
CO2 LASER SEQUENCE LINES.
IEEE J.QUANT.ELECTRON. QE-13,892(1977).

[524] LIN,M.C.:
PHOTOEXCITATION AND PHOTODISSOCIATION LASERS - PART I:
NITRIC OXIDE LASER EMISSIONS RESULTING FROM
C(2PI)-A(2SIGMA+) AND D(2SIGMA+)-A(2SIGMA+) TRANSITIONS.
IEEE J.QUANT.ELECTRON. QE-10,516-521(1974).

[525] ZIEGLER,G. AND U.DUERR:
SUBMILLIMETER LASRR ACTION OF CW PUMPED CD2CL,
CH2DOH, AND CHD2OH.
IEEE J.QUANT.ELECTRON. QE-14,708(1978).

[526] DANIELEWICZ,J. AND C.O.WEISS:
NEW EFFICIENT CW FAR-INFRARED OPTICALLY PUMPED CH2F2 LASER.
IEEE J.QUANT.ELECTRON. QE-14,705-707(1978).

[527] NERESON,N.G. AND H.FLICKER:
WAVENUMBER MEASUREMENT OF WEAK CO2 LASER LINES
AROUND 10.6 MICROMETERS.
OPT.COMMUN.23,171-176(1977).

[528] KOMINE,H. AND R.L.BYER:
OPTICALLY PUMPED ATOMIC PHOTODISSOCIATION LASER.
J.APPL.PHYS.48,2505-2508(1977).

[529] ROSZA,K., M.JANOSSY, J.BERGOU, AND L.CSILLAG:
NOBLE GAS MIXTURE CW HOLLOW CATHODE LASER
WITH INTERNAL ANODE SYSTEM.
OPT.COMMUN.23,15-18(1977).

[530] WEXLER,B.L., T.J.MANUCCIA, AND R,W.WAYNANT:
CW AND IMPROVED PULSED OPERATION OF THE
14- AND 16-MICROMETER CO2 LASERS.
APPL.PHYS.LETT.31,730-732(1977).

[531] MONCHALIN,J.-P., M.J.KELLY, J.E.THOMAS,
N.A.KURNIT, AND A.JAVAN:
ACCURATE WAVELENGTH MEASUREMENT OF P-BRANCH TRANSITIONS
OF THE 011-110 BAND OF C(12)O(16)2 AND
DETERMINATION OF BAND PARAMETERS.
J.MOL.SPECTROSC.64,491-494(1977).

[532] FAHLEN,T.S.:
HIGH-PULSE-RATE 10-W KRF LASER.
J.APPL.PHYS.49,455-456(1978).

[533] PROSNITZ,D., R.R.JACOBS, W.K.BISCHEL, AND C.V.RHODES:
STIMULATED EMISSION AT 9.75 MICROMETERS
FOLLOWING TWO-PHOTON EXCITATION OF METHYL-FLUORIDE.
APPL.PHYS.LETT.32,221-223(1978).

[534] TANG,K.Y., R.O.JUNTER,JR., J.OLDENETTEL, C.HOWTON,
D.HUESTIS, D.ECKSTROM, B.PERRY, AND M.MCCUSKER:
ELECTRON-BEAM-CONTROLLED DISCHARGE HGCL LASER.
APPL.PHYS.LETT.32,226-228(1978).

[535] DEROCHE,J.-C.:
ASSIGNMENTS OF SUBMILLIMETER LASER LINES IN METHYL CHLORIDE.
J.MOL.SPECTROSC.69,19-24(1978).

[536] CHOU,M.S. AND G.A.ZAWADZKAS:
OBSERVATION OF NEW ATOMIC LASER TRANSITION AT 9064 A.
OPT.COMMUN.26,92(1978).

[537] WARNER,B.E., D.C.GERSTENBERGER, R.D.REID, J.R.MCNEIL,
R.SOLANKI, K.B.PERSSON, AND G.J.COLLINS:
1W OPERATION OF SINGLY IONIZED SILVER AND COPPER LASERS.
IEEE J.QUANT.ELECTRON. QE-14,568-570(1978).

[538] MIKHEEV,L.D., A.P.SHIROKIKH, A.V.STARTSEV, AND V.S.ZUEV:
OPTICALLY PUMPED MOLECULAR-IODINE LASER ON THE 342-NM BAND.
OPT.COMMUN.26,237-239(1978).

[539] DANIELEWICZ,E.J. AND C.O.WEISS:
NEW CW FAR-INFRARED D2O, 12CH3F AND 14NH3 LASER LINES.
OPT.COMMUN.27,98-100(1978).

[540] PLATONOV,A.V., A.N.SOLDATOV, AND A.G.FILONOV:
PULSED STRONTIUM VAPOR LASER.
SOV.J.QUANT.ELECTRON.8,120-121(1978).

[541] PACHEVA,Y., M.STEFANOVA, AND P.PRAMATAROV:
CW LASER OSCILLATIONS ON THE KRII 4694 A AND KRII 4318 A
LINES IN A HOLLOW-CATHODE HE-KR DISCHARGE.
OPT.COMMUN.27,121-122(1978).

[542] RICHARDSON,R.J. AND C.E.WISWALL:
COMBUSTOR-DRIVEN CO CHEMICAL LASER.
APPL.PHYS.LETT.33,296-298(1978).

[543] YOSHIDA,T., N.YAMABASHI, K.MIYAZIKI, AND K.FUJISAWA:
INFRARED AND FAR-INFRARED LASER EMISSIONS.
FROM A TE CO2 LASER PUMPED NH3 GAS.
OPT.COMMUN.26,410-414(1978).

[544] WOODWORTH,J.R. AND J.K.RICE:
AN EFFICIENT, HIGH-POWER F2 LASER NEAR 157 NM.
J.CHEM.PHYS.69,2500-2504(1978).

[545] HOLMES,N.C. AND A.E.SIEGMAN:
THE OPTICALLY PUMPED MERCURY VAPOR LASER.
APPL.PHYS.49,3155-3170(1978).

[546] FELDMAN,D.W., C.S.LIU, J.L.PACK, AND L.A.WEAVER:
LONG-LIVED LEAD-VAPOR LASERS.
J.APPL.PHYS.49,3679-3683(1978).

[547] SLADE,P.D.:
GAS PURITY IN THE LARGE-SCALE XENON LASER.
IEEE J.QUANT.ELECTRON. QE-14,637-638(1978).

[548] DELDALLE,A., D.DANGOISSE, J.P.SPLINGARD, AND J.BELLET:
ACCURATE MEASUREMENTS OF CW OPTICALLY PUMPED FIR LASER
LINES OF FORMIC ACID MOLECULES AND ITS ISOTOPIC
SPECIES HC[13]OOH, HCOOD AND DCOOD.
OPT.COMMUN.22,333-336(1977).

[549] JONES,C.R., M.I.BUCHWALD, M.GUNDERSEN, AND H.H.BUSHNELL:
AMMONIA LASER OPTICALLY PUMPED WITH AN HF LASER.
OPT.COMMUN.24,27-30(1978).

[550] GRINDA,M. AND C.O.WEISS:
NEW FAR INFRARED LASER LINES FROM CD3OH.
OPT.COMMUN.26,91(1978).

[551] WELLEGEHAUSEN,B., D.FRIEDE, AND G.STEGER:
OPTICALLY PUMPED CONTINUOUS BI2 AND TE2 LASERS.
OPT.COMMUN.26,391-395(1978).

[552] RUTT,H.N. AND J.M.GREEN:
OPTICALLY PUMPED LASER ACTION IN DIDEUTEROACETYLENE.
OPT.COMMUN.26,422-426(1978).

[553] SCHIMITSCHEK,E.J. AND J.E.CELTO:
MERCURIC BROMIDE DISSOCIATION LASER IN AN
ELECTRIC DISCHARGE.
OPT.LETT.2,64-66(1978).

[554] MARKOVA,S.V., G.G.PETRASH, AND V.M.CHEREZOV:
ULTRAVIOLET-EMITTING GOLD VAPOR LASER.
SOV.J.QUANT.ELECTRON.8,904-906(1978).

[555] HENNINGSEN,J.O.:
NEW FIR LASERLINES FROM OPTICALLY PUMPED CH3OH:
MEASUREMENTS AND ASSIGNMENTS.
IEEE J.QUNAT.ELECTRON. QE-14,958-962(1978).

[556] DANIELEWICZ,E.J. AND F.KEILMANN:
HIGH-POWER FAR-INFRARED EMISSION FROM
CO2 LASER-PUMPED D2S AND HDS.
IEEE J.QUANT.ELECTRON. QE-15,8-11(1979).

[557] BOKHAN,P.A., V.M.KLIMKIN, V.E.PROKOPIEV, AND V.I.SOLOMOMOV:
INVESTIGATION OF A LASER UTILIZING SELF-TERMINATING
TRANSITIONS IN EUROPIUM ATOMS AND IONS.
SOV.J.QUANT.ELECTRON. 7,81-82(1977).

[558] BOKHAN,P.A., V.M.KLIMKIN, AND V.E.PROKOP'EV:
COLLISION GAS-DISCHARGE LASER UTILIZING EUROPIUM VAPOR. I:
OBSERVATION OF SELF-TERMINATING OSCILLATIONS AND TRANSI-
TIONS FROM CYCLIC TO QUASICONTINUOUS OPERATION.
SOV.J.QUANT.ELECTRON. 4,752-754(1974).

[559] DOMNIN,YU.S., V.M.TATARENKOV, AND P.S.SHUMYATSKII:
NEW EMISSION LINES OF A CH3OH LASER PUMPED
BY CO2 LASER RADIATION.
SOV.J.QUANT.ELECTRON. 4,401-402(1974).

[560] LINEVSKY,M.J. AND T.W.KARRAS:
AN IRON-VAPOR LASER.
APPL.PHYS.LETT.33,720-721(1978).

[561] TRAINOR,D.W. AND S.A.MANI:
ATOMIC CALCIUM LASER
PUMPED VIA COLLISION-INDUCED ABSORPTION.
APPL.PHYS.LETT.33,648-650(1978).

[562] PIPER,J.A. AND D.F.NEELY:
CW LASER OSCILLATION ON TRANSITIONS OF CU+
IN HE-CU HALIDE GAS DISCHARGES.
APPL.PHYS.LETT.33,621-623(1978).

[563] TRAINOR,D.W. AND S.A.MANI:
IRON PENTACARBONYL PHOTODISSOCIATION LASER.
APPL.PHYS.LETT.33,31-33(1978).

[564] POWELL,H.T. AND J.J.EWING:
PHOTODISSOCIATION LASERS USING FORBIDDEN TRANSITIONS
OF SELENIUM ATOMS.
APPL.PHYS.LETT.33,165-167(1978).

[565] WEST,W.P. AND H.P.BROIDA:
OPTICALLY PUMPED VAPOR PHASE BI2 LASER.
CHEM.PHYS.LETT.56,283-285(1978).

[566] MARLING,J.B.
PRIVATE COMMUNICATION.

[567] BADCOCK,C.C., W.C.HWANG, J.F.KALSCH, AND R.F.KAMADA:
ISOTOPIC HCL TRANSFER LASER.
APPL.PHYS.LETT.32,363-364(1978).

[568] CHANG,T.Y., T.C.DAMEN, V.T.NGUYEN, J.D.MCGEE,
AND T.J.BRIDGES:
DYNAMIC STARK-TUNED FIR LASERS IN RB.
APPL.PHYS.LETT.32,633-635(1978).

[569] JAIN,K.:
NEW ION LASER TRANSITIONS IN COPPER, SILVER, AND GOLD.
APPL.PHYS.LETT34,398-399(1979).

[570] SOLANKI,R., E.L.LATUSH, W.M.FAIRBANK,JR., AND G.J.COLLINS:
NEW INFRARED LASER TRANSITIONS IN COPPER AND SILVER
HOLLOW CATHODE DISCHARGES.
APPL.PHYS.LETT.34,568-570(1979).

[571] BISCHEL,W.K., H.H.NAKANO, D.J.ECKSTRUM, R.M.HILL,
D.C.HUESTIS, AND D.C.LORENTS:
A NEW BLUE-GREEN EXCIMER LASER IN XEF.
APPL.PHYS.LETT.34,565-567(1979).

[572] REDON,M., C.GASTAUD, AND M.FOURIER:
FAR INFRARED EMISSIONS IN NH3 USING
'FORBIDDEN' TRANSITIONS PUMPED BY A CO2 LASER.
OPT.COMMUN.30,95-96(1979).

[573] BISCHEL,W.K., J.BOKOR, AND C.K.RHODES:
GENERATION OF 16-MICRON RADIATION IN 14NH3 BY
TWO-PHOTON EXCITATION OF THE 2NUE-2(7,5)STATE.
J.APPL.PHYS.50,3867-3870(1979).

[574] MAKI,A.G.:
FURTHER ASSIGNMENTS FOR THE FAR-INFRARED LASER
TRANSITIONS OF HCN AND HC[15]N.
J.APPL.PHYS.49,7-11(1978).

[575] DANGOISSE,D., E.WILLEMOT, A.DELDALLE, AND J.BELLET:
ASSIGNMENT OF HCOOH CW-SUBMILLIMETER LASER.
OPT.COMMUN.28,111-116(1979).

[576] BARANOV,V.YU., S.A.KAZAKOV, V.D.PIS'MENNY, A.I.STORODUBTSEV, E.P.VELIKHOV, YU.A.GORUKHOV, V.S.LETOKHOV, A.P.DYADKIN, A.Z.GRASIUK, AND B.I.VASIL'YEV:
MULTIWATT OPTICALLY PUMPED AMMONIA LASER OPERATION IN THE 12-13 MICROMETER BAND.
APPL.PHYSICS,17,317-320(1978).

[577] PUMMER,H. K.HOHLA, M.DIEGELMANN, AND J.P.REILLY:
DISCHARGE PUMPED F2 LASER AT 1580 A.
OPT.COMMUN.28,104-106(1979).

[578] POWELL,H.T., D.PROSNITZ, AND B.R.SCHLEICHER:
SULFUR 1S0-1D2 LASER BY OCS PHOTODISSOCIATION.
APPL.PHYS.LETT.34,571-573(1979).

[579] POWELL,H.T., J.R.MURRAY, AND C.K.RHODES:
COLLISION-INDUCED AURORAL LINE LASERS.
IEEE J.QUANT.ELECTRON. QE-11,27D(1975).

[580] LUND,M.W. AND J.A.DAVIS:
NEW CW FAR-INFRARED LASER LINES FROM CO2 LASER-PUMPED CH3OD.
IEEE J.QUANT.ELECTRON. QE-15,537-538(1979).

[581] JAIN,K.:
CW LASER OSCILLATION AT 8096 A IN CU II IN A HOLLOW CATHODE DISCHARGE.
OPT.COMMUN.28,207-208(1979).

[582] GRUZEVA,M.G., N.V.SABOTINOW, AND N.K.VUCHKOV:
CW LASER GENERATION ON TI II IN A HOLLOW-CATHODE NE-TI DISCHARGE.
OPT.COMMUN.29,339-340(1979).

[583] SOLANKI,R., G.J.COLLINS, W.M.FAIRBANK,JR.:
IR LASER TRANSITION IN A NICKEL HOLLOW CATHODE DISCHARGE.
IEEE J.QUANT.ELECTRON. QE-15,525(1979).

[584] JAIN,K.:
A NICKEL-ION LASER.
APPL.PHYS.LETT. 34,845(1979).

[585] HEMMATI,H. AND G.J.COLLINS:
ATOMIC GALLIUM PHOTODISSOCIATION LASER.
APPL.PHYS.LETT.34,844-845(1979).

[586] SUMIDA,S., M.OBARA, AND T.FUJIOKA:
NOVEL NEUTRAL ATOMIC FLUORINE LASER LINES IN A HIGH PRESSURE MIXTURE OF F2 AND HE.
J.APPL.PHYS.50,3884-3887(1979).

[587] MCDOWELL,R.S., CH.W.PATTERSON, C.R.JONES, M.I.BUCHWALD, AND J.M.TELLE:
SPECTROSCOPY OF THE CF4 LASER.
OPT.LETT.4,274-276(1979).

[588] BARANOV V.YU., B.I.VASIL'EV, E.P.VELIKHOV, YU.A.GOROKHOV, A.Z.GRASYUK, A.P.DYADKIN,S.A.KAZAKOV, V.S.LETOKHOV, V.D.PIS'MENNY, AND A.I.STARODUBTSEV:
PULSE-PERIODIC OPERATION OF AN OPTICALLY PUMPED CF4 LASER WITH AN AVERAGE OUTPUT POWER OF 0.2 W.
SOV.J.QUANT.ELECTRON.8,544-546(1978).

[589] KNYAZEV,J.N., V.S.LETOKHOV, AND LOBKO:
WEAKLY FORBIDDEN VIBRATION-ROTATION TRANSITIONS AR NOT EQUAL 0 IN CF4 LASER.
OPT.COMMUN.29,73-76(1979).

[590] SCALABRIN,A. AND K.M.EVENSON:
ADDITIONAL CW FIR LASER LINES FROM OPTICALLY PUMPED CH2F2.
OPT.LETT.4,277-279(1979).

[591] DANIELEWICZ,E.J., F.A.GALATOWICZ, F.B.FOOTE, R.D.REEL, AND D.T.HODGES:
OPT.LETT.4,280-282(1979).

[592] SPENCER,D.J. AND C.WITTIG:
ATOMIC BROMINE ELECTRONIC-TRANSITION CHEMICAL LASER.
OPT.LETT.4,1-3(1979).

[593] INGUSCIO,M., F.STRUMIA, K.M.EVENSON, D.A.JENNINGS, A.SCALABRIN, AND S.R.STEIN:
FAR-INFRARED CH3F STARK LASER.
OPT.LETT.4,9-11(1979).

[594] SILFVAST,W.T., L.H.SZETO, AND O.R.WOOD II:
RECOMBINATION LASERS IN ND AND CO2 LASER-PRODUCED PLASMAS.
OPT.LETT.4,271-273(1979).

[595] HAYS,A.K.:
LASER OSCILLATION ON THE 2580 A BAND SYSTEM OF MOLECULAR CHLORINE.
OPT.COMMUN.28,209-212(1979).

[596] DIEGELMANN,M., K.HOHLA, AND K.L.KOMPA:
INTERHALOGEN UV LASER ON THE 285 NM BAND OF CLF.
OPT.COMMUN.29,334-335(1979).

[597] HERMAN,H. AND B.E.PREWER:
NEW FIR LASER LINES FROM OPTICALLY PUMPED METHANOL ANALOGUES.
APPL.PHYSICS,19,241-242(1978).

[598] REDON,M., C.GASTAUD, AND M.FOURRIER:
NEW CW FAR-INFRARED LASING IN 14NH3 USING STARK TUNING.
IEEE J.QUANT.ELECTRON. QE-15,412-414(1979).

[599] PONTNAU,J., J.M.LOURTIAU, AND C.MEYER:
SUBMILLIMETER LASER-ACTION OF CW OPTICALLY PUMPED CF3BR.
IEEE J.QUANT.ELECTRON. QE-15,1088-1090(1979).

[600] SOLANKI,R., E.L.LATUSH, D.C.GERSTENBERGER, W.M.FAIRBANK,JR., AND G.J.COLLINS:
HOLLOW-CATHODE EXCITATION OF ION LASER TRANSITIONS IN NOBLE-GAS MIXTURES.

[601] EICHLER,H.J., H.J.KOCH, J.SALK, AND G.SCHAEFER:
PERFORMANCE OF CU '' LASERS WITH ZYLINDRICAL
HOLLOW CATHODES.
IEEE J.QUANT.ELECTRON. QE-15,908-912(1979).

[602] POPP,H.-P. AND E.SCHMIDT:
A WHITE-LIGHT HESE+ LASER.
IEEE J.QUANT.ELECTRON. QE-15,840-842(1979).

[603] CW OSCILLATION ON A NEW ARGON ION LASER LINE AT 5062 A
AND RELATION TO LASER RAMAN SPECTROSCOPY.
IEEE J.QUANT.ELECTRON.QE-15,842-843(1979).

[604] EHRLICH,D.J. AND R.M.OSGOOD,JR.:
ALKALI-METAL RESONANCE-LINE LASERS BASED
ON PHOTODISSOCIATION.
APPL.PHYS.LETT.34,655-658(1979).

[605] LEE,W., D.KIM, E.MALK, AND J.LEAP:
HOT-BAND LASING IN NH3.
IEEE J.QUANT.ELECTRON. QE-15,838-839(1979).

[606] PENCE,W.H. AND S.R.LEONE:
CASCADE LASING AND COLLISIONAL EXCITATION TRANSFER
IN OPTICALLY PUMPED CALCIUM ATOMS.
IEEE J.QUANT.ELECTRON. QE-15,900-910(1979).

[607] BEAN,B.L. AND S.PERKOWITZ:
COMPLETE FREQUENCY COVERAGE FOR SUBMILLIMETER
LASER SPECTROSCOPY WITH OPTICALLY PUMPED
CH3OD, CD3OD, AND CH2F2.
OPT.LETT.1,202-204(1979).

[608] DALE,R.M., M.HERMAN, J.W.C.JOHNS, A.R.W.MCKELLAR,
S.NAGLER, AND I.K.M.STRATHY:
IMPROVED LASER FREQUENCIES AND DUNHAM COEFFICIENTS
FOR ISOTOPICALLY SUBSTITUTED CARBON MONOXIDE.
CAN.J.PHYS.57,677-686(1979).

[609] PUERTA,J. W.HERMANN, G.BOURAUEL, AND W.URBAN:
EXTENDED SPECTRAL DISTRIBUTION OF LASING TRANSITIONS
IN A LIQUID NITROGEN COOLED CO-LASER.
APPL.PHYS.19,439-440(1979).

[610] WEXLER,B.L. AND R.W.WAYNANT:
HIGH-AVERAGE-POWER 16-MICROMETER GASDYNAMIC CO2-LASER
USING A MULTIPASS CAVITY.
APPL.PHYS.LETT.34,674-677(1979).

[611] DAVIES,P.B. AND H.JONES:
NEW CW FAR INFRARED MOLECULAR LASERS FROM
CLO2, HCCF, FCN, CH3NC, CH3F, AND PROPYNAL.
APPL.PHYS.22,53-55(1980).

[612] KOFFEND,B.J., R.BACIS, AND R.W.FIELD:
CW OPTICALLY PUMPED IODINE LASER. II. SPECTROSCOPY AND
LONG RANGE ANALYSIS OF THE X1SIGMA+G GROUND STATE OF I2.
J.MOL.SPECTROSC.77,202-212(1979).

[613] WEI,J. AND J.TELLINGHUISEN:
PARAMETRIZING DIATOMIC SPECTRA:
"BEST" SPECTROSCOPIC CONSTANTS FOR THE I2 B-X TRANSITION.
J.MOL.SPECTROSC.50,317-332(1974).

[614] FIELD,R.W.:
PRIVATE COMMUNICATION.

[615] GERSTENKORN,S. AND P.LUC:
ASSIGNMENTS OF SEVERAL GROUPS OF IODINE (I2) LINES
IN THE B-X SYSTEM.
J.MOL.SPECTROSC.77,310-321(1979).

[616] KASNER,W.H. AND L.D.PLEASANCE:
LASER EMISSION FROM THE 13.9-MICRON 100-010 CO2 TRANSITION
IN PULSED ELECTRICAL DISCHARGES.
APPL.PHYS.LETT.31,82-84(1977).

[617] HENNINGSEN,J.O., J.C.PETERSEN, F.R.PETERSEN,
D.A.JENNINGS, AND K.M.EVENSON:
HIGH RESOLUTION SPECTROSCOPY OF VIBRATIONALLY EXCITED
13CH3OH BY FREQUENCY MEASUREMENT OF FIR LASER EMISSION.
J.MOL.SPECTROSC.77,298-309(1979).

# Subject Index

| | | |
|---|---|---|
| THALLIUM | TL | 18 |
| THULIUM | TM | 29 |
| TIN | SN | 20 |
| TRIFLUOROBROMO METHANE | CF3BR | 113 |
| TRIFLUOROIODO METHANE | CF3I | 114 |
| TRIOXANE | [H2CO]3 | 115 |
| VANADIUM | V | 23 |
| VINYL BROMIDE | C2H3BR | 136 |
| VINYL CHLORIDE | C2H3CL | 135 |
| VINYL CYANIDE | C2H3CN | 137 |
| WATER | H2O | 104 |
| XENON | XE | 7 |
| YTTERBIUM | YB | 30 |
| ZINC | ZN | 15 |

# Dye Lasers

Editor: F. P. Schäfer

2nd, revised edition 1977. 114 figures. XI, 299 pages
(Topics in Applied Physics, Volume 1)
ISBN 3-540-08470-3

**Contents:**
*F. P. Schäfer:* Principles of Dye Lasers Operation. – *B. B. Snavely:* Continuous-Wave Dye Lasers. – *C. V. Shank, E. P. Ippen:* Mode-Locking of Dye Lasers. – *K. H. Drexhage:* Structure and Properties of Laser Dyes. – *Th. W. Hänsch:* Application of Dye Lasers.

# Eximer Lasers

Editor: C. K. Rhodes

1979. 59 figures, 29 tables. XI, 194 pages
(Topics in Applied Physics, Volume 30)
ISBN 3-540-09017-7

**Contents:**
*P. W. Hoff, C. K. Rhodes:* Introduction. – *M. Krauss, F. H. Mies:* Electronic Structure and Radiative Transitions of Excimer Systems. – *M. V. McCusker:* The Rare Gas Excimers.– *C. A. Brau:* Rare Gas Halogen Excimers. – *A. Gallagher:* Metal Vapor Excimers. – *C. K. Rhodes, P. W. Hoff:* Applications of Excimer Systems.

# Laser Monitoring of the Atmosphere

Editor: E. D. Hinkley

1976. 84 figures, XV, 380 pages
(Topics in Applied Physics, Volume 14)
ISBN 3-540-07743-X

**Contents:**
*E. D. Hinkley:* Introduction. – *S. H. Melfi:* Remote Sensing for Air Quality Management. – *V. E. Zuev:* Laser-Light Transmission through the Atmosphere. – *R. T. H. Collis, P. B. Russell:* Lidar Measurement of Particles and Gases by Elastic Backscattering and Differential Absorption. – *H. Inaba:* Detection of Atoms and Molecules by Raman Scattering and Resonance Fluorescence. – *E. D. Hinkley, R. T. Ku, P. L. Kelley:* Techniques for Detection of Molecular Pollutants by Absorption of Laser Radiation. – *R. T. Menzies:* Laser Heterodyne Detection Techniques.

# Raman Spectroscopy of Gases and Liquids

Editor: A. Weber

1979. 113 figures, 25 tables. XI, 318 pages
(Topics in Current Physics, Volume 11)
ISBN 3-540-09036-3

**Contents:**
*A. Weber:* Introduction. – *S. Brodersen:* High-Resolution Rotation-Vibrational Raman Spectroscopy. – *A. Weber:* High-Resolution Rotational Raman Spectra of Gases. – *H. W. Schrötter, H. W. Klöckner:* Raman Scattering Cross Sections in Gases and Liquids. – *R. P. Srivastava, H. R. Zaidi:* Intermolecular Forces Revealed by Raman Scattering. – *D. L. Rousseau, J. M. Friedman, P. F. Williams:* The Resonance Raman Effect. – *J. W. Nibler, G. V. Knighten:* Coherent Anti-Stokes Raman Spectroscopy.

Springer-Verlag
Berlin
Heidelberg
New York